AF342457

THE GEOMETRY CENTER

Mathematical Sciences Research Institute
Publications

11

Editors

S.S. Chern
I. Kaplansky
C.C. Moore
I.M. Singer

D. Drasin   C.J. Earle   F.W. Gehring
I. Kra   A. Marden
Editors

Holomorphic Functions and Moduli II

Proceedings of a Workshop held
March 13-19, 1986

With 29 Illustrations

Springer-Verlag
New York Berlin Heidelberg London Paris Tokyo

D. Drasin
Department of Mathematics
Purdue University
West Lafayette, IN 47909
USA

C.J. Earle
Department of Mathematics
Cornell University
Ithaca, NY 14853
USA

F.W. Gehring
Department of Mathematics
University of Michigan
Ann Arbor, MI 48109
USA

I. Kra
Department of Mathematics
State University of
   New York at Stony Brook
Stony Brook, NY 11794
USA

A. Marden
Department of Mathematics
University of Minnesota
Minneapolis, MN 55455
USA

Mathematical Sciences Research Institute
1000 Centennial Drive
Berkeley, CA 94720
USA

The Mathematical Sciences Research Institute wishes to acknowledge support by the National Science Foundation.

Mathematics Subject Classification (1980): 30XX, 32XX

Library of Congress Cataloging-in-Publication Data
Holomorphic functions and moduli / D. Drasin...[et al.], editors.
    p. cm. -- (Mathematical Sciences Research Institute
  publications ; 10-<11 >)
  Includes bibliographies.
  ISBN 0-387-96766-4 (v. 1). ISBN 0-387-96786-9 (v. 2)
  1. Holomorphic functions. 2. Teichmuller spaces.
3. Quasiconformal mappings. 4. Riemann surfaces. I. Drasin, D.
(David) II. Series: Mathematical Sciences Research Institute
publications ; 10, etc.
QA331.H66 1988   88-12381
515.9'8--dc19

Camera-ready text prepared by the Mathematical Sciences Research Institute using PC T_EX.
Printed and bound by Edward Brothers, Inc., Ann Arbor, Michigan.
Printed in the United States of America.

9 8 7 6 5 4 3 2 1

ISBN 0-387-96786-9 Springer-Verlag New York Berlin Heidelberg
ISBN 3-540-96786-9 Springer-Verlag Berlin Heidelberg New York

PREFACE

The Spring 1986 Program in Geometric Function Theory (GFT) at the Mathematical Sciences Research Institute (MSRI) brought together mathematicians interested in Teichmüller theory, quasiconformal mappings, Kleinian groups, univalent functions and value distribution. It included a large and stimulating Workshop, preceded by a mini-conference on String Theory attended by both mathematicians and physicists. These activities produced interesting results and fruitful interactions among the participants. These volumes represent only a portion of the papers that will eventually result from ideas developed in the offices and corridors of MSRI's elegant home.

The Editors solicited contributions from all participants in the Program—whether or not they gave a talk at the Workshop. Papers were also submitted by mathematicians invited but unable to attend. All manuscripts were refereed. The articles included here cover a broad spectrum, representative of the activities during the semester. We have made an attempt to group them by subject, for the reader's convenience.

The Editors take pleasure in thanking all participants, authors and referees for their work in producing these volumes.

We are also grateful to the Scientific Advisory Council of MSRI for supporting the Program in GFT. Finally thanks are due to the National Science Foundation and those Universities (including Cornell, Michigan, Minnesota, Rutgers Newark, SUNY Stony Brook) who gave released time to faculty members to participate for extended periods in this program.

The staff at MSRI, the beautiful surroundings and fine weather were instrumental in significantly increasing the usual pleasures of doing mathematics. We look forward to another program at MSRI.

D. Drasin
C.J. Earle
F.W. Gehring
I. Kra
A. Marden

Holomorphic Functions and Moduli

TABLE OF CONTENTS – VOLUME 2

TEICHMÜLLER SPACES

Holomorphic Functions and Moduli

TABLE OF CONTENTS – VOLUME 1

RIEMANN SURFACES

R. Brooks
ISOSPECTRAL POTENTIALS ON A SURFACE OF GENUS 3

Y. Kusunoki
INTEGRABLE HOLOMORPHIC QUADRATIC DIFFERENTIALS
WITH SIMPLE ZEROS

H. Masur
LOWER BOUNDS FOR THE NUMBER OF SADDLE
CONNECTIONS AND CLOSED TRAJECTORIES OF
A QUADRATIC DIFFERENTIAL

M. Sheingorn
RATIONAL SOLUTIONS OF $\sum_{i=1}^{3} a_i x_i^2 = d x_1 x_2 x_3$
AND SIMPLE CLOSED GEODESICS ON
FRICKE SURFACES

M. Shiba
THE PERIOD MATRICES OF COMPACT
CONTINUATIONS OF AN OPEN RIEMANN
SURFACE OF FINITE GENUS

Mostow rigidity on the line: A survey

BY STEPHEN AGARD

§1. Introduction.

G.D. Mostow's celebrated Rigidity Theorem has taken some curious forms on the real line. All assume that f (a continuous strictly increasing real valued function of a real variable) is the "boundary mapping" of an isomorphism between Fuchsian groups. Sample forms of the Rigidity Theorem are the following:

(D1) Either f is absolutely continuous, or f and f^{-1} are *singular**.

(D1') Either f^{-1} is absolutely continuous or f and f^{-1} are singular.

(P2) If f is absolutely continuous, then f is Möbius.

(D3) Either f is Möbius, or it is singular.

(D4) Either f is Möbius, or it is *purely* singular*.

Due to occasional slight misstatement, the development and present status of these propositions is not completely clear. We shall try to bring some light into this forest, and at the same time to investigate the extent to which Dennis Sullivan's comparatively recent and extremely important articles [S1,S2] can be applied to the line.

Since this article is directed toward nonspecialists in these matters, I have included many details which can be skipped over by experts. However, I feel that these are necessary for the intended audience.

§2. Terminology.

We let $\mathbf{M}$ be the group of Möbius transformations fixing the upper half plane $\mathbf{U}$, (and hence as well the real axis $\mathbf{R}$). We assume $\Gamma \subseteq \mathbf{M}$ is a subgroup, and that f is a continuous strictly increasing real valued function of a real variable with the *compatiblity* property: for each $\gamma \in \Gamma$, there exists $\tilde{\gamma} \in \mathbf{M}$ with

$$f(\gamma(x)) \; = \; \tilde{\gamma}(f(x)) \qquad (x \in \mathbf{R}).$$

The author wishes to thank the IMA for financial support which made possible his participation in the conference.
*singular means $f'(x) \, = \, 0$ a.e., whereas *purely* singular means $f'(x) \, = \, 0$ wherever it exists.

The theory of these "boundary mappings" (f) is much built up, and has treated questions of existence, uniqueness, and regularity. Here we mention only the fact that uniqueness never extends beyond the limit set, and therefore confine this note to groups of the first kind.

When Γ is discrete, hence *Fuchsian*, it acts discontinuously in $\mathbf{U}$, and gives rise to a Riemann surface $M = \Gamma\backslash\mathbf{U}$, whose points are the orbits $\Gamma(z)$ ($z \in \mathbf{U}$). The natural covering $z \to \Gamma(z)$ is unlimited, but branched over elliptic fixpoints. With obvious modifications near elliptic fixpoints, coordinates in $\mathbf{U}$ become local coordinates on M. If P denotes the set of parabolic fixpoints, then each $x \in P$ corresponds to a *puncture* on M.

An orbit $\Gamma(z)$ may accumulate near $x \in \mathbf{R}$, and in doing so it may enter every horocycle* at x (horocyclic approach), or even some Stolz angle* at x (conical approach). The manner of approach is independent of the reference orbit $\Gamma(z)$, and we say in these cases that x belongs to the topological, horocyclic, or conical *limit sets*, denoted Λ, Λ_H, Λ_C. The latter is also known as the *radial* set. There is neither horocyclic nor conical approach to points of P, although $P \subseteq \Lambda$. In general, Λ is closed, and its complement $\Omega = \mathbf{R}\backslash\Lambda$ is called the *ordinary* set. Open intervals of Ω, if any, become *border arcs* for M.

A second dichotomy in $\mathbf{R}$ is between the *recurrent* set ($\mathcal{R}$) and the *dissipative* set ($\mathcal{D}$). These are determined up to null sets by the requirements: $\mathcal{D}$ has a measurable fundamental set (one representative from each orbit), and each subset $A \subseteq \mathcal{R}$ with positive measure has $\gamma(A) \cap A$ also with positive measure for infinitely many $\gamma \in \Gamma$.** Since the ordinary set has a fundamental region, we have $\Omega \subseteq \mathcal{D}$. We say the action of Γ is *recurrent* (*conservative* is also used) if $\mathcal{R} = \mathbf{R}$.

In general when Γ acts on a measure space $(\mathbf{X}, \mu)$, we say the action is *ergodic* if each measurable invariant set E has either $\mu(E) = 0$ or $\mu(\mathbf{X}\backslash E) = 0$. We speak at times of Γ acting ergodically on $\mathbf{R}$, or on $\mathbf{R} \times \mathbf{R}$, always with Lebesgue measure.*** Thanks to Fubini's theorem, the latter is evidently the stronger condition. The former, however, is still stronger than recurrent action, for we note that as soon as $\mathcal{D}$ has positive measure, Γ will not act ergodically on $\mathbf{R}$.

---

*A horocycle is a disk in $\mathbf{U}$ tangent to $\mathbf{R}$. A Stolz angle at x is a set $\{\varsigma \in \mathbf{U} : \alpha < arg(\varsigma - x) < \pi - \alpha\}$ ($0 < \alpha < \pi/2$).

**Extensive search and consultation has failed to produce a proof of this folk theorem in the literature. A brief sketch of a proof will be found in the appendix.

***The action on $\mathbf{R} \times \mathbf{R}$ is $(x, y) \to (\gamma(x), \gamma(y))$.

Given $z \in \mathbf{U}$, the Dirichlet fundamental region $\mathcal{F} = \mathcal{F}_z$ is the set $\{\varsigma \in \mathbf{U} : d(\varsigma, z) \leq d(\varsigma, \Gamma(z) \backslash \{z\})\}$. It is, apart from duplicates on its boundary, a measurable fundamental set for the action of Γ in $\mathbf{U}$. Here d is the non-Euclidean distance in $\mathbf{U}$, and we denote by $\mathcal{F}^*$ the set of $x \in \mathbf{R}$ such that the geodesic ray $[z, x)$ lies in $\mathcal{F}$. The qualities of $\mathcal{F}$, for purposes of this discussion, are independent of $z \in \mathbf{U}$.*

§3. Background.

A number of classifications of groups and/or Riemann surfaces have been introduced at various times as being relevant to certain problems. Of course one wants to know about equivalences and inclusions among superficially different classes. For the present discussion, we give the following summary, in which the categories are ordered by proper inclusion, with the largest at the bottom:

Γ	M	Limit sets	$\mathcal{F}$
cocompact in $\mathbf{M}$	compact	$\mathbf{R} = \Lambda_C$	$\mathcal{F}^*$ is void
finitely generated and first kind	finite non-Euclidean area	$\mathbf{R} = \Lambda_C \cup \mathcal{P}$	$\mathcal{F}$ has a finite number of sides and $\mathcal{F}^* \subseteq \mathcal{P}$
ergodic action on $\mathbf{R} \times \mathbf{R}$	without Green's function (O_G)	Λ_C has full measure in $\mathbf{R}$	Q1
ergodic action on $\mathbf{R}$	without bounded nonconstant harmonics (O_{HB})	Q2	Q3
recurrent action on $\mathbf{R}$	Q4	Λ_H has full measure in $\mathbf{R}$	$\mathcal{F}^*$ is null
first kind	without border	$\Lambda = \mathbf{R}$	

Much of the literature has treated the case of groups acting on $\mathbf{U}^{n+1}$ with boundary maps acting on $\mathbf{R}^n$. Ours is the case $n = 1$. In treating the cases $n > 1$, the second line becomes three. Finite non-Euclidean volume now is sufficient but not necessary for $\mathcal{F}$ to have a finite number of

---

*We assume z is not an elliptic fixpoint.

sides (category known as "geometrically finite"), which is in turn sufficient but not necessary for Γ to be finitely generated. Beardon and Maskit's important dichotomy

$$\Lambda \;=\; \Lambda_C \;\cup\; \{\text{cusped parabolic fixpoints}\},$$

[B/M], which applies to geometrically finite groups, takes on for us the simpler form $\mathbf{R} \;=\; \Lambda_C \;\cup\; \mathcal{P}$, because our groups are assumed to be first kind ($\Lambda \;=\; \mathbf{R}$) and because all parabolic fixpoints are cusped in case $n \;=\; 1$. Continuing briefly with general n, one other equivalence might be noted: $\mathcal{M}$ has "finite volume" iff $\Gamma \backslash \mathbf{M}$ carries a *finite* invariant measure (Γ and $\mathbf{M}$ are both unimodular, so $\Gamma \backslash \mathbf{M}$ carries a σ-finite invariant measure in any case) [Ag2]. With respect to common names, groups acting ergodically on $\mathbf{R}^n \times \mathbf{R}^n$ are often known as "divergence type", [Ah1], the corresponding manifolds having been termed the "first class" by Hopf [H].

The equivalences regarding finite area are well known. Those concerning O_G are fully developed in [Ah1]. Those concering O_{HB} involve the interplay via the Poisson transform and radial limits, between the class $HB(\mathcal{M})$ of bounded harmonic functions on $\mathcal{M}$, and bounded invariant functions on $\mathbf{R}$. An example of $M \;\in\; O_{HB} \backslash O_G$ is to be found in [A/S], pages 256-257. We will return to the case of recurrent action in §5.

It would certainly be of interest to provide good descriptions where the entries Q1-Q4 are found in the above table. The author indicates his own unfamiliarity—some of these may be already known. However, I would like to pass on some observations of Prof. Taniguchi regarding Q4: when the dissipative set $\mathcal{D}$ has the measurable fundamental set E, a bounded measurable function with support in E can be made invariant on $\mathcal{D}$ (by making it constant on orbits) and then pushing it through the Poisson transform to give an invariant bounded harmonic function in $\mathbf{U}$, and hence a bounded harmonic function on $\mathcal{M}$. The steps in this passage $L_\infty(E) \;\rightarrow\; L_\infty(\mathcal{D}) \;\rightarrow\; HB(\mathcal{M})$ are injective and it follows that the space $HB(\mathcal{M})$ is infinite dimensional, as is $L_\infty(E)$ *as soon as E has positive measure*. Examples in [K/T] of $\mathcal{M}$ with $dim(HB(\mathcal{M}))$ any finite number show clearly that ergodic action on $\mathbf{R}$ is not equivalent to recurrent action. Prof. Taniguchi further conjectures that the correct equivalence to recurrent action is the class O_{HB}^∞ (see for example [C/C], page 122).

§4. Kuusalo and Tukia.

The dichotomy (D1) was first articulated by Kuusalo [**K**] for surfaces of class O_{HB}. The proof is so simple we recall it: If E is any set of positive measure mapped by f on a null set, then $\Gamma(E)$ becomes invariant with measure $\geq$ meas$(E) > 0$, and therefore $\Gamma(E)$ is a set of full measure. On the other hand, $f(\Gamma(E)) = \cup \tilde{\gamma}(f(E))$ is a countable union of null sets, and null. Thus f is singular, mapping as it does a set of full measure on a null set. So also is f^{-1} in this case, by consideration of the complements. However, the alternative, that no such E exists, actually only shows that f^{-1} is absolutely continuous. In other words, it is (D1') rather than (D1) which Kuusalo proved for $M \in O_{HB}$.*

The next proposition (P2) was also proved by Kuusalo for surfaces of the first class (O_G), by observing that the Möbius identity

$$[\gamma(x) - \gamma(y)]^2 = (x - y)^2 \gamma'(x)\gamma'(y)$$

together with the compatibility, shows that the map

$$(x, y) \to f'(x)f'(y)[(x - y)/(f(x) - f(y))]^2 \quad ((x, y) \in \mathbf{R} \times \mathbf{R})$$

is invariant, and therefore a.e. constant, nonzero in the absolutely continuous case. From this he deduced that f (with appropriate normalization) satisfies the o.d.e. $f'(x) = [f(x)/x]^2$, for which the only normalized absolutely continuous solution is $f(x) \equiv x$.

These results do not quite give the dichotomy (D3) for the first class, though certainly for the case of finite area. The proposition (P2) for compact M was also proved somewhat later by the late Rufus Bowen [**Bo**], though certainly not the main point of that paper. The present author was able to deduce (D3) for the first class as a part of the general case (higher dimensions) [**Ag1**].

The fourth dichotomy (D4) was stated by Mostow ([**M2**], page 178) apparently for surfaces of finite area. However, elementary examples [**Ag3**] show that the presence of a parabolic fixpoint makes (D4) untenable. In establishing (D4) for compact M, (proof due essentially to Sullivan), the present author introduced in [**Ag3**] the notion of the *rigidity set M_r* : $x \in M_r$ if $f'(x) > 0$ implies that f is Möbius.

---

*In recent correspondence, Kuusalo reports that he still has neither proof nor counterexample to (D1) for $M \in O_{HB}$. On the other hand, (D3) and (D1') in conjunction, do give (D1) for $M \in O_G$.

Tukia [**T**] then showed in complete generality that $\Lambda_C \subseteq M_r$, and in view of Beardon and Maskit's aforementioned result $\mathbf{R} = \Lambda_C \cup \mathcal{P}$, the situation is now fully understood for at least finite area: $M_r = \Lambda_C$ and $\mathbf{R} \backslash M_r = \mathcal{P}$. Tukia's beautiful construction also vastly simplifies the proof of (D3) for the first class, for as soon as $E = \{x : f'(x) > 0\}$ has positive measure, it must intersect the radial set Λ_C (full measure) at some $x_0 \in M_r$, and f is Möbius.

§5.   Sullivan.

The equivalences associated to recurrent action are to be found in Sullivan's important paper [**S1**]. Because of their simple form in our case ($n = 1$) we include a brief explication. The horocycle $H(x,s)$ ($x \in \mathbf{R}$, $s > 0$) will be $\{\varsigma \in \mathbf{U} : |\varsigma - (x + is)| < s\}$. We define the (nonzero) point function λ and (totally finite) measure μ by:

$$\lambda(x,\gamma) = (a + bx)^2 + (d + cx)^2 \ (x \in \mathbf{R}, \gamma \in \mathbf{M},$$

$$\gamma(t) = (at + b)/(ct + d), ad - bc = 1)$$

$$\mu(E) = \int_E dx/(1 + x^2) \qquad (E \text{ measurable}, E \subseteq \mathbf{R}).$$

One verifies routinely the following facts:

(a) for $\gamma \in \mathbf{M}$, $\mu(\gamma(E) = \int_E dx/\lambda(x,\gamma)$.

(b) $\gamma(i) \in H(x,s)$ iff $\lambda(x,\gamma^{-1}) < s$, hence given $\Gamma \subseteq \mathbf{M}$,

(c) $x \in \Lambda_H$ iff $0 = \inf\{\lambda(x,\gamma^{-1}) : \gamma \in \Gamma\} = \inf\{\lambda(x,\gamma) : \gamma \in \Gamma\}$.

If E is a measurable fundamental set for $\mathcal{D}$, then

$$\infty > \mu(\mathcal{D}) = \sum_{\gamma \in \Gamma} \mu(\gamma(E)) = \int_E \sum_{\gamma \in \Gamma} (1/\lambda(x,\gamma))dx,$$

hence

(d) $\sum_{\gamma \in \Gamma} (1/\lambda(x,\gamma)) < \infty$ for a.e. $x \in E$, and therefore

(e) $\inf\{\lambda(x,\gamma) : \gamma \in \Gamma\} > 0$ for a.e. $x \in E$.

Hence E and Λ_H have at most a null set in common, which shows that Λ_H and $\mathcal{D}$ have at most a null set in common.

Now the possibilities for non-horocyclic-approach fall into three types which we label $\mathcal{C}$, $\mathcal{N}$, $\mathcal{G}$:

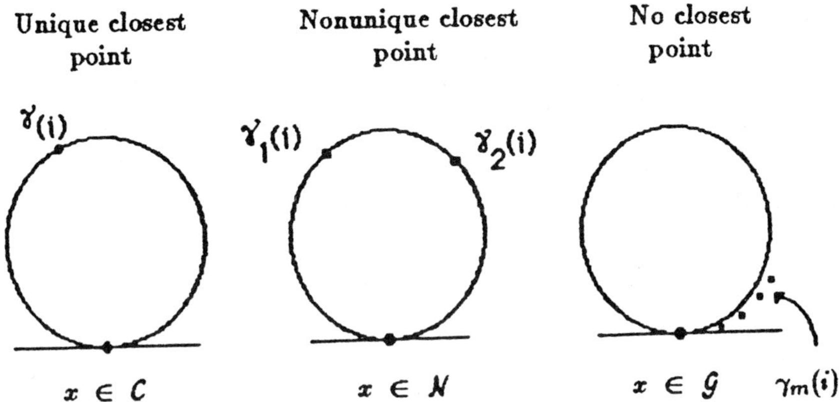

For each $x \notin \Lambda_H$, we have pictured the horocycle $H(x,s)$ with maximal $s = s(x)$ such tht the orbit $\Gamma(i)$ never enters $H(x,s)$. In case $x \in \mathcal{G}$, however, the orbit enters every $H(x,t)$ $(t > s)$. If $\mathcal{G}_\sigma$ is the set $\{x : 2\sigma > s(x) > \sigma\}$, then to prove $\mathcal{G}$ has measure zero*, it suffices to prove $\mathcal{G}_\sigma$ has measure zero. Accordingly, fix $x \in \mathcal{G}_\sigma$ and let $s = s(x)$.

There is an infinity of points $\xi_m + i\eta_m = \gamma_m(i)$ in $\partial H(x,s_m)$ with $s_m \to s > \sigma$. We assume $\xi_m > x$, and to each integer m associate a unique point x_m in the interval (x, ξ_m) determined by the relation $\xi_m + i\eta_m \in \partial H(x_m, \sigma)$. It is clear (Fig. 2 below) that no point of the interval (x_m, ξ_m) can lie in $\mathcal{G}_\sigma$. This is Sullivan's "forbidden zone", and occupies the relative part

$$\frac{\xi_m - x_m}{\xi_m - x} = \sqrt{\frac{\sigma^2 - \eta_m^2}{s_m^2 - \eta_m^2}}$$

of the interval (x, ξ_m), which ratio tends to σ/s as $m \to \infty$ (forcing $\eta_m \to 0$ and $s_m \to s$). Since $\sigma/s > 1/2$, it follows that the set $\mathcal{G}_\sigma$ has length-density at most $3/4$ at an arbitrary point x, and is consequently of measure zero.

Regarding $x \in C$, we have a map $\varphi : C \to \Gamma$ in which $\varphi(x) = \gamma$ iff $\gamma(i)$ is the unique closest orbit point to x. Clearly φ has the invariance $\varphi(\tau(x)) = \tau(\varphi(x))$ $(x \in C, \tau \in \Gamma)$ and consequently the set $E = \{x \in$

---

*The identification $\mathcal{G}$ stands for "Garnett points". Sullivan credits Jon and/or Lucy Garnett with their discovery and with showing that they are negligible.

$C : \varphi(x) = id\}$ is a measurable fundamental set for C. Thus the disjoint union

$$\mathbf{R} = \Lambda_H \cup C \cup \mathcal{N} \cup \mathcal{G}$$

in which $\mathcal{N}$ is countable, $\mathcal{G}$ is null, and C has the measurable fundamental set E, shows that up to the null sets $\mathcal{N}$ and $\mathcal{G}$, Λ_H *is* $\mathcal{R}$ and C *is* $\mathcal{D}$.

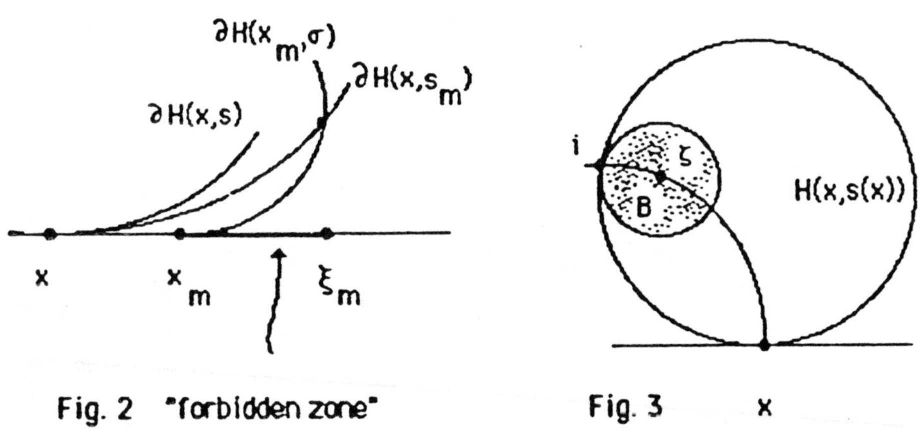

Fig. 2 "forbidden zone" Fig. 3

Finally we note (Fig. 3) that if $x \in E$, then for any ς on the geodesic ray $[i, x)$, it is clear that the set $B = \{z \in \mathbf{U} : d(\varsigma, i) > d(\varsigma, z)\}$ lies entirely in $H(x, s)$ and hence $\varsigma \in \mathcal{F}_i$. It follows that $x \in \mathcal{F}^*$, hence $E \subseteq \mathcal{F}^*$, and we also easily see $\mathcal{F}^* \subseteq E \cup \mathcal{N}$. In summary, then,

$$\Lambda_H \text{ is full measure} \iff \Gamma(E) \text{ is null} \iff \mathcal{F}^* \text{ is null}.$$

§6. Sullivan and Ahlfors.

The existence of non-Möbius boundary mappings f is a reflection of the richness of Teichmüller spaces of Riemann surfaces. It is the content of Mostow's Rigidity Theorem [**M1**] that there are none as soon as $n > 1$, if $\mathcal{M}$ has finite volume. Extensions of this theorem to groups of the other classes is an area of study in itself. In particular, the aforementioned Sullivan article [**S1**] has made a startling advance to groups of recurrent action. In an explication of part of Sullivan's work, Ahlfors [**Ah2**] states and proves this

THEOREM. *There is no nontrivial invariant vector field on the recurrent set.*[*]

By the conventions established in §3, Sullivan's article clearly deals with $n = 2$, whereas Ahlfors's article permits any n. (The context has shifted from $\mathbf{U}^{n+1}$ with boundary $\mathbf{R}^n$ to $\mathbf{B}^n$ with boundary $\mathbf{S}^{n-1}$, adding further confusion to the specific n's such as 1 or 2. Shifting along with Ahlfors, the present focus is $n = 2$, a specific not explicitly *included* in [**Ah2**] however.)

To clarify the terminology, a *vector field* is a mapping

$$\varsigma \;\to\; \varphi(\varsigma) \qquad (|\varsigma| = 1),$$

with $\varphi(\varsigma) \in \mathbf{C}$ and $\perp \varsigma$, hence of the form $\varphi(\varsigma) = i\varsigma\rho(\varsigma)$ with real valued ρ. If $\gamma \in \mathbf{M}^*$ (Möbius fixing $\mathbf{B}^2$), *invariance* is defined to mean

$$\varphi(\gamma(\varsigma)) \;=\; \gamma'(\varsigma)\varphi(\varsigma)/|\gamma'(\varsigma)|.$$

Owing to the general structure of $\gamma \in \mathbf{M}^*$: $\gamma(z) = e^{i\Theta}(z - a)/(1 - \overline{a}z)$ $(|a| < 1)$, one easily verifies that invariance comes to nothing more than ordinary invariance of ρ : $\rho(\gamma(\varsigma)) = \rho(\varsigma)$, and *triviality* amounts to the a.e. vanishing of ρ. Alternatively put, the vector field φ_0 : $\varsigma \to i\varsigma$ is already a *universally invariant nontrivial vector field on* $\mathbf{S}^1$.

We can now be certain that Ahlfors did not intend to include the case $n = 2$, but it is still instructive to see where the proof breaks down: on page 6, §9, with a fixed vector $v' \in \mathbf{R}^{n-1}$ $(|v'| = 1)$ in hand, one needs to know that the sets

$$E_\lambda \;=\; \{u' \in \mathbf{R}^{n-1} : |u'|^2 - (v'{\cdot}u')^2 > \lambda\}$$

expand to include almost all points as $\lambda \downarrow 0$. Considering the case $n = 3$ with v' given by (a,b), u' by (x,y), the inequality defining E_λ becomes

$$x^2 + y^2 - (ax + by)^2 > \lambda.$$

The quadratic form is always rank 1, the case $v' = (1,0)$ is representative, and the set amounts to $\{(x,y) : y^2 > \lambda\}$. On the other hand in case

---

[*]In the later article [**S2**, part II], Sullivan himself explicates this theorem. In part I, he also gives a proof, not unlike Kuusalo's (however not restricted to dimension one), of the rigidity theorem for groups of divergence type.

$n = 2$, with $v'^2 = 1$ and $u' = x$, we have an empty set as soon as $\lambda > 0$.

Thus it appears that the more precise statement of Ahlfors's version of Sullivan's theorem for the case $n = 2$ might read: *invariant vector fields φ on $\mathcal{R}$ in the context $\mathbf{S}^1$ (as boundary of $\mathbf{B}^2$) arise from invariant scalar functions ρ on $\mathcal{R}$ through the connection $\varphi(\varsigma) = i\varsigma\rho(\varsigma)$.* In the ergodic case ($\mathcal{M}$ is O_{HB} and $\mathcal{R}$ is $\mathbf{S}^1$) ρ will be essentially constant.

It is yet a difficult and improperly understood (at least by this author) transition from the above Sullivan Theorem to the Mostow Rigidity Theorem, even in higher dimensions. Whatever implications the present "mini-theorem" has for Mostow Rigidity Theory on the line, if any, remain obscure.

§7. Appendix. A proof of the dissipative/recurrent dichotomy*.

We first observe that since the $\gamma \in \Gamma$ are homeomorphisms, the property of *invariance* ($\gamma(A) = A$ for all $\gamma \in \Gamma$) is preserved under unions, intersections, and differences. Further it is clear that an invariant subset of a recurrent set is recurrent, and it is not hard to see that an invariant subset of a dissipative set is dissipative. Finally a set which is both dissipative and recurrent is null. Therefore the uniqueness of the decomposition $\mathbf{R}^n = \mathcal{R} \cup \mathcal{D}$ is clear, since for a second decomposition $\mathbf{R}^n = \mathcal{R}_1 \cup \mathcal{D}_1$, the sets $\mathcal{R}\backslash\mathcal{R}_1 = \mathcal{R} \cap \mathcal{D}_1$ and $\mathcal{D}\backslash\mathcal{D}_1 = \mathcal{D} \cap \mathcal{R}_1$ are null.

Next we observe that if $\mathcal{D}_1$ and $\mathcal{D}_2$ are dissipative with measurable fundamental sets (MFS) E_1 and E_1, then $E_1 \cup (E_2\backslash\mathcal{D}_1)$ is an MFS for $\mathcal{D}_1 \cup \mathcal{D}_2$. In other words, the union of two dissipative sets is dissipative and we *may assume* that the MFS for the union contains the MFS for the first set.

We now assume that the measure μ is finite. (We can use the measure induced by stereographic projection and Lebesgue measure on $\mathbf{S}^n$, which is precisely the μ introduced for $\mathbf{R}$ in §5.) We then let $\{\mathcal{D}_k\}$ be a sequence of dissipative sets with $\lim \mu(\mathcal{D}_k) = \sup\{\mu(\mathcal{D}) : \mathcal{D}$ is dissipative$\}$. We may assume $\mathcal{D}_k \subseteq \mathcal{D}_{k+1}$ and that the MFS's E_k are also increasing. Then clearly $\cup E_k$ is an MFS for $\mathcal{D} = \cup \mathcal{D}_k$. Clearly the complement $\mathcal{S} = \mathbf{R}^n\backslash\mathcal{D}$ can have no dissipative part of positive measure. It remains to show that $\mathcal{S}$ is recurrent.

Accordingly, assume that $A \subseteq \mathcal{S}$ is a measurable set of positive measure. If the set $\Gamma_0 = \{\gamma \in \Gamma : \mu(\gamma(A) \cap A) > 0, \gamma \neq id\}$ is finite, then

---

*We purport no great generality. Γ is a discrete Möbius group.

define the measurable sets

$$C = \cup \{\gamma(A) : \gamma \in \Gamma_0\}, \ D = \cup \{\gamma(A) : \gamma \notin \Gamma_0, \gamma \neq id\}$$

and consider $B = (A \backslash C) \backslash D$. Clearly $\gamma(B) \cap B = \emptyset$ for all $\gamma \neq id$, hence B is an MFS for $\Gamma(B)$, and B is null. The latter deleted part $D \cap A$ being null, this means that $A \subseteq C$ up to null sets, and by deletion of null sets from A, we may assume in fact $A \subseteq C$ and $D \cap A = \emptyset$.

Now let x_0 be a point of μ-density for A, which is also not fixed for any $\gamma \in \Gamma_0$. There is a small ball E centered at x_0 which is mapped by Γ_0 outside itself.* Take $F = A \cap E$. It is now clear that $\gamma(F) \cap F = \emptyset$ for all $\gamma \neq id$. Hence F is null. This forces us to reject the hypotheses that Γ_0 is finite. Hence S is recurrent.

University of Minnesota, Minneapolis, Minnesota

---

*It is important to remember that Γ may not act discontinuously. Therefore the set Γ_0 must be finite at this point in the argument. The fixpoints of course form a null set.

REFERENCES

[**Ag1**] Agard, S., *A geometric proof of Mostow's rigidity theorem for groups of divergence type*, Acta. Math. **151** (1983), 231–252.

[**Ag2**] _________, *Elementary properties of Möbius transformations in $\mathbb{R}^n$ with applications to rigidity theory*, University of Minnesota Math. Report 82-110, (1982), Minneapolis.

[**Ag3**] _________, *Remarks on the boundary mapping for a Fuchsian group*, Ann. Acad. Sci. Fenn. A.I. Math **10** (1985), 1–13.

[**Ah1**] Ahlfors, L., "Möbius Transformations in Several Variables," Ordway Professorship Lectures in Mathematics, University of Minnesota, Minneapolis, 1981.

[**Ah2**] _________, *Ergodic properties of Möbius transformations*, Analytic functions Kozubnik 1979, Lecture Notes in Math. **798**; Springer (1980), 1–9.

[**A/S**] Ahlfors, L., and Sario, L., "Riemann Surfaces," Princeton, 1960.

[**B/M**] Beardon, A.F., and Maskit, B., *Limit points of Kleinian groups and finite sided fundamental polyhedra*, Acta. Math. **132** (1974), 1–12.

[**Bo**] Bowen, R., *Hausdorff dimension of quasicircles*, Inst. Hautes Etudes Sci. Publ. Math **50** (1979), 11–25.

[**C/C**] Constantinescu, C., and Cornea, A., "Ideale Ränden Riemannschen Flächen," Springer, 1963.

[**H**] Hopf, E., *Ergodic theory and the geodesic flow on surfaces of constant negative curvature*, Bull. Amer. Math. Soc. **77** (1971), 863–877.

[**K**] Kuusalo, T., *Boundary mappings of geometric isomorphisms of Fuchsian groups*, Ann. Acad. Sci. Fenn. A.I. Math **545** (1973), 1–7.

[**K/T**] Kusunoki, Y., and Taniguchi, M., *Remarks on Fuchsian groups associated with open Riemann surfaces*, Riemann surfaces and related topics: Proceedings of the 1978 Stony Brook conference, edited by I. Kra and B. Maskit, Ann. of Math. Studies **97** (1981); Princeton, 377–390.

[**M1**] Mostow, G.D., *Quasiconformal mappings in n-space and the rigidity of hyperbolic space forms*, Inst. Hautes Etudes Sci. Publ. Math. **34** (1968), 53–104.

[**M2**] _________, "Strong Rigidity of Locally Symmetric Spaces," Ann. Math. Studies **78** (1973), Princeton.

[**S1**] Sullivan, D., *On the ergodic theory at infinity of an arbitrary discrete group of hyperbolic motions*, Riemann surfaces and related topics: Proceedings of the 1978 Stony Brook conference, edited by I. Kra and B. Maskit, Ann. of Math. Studies **97** (1981); Princeton, 465–496.

[**S2**] _________, *Discrete conformal groups and measurable dynamics*, Bulletin (New Series) AMS 6 (No. 1) (1982), 57–73.

[**T**] Tukia, P., *Differentiability and rigidity of Möbius groups*, Invent. Math. **82** (1985); No. 3, 557–578.

Fuchsian groups and n^{th} roots of parabolic generators

BY A.F. BEARDON

1. Introduction.

We consider those subgroups of $PSL(2,\mathbb{R})$ which are generated by two parabolic elements, say g and h, and our main objective is to understand precisely when $\langle g^{1/n}, h^{1/m}\rangle$ is discrete. In addition, and as a by-product of our work, we shall sharpen and clarify some of the results concerning $\langle g^{1/n}, h^{1/m}\rangle$ given in [2].

As usual, $SL(2,\mathbb{R})$ is the group of real 2×2 matrices with unit determinant and $PSL(2,\mathbb{R})$ the group of Möbius transformations preserving the upper half-plane $H = \{x + iy : y > 0\}$. Each of the matrices A and $-A$ in $SL(2,\mathbb{R})$ project to the same g in $PSL(2,\mathbb{R})$ and we denote by $\mathrm{tr}^2(g)$ the square of the trace of A (or of $-A$). Note that the trace of the commutator

$$[g,h] = ghg^{-1}h^{-1}$$

is independent of the choice of matrices representing g and h.

A Möbius transformation g is *parabolic* if it has a unique fixed point on the complex sphere; equivalently (excluding the identity) if $\mathrm{tr}^2(g) = 4$. Given a parabolic g, consider the group G of parabolic elements with the same fixed point as g. It is well known that there is a unique isomorphism $\theta : (\mathbb{R}, +) \to G$ with $\theta(1) = g$ and we define the transformation g^a to be $\theta(a)$. Of course, g^{-1}, g^2 and so on have their usual interpretations and, quite generally,

$$f(g^a)f^{-1} = (fgf^{-1})^a.$$

Given any parabolic elements g and h with distinct fixed points, we may (up to conjugation by the same transformation) assume that

$$g(z) = z + t;$$

$$h(z) = z/(sz + 1).$$

It is convenient to refer to this by saying that these g and h are in *canonical form* and that the original g and h are *jointly conjugate* to these. We then have

$$g^a(z) = z + at;$$

$$h^b(z) = z/(bsz + 1)$$

and so

$$(1.1) \qquad \mathrm{tr}[g,h] - 2 = (st)^2;$$

$$(1.2) \qquad \mathrm{tr}[g^a, h^b] - 2 = (ab)^2(\mathrm{tr}[g,h] - 2).$$

Note that as the terms in (1.2) are invariant under joint conjugation, (1.2) holds for any pair of parabolic elements g and h (not necessarily in canonical form). Moreover, neither side of (1.1) or (1.2) changes if we replace either (or both) of the generators g and h by their inverse.

Further, taking g and h in canonical form, we can conjugate $\langle g^a, h^b \rangle$ by $f : z \mapsto bz$ and obtain the conjugate group $\langle g^{ab}, h \rangle$. Thus, if we wish, we may confine our attention to groups of the form $\langle g^a, h \rangle$.

In [2], the authors are interested in the trace of the composition gh and, to avoid ambiguity, they work with $\mathrm{tr}^2(gh)$. It is better, however, to consider instead the matrix invariant form

$$|T(g,h) - 2|$$

where

$$T(g,h) = 4\mathrm{tr}(gh)/\mathrm{tr}(g)\mathrm{tr}(h)$$

and where we use the same matrices in computing the numerator and the denominator (of course, the denominator is -4 or 4). The advantage of using $|T(g,h) - 2|$ is simply that it is unchanged if either (or both) of g and h are replaced by their inverse: this is not true of $\mathrm{tr}^2(gh)$. Such a replacement is often convenient, yet it cannot be achieved by conjugation as g and g^{-1} are never conjugate within $SL(2,\mathbb{R})$.

Taking g and h in canonical form, another simple computation leads to

$$(1.3) \qquad |T(g,h) - 2| = \{\mathrm{tr}[g,h] - 2\}^{1/2}$$

(both sides are $|st|$) and this, too, is invariant under joint conjugation and so holds for any pair g and h.

Before stating our main results (Theorems 1 and 2), let us dispose of one simple case. With g and h in canonical form it is easy to see that $\langle g,h \rangle$ is discrete and free if and only if the straight lines $x = t/2$ and $x = -t/2$ do not meet the isometric circles of h (in H). This condition is simply $|st| > 4$ or, from (1.1), $\mathrm{tr}[g,h] - 2 > 16$. Using (1.2) we find that

$$\langle g^a, h^b \rangle \text{ is discrete and free if and only if}$$

$$|ab| \geq 4(\mathrm{tr}[g,h] - 2)^{-1/2} :$$

this holds regardless of whether $\langle g, h \rangle$ is discrete or not.

In order to state Theorem 1, we need to introduce the groups

$$\Gamma_n = \langle z \mapsto z + n/2, z \mapsto z/(2z+1) \rangle$$

where $n = 1, 2, 3$ and 4. It is easily seen that these Γ_n are discrete and of the first kind: in fact, the signatures of $\Gamma_1, \Gamma_2, \Gamma_3$ and Γ_4 are

$$(0; 2, 3, \infty), \quad (0; 2, \infty, \infty), \quad (0; 3, \infty, \infty), \quad (0; \infty, \infty, \infty)$$

respectively.

Note that the region enclosed by the geodesics

$$x = -n/2, \quad x = n/2, \quad |2z+1| = 1, \quad |2z-1| = 1$$

is a fundamental region for Γ_n when $n = 2, 3$ or 4: when $n = 1$ the region contains two copies of a fundamental region for Γ_1 (which is conjugate to the Modular group).

We can now state Theorems 1 and 2: in §5, we shall compare Theorem 2 with the results in [2].

THEOREM 1. *Suppose that g and h are parabolic elements with distinct fixed points, that the groups $\langle g, h \rangle$ and $\langle g^a, h^b \rangle$ are discrete and of the first kind and that a and b are rational. Then, to within conjugacy, the pair $\langle g, h \rangle$ and $\langle g^a, h^b \rangle$ is one of the six cases*

$$\Gamma_p, \quad \Gamma_q, \quad a = p/q, \quad b = 1$$

where $p < q$ and $p, q \in \{1, 2, 3, 4\}$.

THEOREM 2. *Suppose that $G = \langle g, h \rangle$ and $G_n = \langle g^{1/n}, h \rangle$ are discrete where g and h are parabolic elements with distinct fixed points and where $n \geq 2$.*

(i) *If G_n (and hence G) is of the second kind, then*

$$|T(g, h) - 2| > 8;$$

(ii) *If G_n is of the first kind but G is of the second kind, then*

$$|T(g, h) - 2| \geq 5;$$

(iii) *If G_n and G are of the first kind, then*

$$|T(g, h) - 2| = 2, \ 3 \text{ or } 4.$$

2. Existence of elliptic elements.

The analysis above shows that, irrespective of discreteness, $\langle g, h \rangle$ contains elliptic elements precisely when $\mathrm{tr}[g, h] < 18$. We shall assume now that this is so and proceed to characterize the set of values of $\mathrm{tr}[g, h]$ for which $\langle g, h \rangle$ is discrete. Although there are results of this type in the literature, we need a more explicit form, the details of which play an important role in the subsequent discussion. For this reason, we include a short proof of the next result.

THEOREM 3. *Suppose that g and h are parabolic with distinct fixed points and that $\mathrm{tr}[g, h] < 18$. Then $\langle g, h \rangle$ is discrete if and only if for some integer q $(3 \leq q \leq +\infty)$,*

$$\mathrm{tr}[g, h] = 2 + 16 \cos^4(\pi/q).$$

Moreover, if this holds, then either

   (i) *q is even and $\langle g, h \rangle$ has signature $(0; q/2, \infty, \infty)$ or*
   (ii) *q is odd and $\langle g, h \rangle$ has signature $(0; 2, q, \infty)$.*

PROOF: We take g and h in canonical form and the proof depends on a discussion of the hyperbolic polygon P bounded by the geodesics

$$x = -t/2, \quad x = t/2, \quad |sz + 1| = 1, \quad |sz - 1| = 1$$

where we assume that $s > 0$ and $t > 0$ (for we may replace either of g and h by their inverse as necessary).

The assumption that $\mathrm{tr}[g, h] < 18$ is equivalent to $st < 4$ and this, in turn, means

   (1) $x = t/2$ meets $|sz - 1| = 1$ at w_1 (in H) say;
   (2) $x = -t/2$ meets $|sz + 1| = 1$ at w_2

where the interior angles of P at the w_j are, say, θ [the reader is urged to draw a diagram].

An elementary calculation shows that

$$st = 2 + 2 \cos \theta$$

$$= 4 \cos^2(\theta/2)$$

so

$$\mathrm{tr}[g, h] = 2 + (st)^2$$

$$= 2 + 16 \cos^4(\theta/2).$$

First, we shall show that the possibilities in (i) and (ii) can occur. If θ is of the form π/r, then Poincaré's Theorem ensures that $\langle g, h \rangle$ is discrete and that P is a fundamental region for $\langle g, h \rangle$. In this case, $\langle g, h \rangle$ has signature $(0; r, \infty, \infty)$ and this is Theorem 3,(i) with $q = 2r$.

If θ is of the form $2\pi/r$ (where r is odd else the previous paragraph applies), we observe that $h(w_2) = w_1$ and, by considering g and h as compositions of reflections, $f(= gh^{-1})$ is a rotation of angle 2θ about w_1. As $2\theta = 4\pi/r$, and as r is odd, there is some power of f, say f^q, which is a rotation of angle $2\pi/r$ about w_1. It is easy to see that $f^q g$ interchanges w_1 and w_2 so $f^q g$ is of order two and fixes the mid-point of the hyperbolic segment $[w_1, w_2]$. This means that $\langle g, f^q g \rangle$ is the Hecke group with signature $(0; 2, r, \infty)$ and the same is true of $\langle g, h \rangle$ as

$$\langle g, h \rangle = \langle g, f \rangle = \langle g, f^q \rangle.$$

This is Theorem 3,(ii) with $q = r$.

The remainder of the proof consists in showing that if $\langle g, h \rangle$ is discrete, then we are in one of these two cases. Assume, then, that $\langle g, h \rangle$ is discrete: the angle θ in P is then of the form $k\pi/r$ where k and r are coprime positive integers and we want to show that $k = 1$ or 2.

We assume that $k \geq 2$. As before, $\langle g, h \rangle$ contains rotations f_1 and f_2, each of order r, which fix w_1 and w_2 respectively. The isometric circle for f_1 meets the line $x = t/2$ at an angle π/r at the point w_1 and similarly for f_2. As $k \geq 2$ and $\theta = k\pi/r$, we see that these isometric circles contain the hyperbolic segment $[w_1, w_2]$ (which must bisect each angle at w_1 and w_2). Thus the triangle bounded by $x = t/2$, $x = -t/2$ and $[w_1, w_2]$ contains a fundamental region F for $\langle g, h \rangle$ and we deduce that

$$\text{area}(F) \leq \pi - (0 + \theta/2 + \theta/2)$$

$$= \pi(1 - k/r).$$

If G is any Fuchsian group with parabolic elements and with a fundamental domain F of area at most π, then, by writing down the area of F in terms of the signature of G, we find that G has signature, say, $(0; u, v, \infty)$. Applying this to $\langle g, h \rangle$, we find that

$$2(1 - u^{-1} - v^{-1}) \leq 1 - k/r.$$

Now f_1 is of order r so r divides u, say: hence $u \geq r$ and $v \geq 2$. This yields

$$2(1 - 1/r - 1/2) \leq 2(1 - 1/u - 1/v) \leq 1 - k/r$$

which implies that $k \leq 2$. The proof is now complete.

3. The proof of Theorem 1.

We are assuming that both groups

$$\langle g, h \rangle, \quad \langle g^a, h^b \rangle$$

are discrete and of the first kind. We know that to within joint conjugacy, we may replace the pair a, b by $\alpha, 1$ where $\alpha = ab$ and, as a and b are rational, so is α. If $\alpha > 1$ we may write $f = g^\alpha$ (so $g = f^{1/\alpha}$) and so we may further confine our attention to the case $0 < \alpha < 1$. Thus we assume that

$$\alpha = m/n, \quad (m, n) = 1, \quad 1 \leq m < n.$$

With these assumptions, Theorem 3 and (1.2) imply that for some q (with $3 \leq q \leq +\infty$),

$$\mathrm{tr}[g, h] - 2 = 16 \cos^4(\pi/q)$$

(where we allow $q = +\infty$ corresponding to $\mathrm{tr}[g, h] = 18$) and for some p (with $3 < p < +\infty$),

$$\alpha^2 (\mathrm{tr}[g, h] - 2) = \mathrm{tr}[g^\alpha, h] - 2$$

$$= 16 \cos^4(\pi/p).$$

From this we obtain

$$m \cos^2(\pi/q) = n \cos^2(\pi/p)$$

and we seek all integral solutions of this equation in the restricted integer variables m, n, p and q. The solutions are known, [1] p. 238, and are as follows.

CASE 1: $\cos^2(\pi/4) = 2\cos^2(\pi/3) = 1/2$.
Here, $n = 2$, $m = 1$, $p = 3$, $q = 4$ and $\alpha = 1/2$. Using Theorem 1, we see that in this case, the two groups $\langle g, h \rangle$ and $\langle g^{1/2}, h \rangle$ have signatures $(0; 2, \infty, \infty)$, $(0; 2, 3, \infty)$ and are jointly conjugate to Γ_2 and Γ_1 respectively.

CASE 2: $\cos^2(\pi/6) = 3\cos^2(\pi/3) = 3/4$.

Here, $n = 3$, $m = 1$, $p = 3$, $q = 6$ and $\alpha = 1/3$. Using Theorem 1, we see that in this case, the two groups $\langle g, h \rangle$ and $\langle g^{1/3}, h \rangle$ have signatures $(0; 3, \infty, \infty)$, $(0; 2, 3, \infty)$ and are jointly conjugate to Γ_3 and Γ_1 respectively.

CASE 3: $2\cos^2(\pi/6) = 3\cos^3(\pi/4) = 3/2$.

Here $n = 3$, $m = 2$, $p = 4$, $q = 6$ and $\alpha = 2/3$. Using Theorem 1, we see that in this case, the two groups $\langle g, h \rangle$ and $\langle g^{2/3}, h \rangle$ have signatures $(0; 3, \infty, \infty)$, $(0; 2, \infty, \infty)$ and are jointly conjugate to Γ_3 and Γ_2 respectively.

CASE 4: $4\cos^2(\pi/3) = 1$.

Here $n = 4$, $m = 1$, $p = 3$, $q = +\infty$ and $\alpha = 1/4$. Using Theorem 1, we see that in this case, the two groups $\langle g, h \rangle$ and $\langle g^{1/4}, h \rangle$ have signatures $(0; \infty, \infty, \infty)$, $(0; 2, 3, \infty)$ and are jointly conjugate to Γ_4 and Γ_1 respectively.

CASE 5: $2\cos^2(\pi/4) = 1$.

Here $n = 2$, $m = 1$, $p = 4$, $q = +\infty$ and $\alpha = 1/2$. Using Theorem 1, we see that in this case, the two groups $\langle g, h \rangle$ and $\langle g^{1/2}, h \rangle$ have signatures $(0; \infty, \infty, \infty)$, $(0; 2, \infty, \infty)$ and are jointly conjugate to Γ_4 and Γ_2 respectively.

CASE 6: $4\cos^2(\pi/6) = 3$.

Here $n = 4$, $m = 3$, $p = 6$, $q = +\infty$ and $\alpha = 3/4$. Using Theorem 1, we see that in this case, the two groups $\langle g, h \rangle$ and $\langle g^{3/4}, h \rangle$ have signatures $(0; \infty, \infty, \infty)$, $(0; 3, \infty, \infty)$ and are jointly conjugate to Γ_4 and Γ_3 respectively.

The proof of Theorem 1 is completed.

4. The proof of Theorem 2.

First, of all g and h we have (from (1.2) and (1.3))

$$|T(g, h) - 2| = (\mathrm{tr}[g, h] - 2)^{1/2}$$

$$= n(\mathrm{tr}[g^{1/n}, h] - 2)^{1/2}.$$

If G_n is of the second kind, then

$$\mathrm{tr}[g^{1/n}, h] > 18, \quad n \geq 2$$

and so (i) holds.

If G_n is of the first kind but G is of the second kind, then for some q, we have

$$16 < \ \mathrm{tr}[g,h] - 2$$

$$= n^2(\mathrm{tr}[g^{1/n}, h] - 2)$$

$$= 16n^2 \cos^4(\pi/q)$$

and so $n \geq 2$, $q \geq 3$ and

$$n \cos^2(\pi/q) > 1.$$

Subject to these constraints, $n \cos^2(\pi/q)$ has a minimum value of $5/4$ (when $n = 5$ and $q = 3$) so in (ii) we have

$$|T(g,h) - 2| = (\mathrm{tr}[g,h] - 2)^{1/2} \geq 5.$$

Finally, to verify (iii) it is only necessary to check the possibilities listed in Theorem 1: we omit the details.

5. The results in [2].

Discrete groups generated by two parabolic elements are discussed in [2] and, as a by-product of our results, we obtain more detailed information than is given (or needed) there. Specifically, we shall consider the Proposition (p. 306) and Lemmas 1 and 2 (p. 311) of [2] and it is apparent that these are not stated in the most appropriate form for a sharp result. The Proposition states that there is a (universal) positive constant ε such that if

$$4 < \ \mathrm{tr}^2(gh) < 4 + \varepsilon$$

then no group of the form $\langle g^{1/n}, h \rangle$, $n \geq 2$, is discrete. Theorem 2 shows that the best possible value of ε is 5 but, of course, Theorem 2 is a better result as we have chosen to state it in terms of $|T(g,h) - 2|$ rather than $\mathrm{tr}^2(gh)$. Similar methods can be used to analyse Lemma 1 but these are left to the interested reader.

Finally, we comment on Lemma 2, [2], namely that if

$$g(z) = z + 2, \quad h(z) = z/(-2z + 1)$$

(so $\langle g, h \rangle = \Gamma_4$) then the only discrete groups of the form $\langle g^{1/n}, h \rangle$ are with $n = 2$ and $n = 4$. We need only remark that the preceeding discussion contains a complete list of such situations and that Lemma 2 contains two of the four possibilities of this type (namely the Cases 4 and 5, the other two Cases being 1 and 2).

D.P.M.M.S., University of Cambridge, England

REFERENCES

1. Conway, J.H. and Jones, A.J., *Trigonometric diophantine equations*, Acta Arithmetica **30** (1976), 229–240.
2. Jorgensen, T., Marden, A., and Pommerenke, C., *Two examples of covering surfaces, Riemann surfaces and related topics*, in "Riemann Surfaces and Related Topics: Proceedings of the 1978 Stony Brook Conference," pp. 305–317, Annals of Maths Studies **97**, Princeton University Press, 1980.

On the existence of elliptics in subgroups of $PSL(2,\mathbb{R})$: A graphical picture

BY JANE GILMAN

Let F be a non-elementary subgroup of $PSL(2,\mathbb{R})$. The purpose of this paper is to elucidate the nature and the extent of elliptic elements in F.

If M is the matrix of an elliptic Möbius transformation, so that $|trM| < 2$, papers in the literature usually distinguish between M being of finite order or infinite order by asking whether or not M corresponds to a rotation by a rational multiple of π. This is equivalent to asking whether or not $\arccos((trM)/2)$ is a rational multiple of π. While this condition definitely differentiates between finite and infinite order elliptics, it is not clear that one could ever use this to determine in a finite number of steps that a given elliptic element is *not* of finite order. Unless one can produce an algorithm that tells in a finite number of steps that a given M with $|trM| < 2$ is *not* of finite order, any necessary and sufficient conditions for a non-elementary F to be discrete with torsion will be of limited value. I have not seen this question addressed anywhere in the literature.

This paper show how to partially avoid the issue. First, we produce a necessary and sufficient condition for F to have elliptic elements. This condition is an inequality and only involves the hyperbolic elements of F. We then display a graphical picture (Figure I) that gives some indication of the nature and extent of elliptics within the group. The picture explains why it is that in a two generator group elliptic elements occur "soon" (Corollary 2 and Remark 2). We obtain some partial results that begin to test for discreteness that avoid the algorithm problem when F has elliptics. Finally, as a by-product of our method we also give a simplification of Theorem 6.1 of [G].

Notation.

If g and h are hyperbolic fractional linear transformations with attracting fixed points v_g and v_h, respectively, and repelling fixed points w_g and w_h, respectively, let $C(g,h) = ((v_g - v_h)(w_g - w_h)/((v_g - w_h)(w_g - v_h)))$ be *the cross ratio of g and h*. Let $C^\delta = C^\delta(g,h) = C(g,h)$ when $0 < C(g,h) < 1$ and $C^\delta = 1/C(g,h)$ when $C(g,h) > 1$. Note that $C^\delta(g^{\pm 1}, h^{\pm 1}) = C^\delta(h,g)$.

Let $\tilde{R}$ and $\tilde{K}$ be the multipliers of g and h, respectively. Let K be chosen so that $K > 1$ and $K = \tilde{K}$ or $1/\tilde{K}$. Choose R similarly. Set

$$Q^2_{M,N}(R,K) = \left(\frac{\sqrt{R^M} + \sqrt{K^N}}{\sqrt{R^M K^N} + 1}\right)^2 \quad \text{and} \quad S^2_{M,N}(R,K) = \left(\frac{\sqrt{R^M} - \sqrt{K^N}}{\sqrt{R^M K^N} - 1}\right)^2.$$

Finally, let $Q^2 = Q^2_{1,1}(g,h)$ and $S^2 = S^2_{1,1}(g,h)$.

PROPOSITION. *Let F be a non-elementary subgroup of $PSL(2,\mathbb{R})$. Then F contains an elliptic element if and only if $\exists$ g and h hyperbolic with disjoint axes in F with $\langle g,h \rangle$ non-elementary and*

$$S^2(g,h) < C^\delta(g,h) < Q^2(g,h).$$

PROOF: If F contains an elliptic e, it is well-known that $e = gh$ where g and h are hyperbolic (see Lemma 6.2 of [G], for example). If the axes of g and h intersect, one can compute that $tr g^{\pm 1} h > 2$ to conclude that the axes of g and h must be disjoint. Alternately, one can compute that $tr[g,h] = tr[e,h]$ and that whenever e is elliptic, $tr[e,h] > 2$. Here $[\,,\,]$ denotes the commutator. Thus $tr[g,h] > 2$, which implies that the axes of g and h are disjoint. To complete the proof now apply Lemma 6.3 of [G].

Combining this with Theorem 6.1 of [G], we have actually shown:

COROLLARY 1. *Let F be a non-elementary subgroup of $PSL(2,\mathbb{R})$. Then F is disctete without elliptics if and only if for each pair g and h of hyperbolics with disjoint axes such that $G = \langle g,h \rangle$ is non-elementary either*

$$(i) \quad C^\delta(g,h) \geq Q^2(g,h) \qquad \text{or} \qquad (ii) \quad C^\delta(g,h) \leq S^2(g,h).$$

While the proposition shows that whether or not F contains elliptics is governed by when $S^2(g,h) < C^\delta(g,h) < Q^2(g,h)$ for some disjoint hyperbolics g and h, we want to see that the relation between C^δ, Q^2 and S^2 determines how many elliptics F contains and gives some information about their nature. To this end we study the Q^2 and S^2 functions more carefully. The following facts are easy to verify:

$$\lim_{N \to \infty} Q^2_{N,M}(R,K) = \frac{1}{K^M}$$

$$\lim_{N \to \infty} S^2_{N,M}(R,K) = \frac{1}{K^M}$$

$$\lim_{m \to \infty} Q^2_{N,M}(R,K) = \frac{1}{R^N}$$

$$\lim_{m \to \infty} S^2_{N,M}(R,K) = \frac{1}{R^N}$$

Fixing one variable or the other we see that $Q^2_{N,M}(R,K)$ is decreasing as a function of K or as a function of R.

As a function of R with K fixed, $S^2_{N,M}(R,K)$ is increasing for $R > K$ and decreasing for $R < K$.

As a function of K with R fixed, $S^2_{N,M}(R,K)$ is increasing for $R < K$ and decreasing for $R > K$.

If $R > K$, $S^2_{1,1} < \frac{1}{K}$.

If $K > R$, $S^2_{1,1} < \frac{1}{R}$.

Let N_0 be an integer satisfying $K^{N_0} < R < K^{N_0+1}$. If $R > K$, $N_0 \geq 0$, while if $R < K$, $N_0 \leq 0$.

The graphs of S^2, Q^2 are pictured in Figure I for the case $R > K$. The case $K > R$ is similar.

One can compute that $S^2_{1,1} < \frac{1}{R}$ if $K > R$ and $S^2_{1,1} < \frac{1}{K}$ if $K < R$. Also if $R > K$, then either $\frac{1}{R} < S^2_{1,1} < \frac{1}{K}$ or $S^2_{1,1} < \frac{1}{R}$. We thus obtain two possibilities in Figure I. The case $\frac{1}{R} < S^2_{1,1} < \frac{1}{K}$ is given by the dotted line.

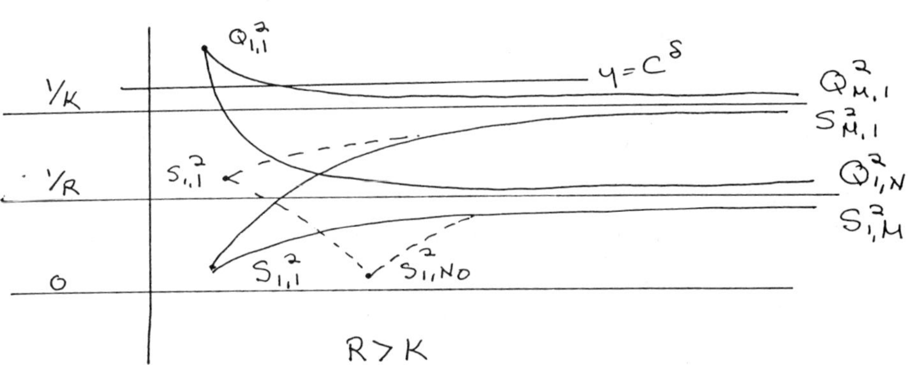

Figure I

The Proposition shows that where the line $y = C^\delta$ lies determines the existence of elliptics in the group. If $S^2_{M,N} < C^\delta(g,h) < Q^2_{M,N}$ then $g^{\varepsilon M} h^N$ is elliptic where $\varepsilon = +1$ or -1. The Proposition actually shows:

COROLLARY 2. *If* $S^2_{1,1} < C^\delta(g,h) < Q^2_{1,1}$ *then* $g^\varepsilon h$ *is elliptic where either* $\varepsilon = +1$ *or* -1. *Further if* T_0 *is the smallest integer such that* $C^\delta(g,h) \geq Q^2_{T_0,1}$ *or* $C^\delta(g,h) \leq S^2_{T_0,1}$, *then* $g^\varepsilon h, g^{2\varepsilon}h, \ldots, g^{\varepsilon T_0}h$ *are all elliptic elements.*

COROLLARY 3. *If* $T_0 > 3$, *then either* F *is not discrete or some elements in* $\{g^{\varepsilon i}h | i = 1, \ldots, T_0\}$ *are in the same conjugacy class of maximal cyclic subgroups of* F.

PROOF: If F is discrete, then Theorem 2 of [R-2] implies that $\langle g, h \rangle$ has one of four possible presentations when the axes of g and h are disjoint. These presentations show that $G = \langle g, h \rangle$ contains at most three conjugacy classes of maximal elliptic cyclic subgroups.

REMARK 1: One could in fact use Rosenberger's complete classification of all two generator groups [R-2] to give very explicit necessary and sufficient conditions for F to be discrete along the lines of Corollary 3. However, without a test for non-finite order of elliptics it is not clear that it would be worth stating these conditions. Since the conjugacy problem is solveable in three of the four presentations involved and since the fourth presentation forces F to be a triangle group, some statements can be made.

REMARK 2: This picture also gives another proof of the fact that if F is a non-elementary subgroup of $PSL(2, \mathbb{R})$, then for each pair of hyperbolics $\{u, v\}$ that generate a non-elementary subgroup of G, $\exists$ an integer N such that $\langle u^N, v^N \rangle$ is a discrete free group of rank 2. (See [R-3].) Namely, for g and h hyperbolic with disjoint axes, $\exists\, T$ (possibly $T = 1$) such that $C^\delta(g,h) > \left(\frac{1}{K}\right)^T$. Since $C^\delta(g,h) \geq \lim_{m \to \infty} Q^2_{M,T}(R, K)$, $\exists\, P > 0$ such that $C^\delta(g,h) \geq Q^2_{P,T}(R, K)$. Apply Remark 6.5 of [G] to see that $\langle g^P, h^T \rangle$ is a discrete free group of rank 2 and take $N = PT$. If the axes of g and h intersect, since f is a decreasing function of N, $f(C) \geq -4f(K^N)f(R)$ for some integer N (possibly $N = 1$). The proof of Theorem 6.1 in [G] or [P] or [R-1] then shows that $\langle u^N, v \rangle$ is a discrete free group of rank 2 as is then $\langle u^N, v^N \rangle$. (As in [G] f is the function $f(x) = x/(x-1)^2$.)

Math. Dept., Rutgers University, Newark, NJ 07102
and MSRI, Berkeley, CA 94720

References

[**G**] Gilman, J., *Inequalities and discrete subgroups of $PSL(2,\mathbb{R})$*, Canad. J., to appear.

[**P**] Purzitski, N., *Two generator discrete free products*, Math. Z. **126** (1972), 209–223.

[**R-1**] Rosenberger, G., *Fuchssche Gruppen, die freies Produkt zweier zyklischer Gruppen sind, und die Gleichung $x^2 + y^2 + z^2 = xyz$*, Math. Ann. **199** (1972), 213–228.

[**R-2**] Rosenberger, G., *All generating pairs of all two-generator Fuchsian groups*, Arch. Math. **46** (1986), 198–204.

[**R-3**] Rosenberger, G., *On discrete free subgroups of linear groups*, J. London Math. Soc. (2) **17** (1978), 79–85.

The kernel of the Poincaré series operator of weight -2

BY MAKOTO MASUMOTO

Introduction. Let Γ be a Fuchsian group keeping the unit disk Δ invariant. Assume that Γ is of convergence type, that is, $\sum_{\gamma \in \Gamma}(1-|\gamma(0)|) < +\infty$. Then, for any f in the Hardy space $H^1 = H^1(\Delta)$, the Poincaré series

$$\Theta_1 f(z) = \Theta f(z) = \sum_{\gamma \in \Gamma} f(\gamma(z))\gamma'(z)$$

of weight -2 converges absolutely and uniformly on compact subsets of Δ. If Γ is finitely generated and of the second kind, then Θf belongs to the Banach space $B(\Gamma)$ of bounded holomorphic automorphic forms of weight -2 on Δ for Γ (cf. [**5**; §**3**] and Section 1 below).

The purpose of this note is to show the following

THEOREM 1. *Let Γ be a finitely generated Fuchsian group of the second kind acting on the unit disk Δ. Denote by $E^2(\Gamma)$ the set of all $f \in H^2$ such that the function $zf(z)$ is the lift of an abelian integral on Δ/Γ. Then the kernel of the Poincaré series operator $\Theta : H^2 \to B(\Gamma)$ is precisely the orthogonal complement $E^2(\Gamma)^\perp$ of $E^2(\Gamma)$ in H^2.*

For finitely generated Fuchsian groups of the first kind, the kernels of the Poincaré series operators of weight $-2q$, $q = 2, 3, \ldots$, can be characterized in a similar manner in terms of Eichler integrals of order $1 - q$ (cf. [**3**]). Note that Eichler integrals of order 0 are just the lifts of abelian integrals, which will be simply called abelian integrals for Γ.

Theorem 1 will be proved in Section 2. As an application of the theorem we will determine the image $\Theta(H^2)$ in the final section.

The author is grateful to Professor F–Y. Maeda for giving helpful advice. He also thanks the referee for valuable suggestions.

1. Let $B(\Gamma)$ be the complex Banach space of holomorphic automorphic forms φ of weight -2 on Δ for Γ such that $\sup_{z \in \Delta} \lambda(z)^{-1}|\varphi(z)| < +\infty$, where $\lambda(z) = (1 - |z|^2)^{-1}$. Rajeswara Rao [**5**] showed that $\Theta f \in B(\Gamma)$ for $f \in H^1$ if Γ is a finitely generated Fuchsian group of the second kind with neither elliptic nor parabolic transformations. Improving his argument, we can show

---

This work was supported in part by the JMS (Educational Project for Japanese Mathematical Scientists) grant.

PROPOSITION 1. *If Γ is an arbitrary finitely generated Fuchsian group of the second kind, then $\Theta f \in B(\Gamma)$ for $f \in H^1$, and the linear operator $\Theta : H^1 \to B(\Gamma)$ is bounded.*

To prove this we need the following lemma due to Rajeswara Rao [5]. Though he proved it under the additional assumption that Γ contains no elliptic transformations, one can easily show that this assumption is unnecessary.

Let Ω denote the set of discontinuity of Γ. If Γ is finitely generated and of the second kind, then Ω is connected and there is a compact Riemann surface S such that $S - \Omega/\Gamma$ consists of at most a finite number of points.

LEMMA 1. *Let Γ be a Fuchsian group of convergence type. Remove the elliptic fixed points of Γ from Ω and denote the punctured open set by Ω_0. Then,*

(i) $\sup_{z \in \Delta} \sum_{\gamma \in \Gamma} (1 - |\gamma(z)|^2) < +\infty$,

(ii) $H(z) = \sum_{\gamma \in \Gamma - \{\mathrm{id}\}} \log \left| \frac{1 - \gamma(z)\bar{z}}{\gamma(z) - z} \right|$ *converges uniformly on compact subsets of Ω_0, and*

$$\sum_{\gamma \in \Gamma} |f(\gamma(z))||\gamma'(z)| \leqq 2\lambda(z) e^{4H(z)} \{1 + 2H(z)\} \|f\|_1$$

for $f \in H^1$ and $z \in \Delta \cap \Omega_0$, where $\| \cdot \|_1$ denotes the H^1-norm.

PROOF OF PROPOSITION 1: Let $f \in H^1$. By Lemma 1 (ii), the Poincaré series Θf converges absolutely and uniformly on compact subsets of Δ, and thus Θf is a holomorphic automorphic form of weight -2.

Since Γ is finitely generated, we can find a fundamental polygon $P(\subset \Delta)$ with finitely many sides. To prove $\Theta f \in B(\Gamma)$, we have only to show that $\lambda^{-1}|\Theta f|$ is bounded on P. By Lemma 1 (ii), it is clearly bounded near the free sides (= the sides on the unit circle $\partial \Delta$) of P. Let $a \in \partial P \cap \partial \Delta$ be a parabolic fixed point of Γ, and denote by Γ_a the stabilizer of $a : \Gamma_a = \{\gamma \in \Gamma | \gamma(a) = a\}$. Choose a Möbius transformation τ such that $\tau(\Delta) = U$, the upper half plane, and that $\tau \circ \Gamma_a \circ \tau^{-1}$ is the cyclic group $\langle T \rangle$ generated by the translation $T : Z \mapsto Z + 1$. Then Θf has a Fourier expansion of the form

$$(1) \qquad \Theta f(z) = \tau'(z) \sum_{\nu = -\infty}^{\infty} a_\nu(f) e^{2\pi i \nu \tau(z)}$$

on Δ. Now, choose a sequence $\{f_n\}$ in H^∞ that converges to f in H^1. Then $\{\Theta f_n\}$ converges to Θf uniformly on compact subsets of Δ by Lemma 1 (ii), and hence $\lim_{n\to\infty} a_\nu(f_n) = a_\nu(f)$. On the other hand Lemma 1 (i) implies $\Theta f_n \in B(\Gamma)$, and thus $a_\nu(f_n) = 0$ for all non-positive ν. Therefore $a_\nu(f) = 0$ for $\nu \leqq 0$, which means that the restriction of $\lambda^{-1}|\Theta f|$ to P is bounded near the parabolic cusp a. Since P has only finitely many sides, we have shown that $\lambda^{-1}|\Theta f|$ is bounded on P.

Finally, the closed graph theorem together with Lemma 1 (ii) implies the boundedness of the operator $\Theta : H^1 \to B(\Gamma)$. This completes the proof.

REMARK: If f is a rational function with poles in Ω, then Θf vanishes at every cusp (i.e., $a_\nu(f) = 0$ for $\nu \leqq 0$), which means that every puncture of Ω/Γ is a removable singularity of the abelian differential induced by Θf. This is an easy application of Lemma 1 (i). The above proof shows that for $f \in H^1$ the same result holds with Ω replaced by Δ. Note that Γ may be infinitely generated. On the other hand the function $H(z)$ defined in Lemma 1 is not bounded near the parabolic cusps as Rajeswara Rao pointed out.

2. Let $\hat{\mathbb{C}} = \mathbb{C} \cup \{\infty\}$ and $\Delta^* = \hat{\mathbb{C}} - \bar{\Delta}$. Set $z^* = 1/\bar{z}$ for $z \in \Delta$. As is well known, the Riemann surface Δ^*/Γ is the reflection of Δ/Γ. Let $\pi : \Omega \to \Omega/\Gamma$ be the natural projection. To each parabolic fixed point $a \in \partial\Delta$ there corresponds two punctures of Ω/Γ. One comes from the cusp with respect to Δ and the other from the cusp with respect to Δ^*. It is thus convenient to distinguish these cusps. We denote the cusp with respect to Δ by the same letter a, and the cusp with respect to Δ^* by a^*. Let $\tilde{\Omega}$ denote Ω plus the cusps with respect to Ω. The sets $\tilde{\Delta}$ and $\tilde{\Delta}^*$ are defined similarly. Then $\tilde{\Delta} \cap \tilde{\Delta}^* = \phi$ and $\tilde{\Omega} = \tilde{\Delta} \cup \tilde{\Delta}^* \cup (\Omega \cap \partial\Delta)$. The projection π is naturally extended to $\tilde{\Omega}$. Thus, for example, $\pi(a)$ and $\pi(a^*)$ are the punctures of Δ/Γ and Δ^*/Γ, respectively. Also the action of Γ is naturally extended to $\tilde{\Omega}$. Thus $\tilde{\Omega}/\Gamma = \pi(\tilde{\Omega})$. We have in general $\gamma(z^*) = \gamma(z)^*$ for $\gamma \in \Gamma$ and $z \in \tilde{\Delta}$.

Let $1 < p < \infty$ and $1/p + 1/p' = 1$. The conjugate space $(H^p)^*$ of the Hardy space H^p is identified anti-linearly with $H^{p'}$ through the pairing

$$\langle f, g \rangle = \frac{1}{2\pi} \int_0^{2\pi} f(e^{i\theta})\overline{g(e^{i\theta})}\,d\theta, \qquad f \in H^p, \; g \in H^{p'}.$$

Since $f \mapsto \Theta f(\varsigma)$ is a bounded linear functional on H^p for each fixed $\varsigma \in \Delta$, there is a unique $k^\Gamma(\cdot, \varsigma) = k(\cdot, \varsigma) \in H^{p'}$ such that $\Theta f(\varsigma) = \langle f, k(\cdot, \varsigma)\rangle$ for all $f \in H^p$. Setting $\psi_\nu(z) = z^\nu$, we have

$$(2) \qquad k(z, \varsigma) = \sum_{\nu=0}^{\infty} \overline{\Theta\psi_\nu(\varsigma)}\, z^\nu = \sum_{\nu=0}^{\infty} \Big(\sum_{\gamma\in\Gamma} \overline{\gamma(\varsigma)^\nu \gamma'(\varsigma)}\Big) z^\nu.$$

In particular k does not depend on p. Since the double series (2) converges absolutely and uniformly on compact subsets of $\Delta \times \Delta$, we may interchange the order of summation, and obtain

$$(3) \qquad k(z,\varsigma) = \sum_{\gamma \in \Gamma} \frac{\overline{\gamma'(\varsigma)}}{1 - \overline{\gamma(\varsigma)}z}.$$

It follows from (3) that $k(\cdot,\varsigma)$ is extended meromorphically to Ω for each fixed $\varsigma \in \Omega - \cup_{\gamma \in \Gamma}\{\gamma(\infty)\}$ and that $k(z,\cdot)$ is extended anti-meromorphically to Ω for each fixed $z \in \Omega$.

Set

$$k_\ell(z,\varsigma) = \frac{\partial^{\ell-1}k}{\partial \bar{\varsigma}^{\ell-1}}(z,\varsigma) = \sum_{\nu=0}^{\infty} \overline{(\Theta\psi_\nu)^{(\ell-1)}(\varsigma)}z^\nu$$

for $\ell \in \mathsf{N} = \{1,2,3,\dots\}$. Since $f \mapsto (\Theta f)^{(\ell-1)}(\varsigma)$ also defines an element of $(H^p)^*$ for each $\varsigma \in \Delta$, we have $(\Theta f)^{(\ell-1)}(\varsigma) = \langle f, k_\ell(\cdot,\varsigma)\rangle$ for all $f \in H^p$. Note that $k_\ell(\cdot,\varsigma) \in H^{p'}$.

For each parabolic fixed point $a \in \tilde{\Delta} - \Delta$ we fix a Möbius transformation $\tau = \tau_a$ as in the proof of Proposition 1: $\tau(\Delta) = U$ and $\tau \circ \Gamma_a \circ \tau^{-1} = \langle T \rangle$. Then Θf has a Fourier expansion of the form (1) with $a_\nu(f) = 0$ for $\nu \le 0$, and again $f \mapsto a_\ell(f)$ is continuous on H^p. Thus there is a unique $k_\ell(\cdot,a) \in H^{p'}$ such that $a_\ell(f) = \langle f, k_\ell(\cdot,a)\rangle$ for all $f \in H^p$. It is easy to see that

$$k_\ell(z,a) = \int_{\tau^{-1}(C_M)} e^{2\pi i \ell \overline{\tau(\varsigma)}} k(z,\varsigma)\,d\bar{\varsigma},$$

where $M > 0$ and C_M denotes the oriented segment $iM + t$, $0 \le t \le 1$.

LEMMA 2. *Let $E^p(\Gamma)$ be the set of $f \in H^p$ such that $\psi_1 f$ is an abelian integral for Γ. Then $k_\ell(\cdot,\varsigma) \in E^p(\Gamma)$ for $\varsigma \in \tilde{\Delta}$, $\ell \in \mathsf{N}$ and $1 < p < \infty$.*

REMARK: Lemma 2 implies that $E^2(\Gamma)^\perp$ is contained in the kernel of $\Theta : H^2 \to B(\Gamma)$. Therefore, to prove Theorem 1 it is sufficient to show that $E^2(\Gamma) \cap \mathrm{Ker}\,\Theta = \{0\}$. We shall prove a more general result in Theorem 2 below.

PROOF OF LEMMA 2: It is sufficient to show that for $\varsigma \in \Delta$ the function $z \mapsto zk(z,\varsigma)$ is an abelian integral for Γ. It is well known and in fact easy to verify that $L_\varsigma(z) = \sum_{\gamma \in \Gamma} \gamma'(\varsigma)/(z - \gamma(\varsigma))$ is an abelian integral on Ω. Since $zk(z,\varsigma) = \overline{L_\varsigma(z^*)}$, the proof is complete.

Set $K_\ell^\Gamma(z,\varsigma) = K_\ell(z,\varsigma) = zk_\ell(z,\varsigma)$. We see from the above proof that $K_\ell(\cdot,\varsigma)$ is a meromorphic abelian integral on Ω.

We need to consider the behavior of abelian integrals at cusps. Let F be a meromorphic abelian integral on Δ. Let $a \in \tilde{\Delta} - \Delta$. If F is holomorphic in the horodisk $\{\operatorname{Im} \tau_a(z) > M\}$ for sufficiently large $M > 0$, then the representation

$$(4) \qquad F(z) = \sum_{\nu=-\infty}^{\infty} A_\nu \exp(2\pi i \nu \tau_a(z)) + B\tau_a(z)$$

is valid on the horodisk. If $B = 0$ and $A_\nu = 0$ for $\nu < 0$, then F is said to be holomorphic at a. If $B = 0$ and $-\infty < -r = \inf\{\nu | A_\nu \neq 0\} < 0$, then F is said to have a pole of order r at a. The behavior of abelian integrals at $a^* \in \tilde{\Delta}^* - \Delta^*$ is defined similarly by using $-\tau_a$ instead of τ_a. If an abelian integral is holomorphic at every point of a subset $E \subset \tilde{\Omega}$, then it is said to be holomorphic on E.

The next lemma is fundamental.

LEMMA 3. *Abelian integrals in the Hardy space H^1 are holomorphic on $\tilde{\Delta}$.*

PROOF: Let $a \in \tilde{\Delta} - \Delta$. An abelian integral $F \in H^1$ has a Fourier expansion of the form (4). Then $F_0 = F - B\tau_a$ induces a holomorphic function $\tilde{F}_0$ on Δ/Γ_a since $F_0 \circ \gamma = F_0$ for $\gamma \in \Gamma_a$. Identify Δ/Γ_a with the punctured disk $\Delta - \{0\}$. Since F_0 is an $H^{1/2}$-function, so is $\tilde{F}_0$ and thus the origin 0 is a removable singularity of $\tilde{F}_0$. Hence $A_\nu = 0$ for $\nu < 0$. Now $F(z) = o((1 - |z|)^{-1})$ as $|z| \to 1$ by a theorem of Hardy–Littlewood (cf. Duren [1; **Theorem 5.9**]) and so $B = 0$. Hence we have the lemma.

Our next step is to determine the behavior of the abelian integrals $K_\ell(\cdot, \varsigma)$ on $\tilde{\Omega}$. We first consider the simplest case.

EXAMPLE: Here we assume that Γ is generated by a single parabolic transformation with fixed point $a \in \partial\Delta$. Recall that $K_1(\cdot, \varsigma)$ is meromorphic on Ω for each fixed $\varsigma \in \Omega' = \Omega - \cup_{\gamma \in \Gamma}\{\gamma(\infty)\}$, and that $K_1(z, \cdot)$ is anti-meromorphic on Ω for each fixed $z \in \Omega$. It follows from Lemma 3 that $K_1(\cdot, \varsigma)$ is holomorphic on $\tilde{\Delta}$ if $\varsigma \in \Delta$. Therefore

$$(5) \qquad K_1(\gamma(z), \varsigma) = K_1(z, \varsigma) \text{ for } \gamma \in \Gamma, \ z \in \Omega \text{ and } \varsigma \in \Omega'.$$

Let $\tau = \tau_a$. By (5) $K_1(\cdot, \varsigma)$ is expanded in the form

$$K_1(z, \varsigma) = \sum_{\nu=-\infty}^{\infty} A_\nu(\varsigma) e^{-2\pi i \nu \tau(z)}$$

on $\{\mathrm{Im}(-\tau(z)) > M\}$ for sufficiently large $M = M(\varsigma) > 0$. We shall show that $A_\nu(\varsigma) \equiv 0$ for $\nu < 0$. For $\varsigma \in \Delta^* \cap \Omega'$ and $z \in \Delta$ we have

$$K_1(\frac{1}{z}, \varsigma) = \frac{1}{z} \sum_{\gamma \in \Gamma} \frac{\overline{\gamma'(\varsigma)}}{1 - \overline{\gamma(\varsigma)}/z} = -\bar{\varsigma}^{-2} \sum_{\nu=0}^{\infty} \Theta(\frac{\psi_\nu}{\psi_1})(\frac{1}{\bar{\varsigma}}) z^\nu.$$

The last series represents an H^2-function of $z \in \Delta$ since $f \mapsto \Theta(f/\psi_1)(1/\bar{\varsigma})$ is a bounded linear functional on H^2 by Lemma 1 (ii). Thus $K_1(\cdot, \varsigma) \in H^2(\Delta^*)$, and hence $A_\nu(\varsigma) = 0$ for $\nu < 0$ and $\varsigma \in \Delta^* \cap \Omega'$ by Lemma 3. But each A_ν is an anti-holomorphic function of $\varsigma \in \Omega'$. Since Ω' is connected, $A_\nu(\varsigma) \equiv 0$ for $\nu < 0$ by the uniqueness theorem. We have proved that $K_1(\cdot, \varsigma)$ is holomorphic on $\tilde{\Omega} - \cup_{\gamma \in \Gamma}\{\gamma(1/\bar{\varsigma})\}$. It is trivial that $\gamma(1/\bar{\varsigma})$, $\gamma \in \Gamma$, are simple poles of $K_1(\cdot, \varsigma)$.

We shall next determine the behavior of $K_\ell(\cdot, a)$. We may identify Ω/Γ with the punctured plane $\mathbb{C} - \{0\}$, and assume that $\pi(a) = 0$, $\pi(a^*) = \infty$ and $\pi(\Omega \cap \partial\Delta) = \partial\Delta$; explicitly, $\pi(z) = e^{2\pi i \tau(z)}$. Define $\tilde{K}$ by $\tilde{K}(w, \omega)d\bar{\omega} = K_1(z, \varsigma)d\bar{\varsigma}$, $w = \pi(z)$, $\omega = \pi(\varsigma)$. Then there is $c \in \mathbb{C} - \{0\}$ such that

$$\tilde{K}(w, \omega) = c\frac{w - \alpha}{(\bar{\omega}w - 1)(\bar{\omega} - \bar{\beta})},$$

where $\alpha = \pi(0)$ and $\beta = \pi(\infty) = 1/\bar{\alpha}$. Hence

$$K_\ell(z, a) = \int_{|\omega| = \varepsilon} \bar{\omega}^{-\ell}\tilde{K}(w, \omega)d\bar{\omega} \qquad (0 < \varepsilon < \min\{\frac{1}{|w|}, 1\})$$
$$= 2\pi i c\alpha(\alpha^\ell - w^\ell).$$

Therefore $K_\ell(\cdot, a)$ has a pole of order ℓ at a^*, and is holomorphic elsewhere.

Now we return to the general case. Using the above example, we prove the following

LEMMA 4. *Let $a \in \tilde{\Delta}$ and $\ell \in \mathbb{N}$. In the case $a \in \Delta$ assume further that $\ell \equiv 0 \pmod{\mu}$, where $\mu = \mathrm{ord}\,\Gamma_a$. Then the abelian integral $K_\ell(\cdot, a)$ has poles of order ℓ at $\gamma(a^*)$, $\gamma \in \Gamma$, and is holomorphic on $\tilde{\Omega} - \cup_{\gamma \in \Gamma}\{\gamma(a^*)\}$.*

PROOF: (i) First let $a \in \Delta$. Assuming $a \neq 0$, we show that $K_\ell(\cdot, a)$ has poles of order ℓ at $\gamma(a^*)$, $\gamma \in \Gamma$. The case $a = 0$ can be treated similarly.

Now, as $z \to a^*$, we have

$$K_\ell(z,a) = z \sum_{\gamma \in \Gamma_a} \left(\frac{\partial}{\partial \bar{\varsigma}}\right)^{\ell-1} \left(\frac{\overline{\gamma'(\varsigma)}}{1 - \overline{\gamma(\varsigma)}z}\right)\bigg|_{\varsigma=a} + O(1)$$

$$= \frac{(\ell-1)! z^\ell}{(1 - \bar{a}z)^\ell} \sum_{\gamma \in \Gamma_a} \overline{\gamma'(a)}^\ell + O((z - a^*)^{-\ell+1})$$

$$= \frac{(\ell-1)! \mu z^\ell}{(1 - \bar{a}z)^\ell} + O((z - a^*)^{-\ell+1})$$

since $\ell \equiv 0 \pmod{\mu}$. Thus $K_\ell(\cdot,a)$ has a pole of order ℓ at a^* and hence at $\gamma(a^*)$ for each $\gamma \in \Gamma$.

By Lemma 3 $K_\ell(\cdot,a)$ is holomorphic at each cusp with respect to Δ. Let $b^* \in \tilde{\Delta}^* - \Delta^*$, and let $\Gamma = \cup_j \Gamma_b \circ \gamma_j$ be the right coset decomposition of Γ with respect to Γ_b. Then, since

$$K_1(z,\varsigma) = \sum_j K_1^{\Gamma_b}(z,\gamma_j(\varsigma))\overline{\gamma_j'(\varsigma)}$$

and for each fixed ς the series converges uniformly on compact subsets of $\Omega - \cup_{\gamma \in \Gamma}\{\gamma(\varsigma^*)\}$, it follows from the example that $K_1(\cdot,\varsigma)$ is holomorphic at b^*. In other words the representation

$$(6) \qquad K_1(z,\varsigma) = \sum_{\nu=0}^{\infty} A_\nu(\varsigma) \exp(-2\pi i \nu \tau_b(z))$$

is valid on $\{\mathrm{Im}(-\tau_b(z)) > M\}$ for sufficiently large $M > 0$. Observe that $A_\nu(\varsigma)$ are anti-holomorphic functions of ς. Hence

$$K_\ell(z,a) = \sum_{\nu=0}^{\infty} \frac{\partial^{\ell-1} A_\nu}{\partial \bar{\varsigma}^{\ell-1}}(a) \exp(-2\pi i \nu \tau_b(z)),$$

which shows that $K_\ell(\cdot,a)$ is holomorphic at b^*.

(ii) Next, let $a \in \tilde{\Delta} - \Delta$. Clearly $K_\ell(\cdot,a)$ is holomorphic on $\tilde{\Omega} - (\tilde{\Delta}^* - \Delta^*)$. Let $\Gamma = \Gamma_a \cup (\cup_j \Gamma_a \circ \eta_j)$ be the right coset decomposition of Γ with respect to Γ_a. Then,

$$K_\ell(z,a) = K_\ell^{\Gamma_a}(z,a) + \sum_j \int_{\eta_j \circ \tau_a^{-1}(C_M)} K_1^{\Gamma_a}(z,\varsigma) \exp(2\pi i \ell \overline{\tau_a \circ \eta_j^{-1}(\varsigma)})d\bar{\varsigma}.$$

Since each term of the series, which is an abelian integral for Γ_a, is holomorphic at a^*, we conclude from the example that $K_\ell(\cdot,a)$ has a pole of order ℓ at a^*. Finally, let $b^* \in \tilde{\Delta}^* - (\Delta^* \cup (\cup_{\gamma \in \Gamma}\{\gamma(a^*)\}))$. Then for each negative integer ν

$$\int_{C_M} e^{-2\pi i \nu Z} K_\ell((-\tau_b)^{-1}(Z), a)\,dZ$$
$$= \int_{\tau_a^{-1}(C_{M'})} \exp(2\pi i \ell \overline{\tau_a(\varsigma)}) \int_{C_M} e^{-2\pi i \nu Z} K_1((-\tau_b)^{-1}(Z), \varsigma)\,dZ\,d\bar{\varsigma}$$
$$= 0$$

by (6). Thus $K_\ell(\cdot, a)$ is holomorphic at b^*. This completes the proof.

REMARK: If $a \in \Delta$ and $\ell \not\equiv 0 \pmod{\mu}$, then $K_\ell(\cdot, a) = 0$ on Δ.

Let $Z^1(\Gamma, \mathbb{C})$ denote the set of group homomorphisms of Γ into the additive group $\mathbb{C}$. It naturally forms a vector space over $\mathbb{C}$. We need the subspace $PZ^1(\Gamma, \mathbb{C})$ consisting of all $\chi \in Z^1(\Gamma, \mathbb{C})$ such that $\chi(\gamma) = 0$ for all parabolic $\gamma \in \Gamma$. If Γ is finitely generated and of the second kind, then $\dim PZ^1(\Gamma, \mathbb{C}) = g_\Gamma$, the genus of Ω/Γ.

If F is an abelian integral for Γ, then $\gamma \mapsto F \circ \gamma - F$ is an element of $Z^1(\Gamma, \mathbb{C})$, which will be denoted by $pd\,F$ and called the period of F. Note that $pd\,K_\ell(\cdot, a) \in PZ^1(\Gamma, \mathbb{C})$ for $a \in \tilde{\Delta}$ by Lemma 4.

Following Earle–Marden [2], we consider the automorphic form

$$\Phi(z) = \Theta \psi_{-1}(z) = \sum_{\gamma \in \Gamma} \frac{\gamma'(z)}{\gamma(z)}, \quad z \in \Omega.$$

Let $\tilde{\Phi}$ denote the abelian differential on $\tilde{\Omega}/\Gamma$ induced by Φ. The restriction of the real part of $-\tilde{\Phi}$ to Δ/Γ is just the differential of Green's function of Δ/Γ with pole at $\pi(0)$. Thus the singularities of $\tilde{\Phi}$ are simple poles at $\pi(0)$ and $\pi(\infty)$. The zeros of $\tilde{\Phi}$ are located symmetrically with respect to $\pi(\Omega \cap \partial\Delta)$. If Γ is finitely generated and of the second kind, then $\tilde{\Phi}$ has $2g_\Gamma$ zeros in $\tilde{\Omega}/\Gamma$, and half of them are in $\tilde{\Delta}/\Gamma$.

We make use of the next lemma due to Heins. A proof of the lemma can be found in [2].

LEMMA 5. *Let Γ be a finitely generated Fuchsian group of the second kind. Then the zero function is the only meromorphic function $\tilde{F}$ on $\tilde{\Omega}/\Gamma$ such that (i) $\tilde{F}$ is holomorphic in $\tilde{\Delta}/\Gamma$, and (ii) $\tilde{F}\tilde{\Phi}$ is a holomorphic differential on $\tilde{\Omega}/\Gamma$.*

LEMMA 6. *Suppose that Γ is finitely generated and of the second kind. Let $P_1, \ldots, P_s \in \tilde{\Delta}/\Gamma$ be the collection of the zeros of $\tilde{\Phi}$ located in $\tilde{\Delta}/\Gamma$,* and set $\lambda_j = \mathrm{ord}_{P_j} \tilde{\Phi}$ (and hence $\sum_{j=1}^{s} \lambda_j = g_\Gamma$). Choose $a_j \in \pi^{-1}(P_j)$, $j = 1, \ldots, s$, and let V be the vector space generated by

$$K_{\ell\mu_j}(\cdot, a_j), \qquad \ell = 1, \ldots, \lambda_j; \; j = 1, \ldots, s,$$

where $\mu_j = \mathrm{ord}\, \Gamma_{a_j}$ if $a_j \in \Delta$ and $\mu_j = 1$ otherwise. Then the period mapping pd gives an isomorphism between V and $PZ^1(\Gamma, \mathbb{C})$.

PROOF: Since $\dim V = \dim PZ^1(\Gamma, \mathbb{C}) = g_\Gamma$ by Lemma 4, it suffices to show that the period mapping is injective. Suppose that $pd\, K = 0$ for some $K \in V$. Then K induces a meromorphic function $\tilde{K}$ on $\tilde{\Omega}/\Gamma$ which has a zero at $\pi(0)$ and whose only sigularities are poles of order at most λ_j at $\pi(a_j^*)$. Thus the possible singularity of the abelian differential $\tilde{K}\tilde{\Phi}$ is a simple pole at $\pi(\infty)$. Since $\tilde{\Omega}/\Gamma$ is compact, the residue of $\tilde{K}\tilde{\Phi}$ at $\pi(\infty)$ must be zero, and thus $\tilde{K}\tilde{\Phi}$ is holomorphic on $\tilde{\Omega}/\Gamma$. Therefore, by Lemma 5, we have $K \equiv 0$, as desired.

THEOREM 2. Let Γ be a finitely generated Fuchsian group of the second kind acting on the unit disk Δ. Set $N^p(\Gamma) = \{f \in H^p | \Theta f \equiv 0\}$. Then $E^p(\Gamma) \cap N^p(\Gamma) = \{0\}$ for $1 < p < \infty$.

PROOF: Let $f \in E^p(\Gamma) \cap N^p(\Gamma)$. By Lemma 6 there is a unique $K \in V$ such that $pd\, (\psi_1 f) = pd\, K$. Set $k = K/\psi_1$, and observe that k is a linear combination of $k_{\ell\mu_j}(\cdot, a_j)$, $1 \leqq \ell \leqq \lambda_j$, $1 \leqq j \leqq s : k = \sum_{j=1}^{s} \sum_{\ell=1}^{\lambda_j} c_{j\ell} k_{\ell\mu_j}(\cdot, a_j)$. Then, since

$$\Theta k = \Theta(k - f) = \Theta\left(\frac{\psi_1(k - f)}{\psi_1}\right) = \psi_1(k - f)\Phi,$$

the abelian differential on $\tilde{\Delta}/\Gamma$ induced by Θk has zeros of order at least λ_j at $P_j = \pi(a_j)$, $j = 1, \ldots, s$. Hence, we have

$$0 = \langle k, k_{m\mu_i}(\cdot, a_i) \rangle = \sum_{j=1}^{s} \sum_{\ell=1}^{\lambda_j} c_{j\ell} \langle k_{\ell\mu_j}(\cdot, a_j), k_{m\mu_i}(\cdot, a_i) \rangle$$

for $1 \leqq m \leqq \lambda_i$ and $1 \leqq i \leqq s$. This may be viewed as a system of g_Γ linear equations in g_Γ unknowns $c_{j\ell}$. The determinant of the coeficient matrix of this system, which is just Gram's determinant formed by $k_{\ell\mu_j}(\cdot, a_j)$, is not zero since $k_{\ell\mu_j}(\cdot, a_j)$, $1 \leq \ell \leq \lambda_j$, $1 \leq j \leq s$, are linearly independent. Thus $c_{j\ell} = 0$ for all j and ℓ, and hence $f = k - (\Theta k)/\psi_1\Phi = 0$. We have proved the theorem.

As a corollary to this theorem, we obtain Theorem 1, which may be stated as

COROLLARY 1. *If Γ is finitely generated Fuchsian group of the second kind, then we have the orthogonal decomposition*

$$H^2 = E^2(\Gamma) \oplus N^2(\Gamma).$$

COROLLARY 2. *If Γ is finitely generated and of the second kind, then $N^2(\Gamma)$ is the smallest closed subspace of H^2 containing $(\psi_\nu \circ \gamma) \cdot \gamma' - \psi_\nu$, $\gamma \in \Gamma$, $\nu = 0, 1, 2, \ldots$.*

One can prove Corollary 2 by the same method as in the proof of [3; **Corollary to Proposition 5**]. A related result for Poincaré series of weight less than -2 was obtained by Metzger [4].

3. In this section we determine the image $\Theta(H^2)$. First we remark the following

LEMMA 7 (EARLE–MARDEN [2]). *Let $W(\Gamma)$ be the space of the lifts of holomorphic abelian differentials on $\tilde{\Omega}/\Gamma$. Assume that Γ is finitely generated and of the second kind. Then $W(\Gamma) \subset \Theta(H^\infty)$, i.e., for each $\varphi \in W(\Gamma)$ there is $f \in H^\infty$ such that $\varphi = \Theta f$ on Δ.*

Though Earle and Marden proved Lemma 7 under the additional assumption that Γ contains neither elliptic nor parabolic transformations, it is easy to suppress this assumption in their argument.

THEOREM 3. *Let $H_0^2(\Gamma)$ be the space of all $f \in H^2$ that vanish at the origin and are automorphic under Γ, that is, $f \circ \gamma = f$ for all $\gamma \in \Gamma$. Assume that Γ is finitely generated and of the second kind. Then $\Theta(H^2)$ is identical with the direct sum of $\Phi \cdot H_0^2(\Gamma)$ and $W(\Gamma)$.*

PROOF: It follows from Lemma 7 and Theorem 1 that $W(\Gamma) \subset \Theta(H^2) = \Theta(E^2(\Gamma))$ and that the restriction of Θ to $E^2(\Gamma)$ is injective. Set $W_0 = \{f \in E^2(\Gamma) | \Theta f \in W(\Gamma)\}$. Suppose that $f \in W_0$ and $pd\,(\psi_1 f) = 0$. Then $\psi_1 f = (\Theta f)/\Phi$ induces a meromorphic function $\tilde{F}$ on $\tilde{\Omega}/\Gamma$ which satisfies the conditions (i) and (ii) in Lemma 5. Hence $f = 0$. This proves that $pd : \psi_1 \cdot W_0 \to PZ^1(\Gamma, \mathbb{C})$ is injective. Since $\dim(\psi_1 \cdot W_0) = \dim PZ^1(\Gamma, \mathbb{C}) = g_\Gamma$, we obtain $pd\,(\psi_1 \cdot W_0) = PZ^1(\Gamma, \mathbb{C})$ and thus $E^2(\Gamma) = \psi_{-1} \cdot H_0^2(\Gamma) + W_0$. Therefore,

$$\Theta(H^2) = \Theta(E^2(\Gamma)) = \Theta(\psi_{-1} \cdot H_0^2(\Gamma)) + \Theta(W_0) = \Phi \cdot H_0^2(\Gamma) + W(\Gamma).$$

REMARK: Earle–Marden [2] showed that *on the unit circle* $\Theta(H^p) = \Phi \cdot H_0^p(\Gamma) + W(\Gamma)$ for $1 \leqq p \leqq \infty$, where $H_0^p(\Gamma)$ consists of all function in H^p that vanish at the origin and are automorphic under Γ.

M. Masumoto, Department of Mathematics, Hiroshima University, Hiroshima, 730, Japan

REFERENCES

1. Duren, P.L., "Theory of H^p Spaces," Academic Press, New York, 1970.
2. Earle, C.J. and Marden, A., *On Poincaré series with application to H^p spaces on bordered Riemann surfaces*, Illinois J. Math. **13** (1969), 202–219.
3. Masumoto, M., *A characterization of the kernel of the Poincaré series operator*, Trans. Amer. Math. Soc. **300** (1987), 695–704.
4. Metzger, T.A., *The kernel of the Poincaré series operator*, Proc. Amer. Math. Soc. **76** (1979), 289–292.
5. Rajeswara Rao, K.V., *Fuchsian groups of convergence type and Poincaré series of dimension −2*, J. Math. Mech. **18** (1969), 629–644.

Strange actions of groups on spheres, II

BY MICHAEL H. FREEDMAN AND RICHARD SKORA

In [**FS**] we investigated certain topological analogs of Schottky groups, called admissible actions, and their compatibility with various structures on spheres. We constructed an action $\phi : F_2 \times S^3 \to S^3$ which was not topologically conjugate to a uniformly quasiconformal action. Also, there was an example $\psi : (F_r \rtimes \mathbb{Z}_2) \times S^3 \to S^3$, r sufficiently large, which was smooth and uniformly quasiconformal, but not topologically conjugate to a conformal action. And we gave examples of admissible actions on higher dimensional spheres, but did not analyse the structures preserved.

This paper continues the discussion in [**FS**] and familiarity with that paper will be helpful. Each action described here has the property that each homeomorphism is individually topologically conjugate to an element of the Möbius group, but the action (except possibly the action of §4) is not topologically conjugate to a conformal action.

In §1 we recall the necessary definitions. Also, we prove two lemmas. Lemma 1.2 gives sufficient conditions for an action to be uniformly quasiconformal. And Lemma 1.3 gives coordinates to S^n which are used in §5 to turn a particular topological action into a uniformly quasiconformal action.

Let $\varsigma : F_2 \times S^n \to S^n$, $n \geq 4$ be the admissible action described in [**FS**]. In §2 we prove ς is not conjugate to a uniformly quasiconformal action.

In §3 we construct an action $\mu : (\mathbb{Z}^2 \rtimes \mathbb{Z}) \times S^3 \to S^3$ with a single limit point and which is not conjugate to a uniformly quasiconformal action. The construction is based on a closed 3-manifold uniformized by the NIL geometry—a geometry which is homeomorphic to Euclidean space. Using 3-manifold connected sum, we turn the example into an admissible action which is not conjugate to a quasiconformal action.

An admissible action $\eta : F_r \times S^3 \to S^3$ which is uniformly quasiconformal is described in §4. The limit set of η is "more wildly" embedded in S^3 than the limit set of ψ in [**FS**]. This difference is described in detail.

Finally in §5 we construct an admissible action $\psi_n : (F_r \rtimes \mathbb{Z}_{2r}) \times S^n \to S^n$, $n \geq 4$ and r sufficiently large and depending on n, which is smooth and

Research of the first author was supported in part by DARPA Grant No. 86-A227500 and research of the second author was supported in part by NSF Grant DMS-86-01037.

uniformly quasiconformal but not conjugate to a conformal action. As in [**FS**] we distinguish it from a conformal action by examining a link L which arises as a union of fixed sets—L is a non-trivial link of $(n-2)$—spheres such that any pair of components forms the unlink.

1. Definitions and Lemmas.

The free group of rank r is denoted F_r. Given an action $\alpha : G \times S^n \to S^n$ the *set of discontinuity*, denoted Ω_α, is the collection of points of S^n which have neighborhoods N such that all but finitely many translates of N under the action are disjoint with N. The *limit set*, denoted Λ_α, is $S^n - \Omega_\alpha$.

1.1   DEFINITION: Let G be a finitely generated group and $\alpha : G \times S^n \to S^n$ an action. Then α is *admissible* if: (1) Λ_α is a cantor set, (2) α is properly discontinuous on Ω_α and (3) Ω_α/G is compact.

In [**GM**] it is shown that (1) and (2) imply that G is a convergence group.

An action $\alpha : G \times S^n \to S^n$ is *uniformly quasiconformal* if for some K each $g \in G$ is K-quasiconformal.

The following lemmas are used in §§4 and 5. The first lemma generalizes that ψ of [**FS**] is uniformly quasiconformal.

1.2   LEMMA (CRITERION FOR QUASICONFORMALITY). *Let $G \times S^n \to S^n$ be properly discontinuous on the domain of discontinuity. If there is some collection $\{g_i\}_i$ of generators of G, some neighborhood N of the limit set and some $K < +\infty$ such that each g_i is conformal on N and K-quasiconformal on S^n, then $G \times S^n \to S^n$ is uniformly quasiconformal.*

PROOF: Define the compact set $C = \overline{S^n - N}$. By hypothesis there is a β such that $C \cap f(C) \neq \phi$ for at most β choices of $f \in G$. We claim the action is K^β-quasiconformal.

Fix $f \in G$ and x in the domain of discontinuity. It suffices to show the dilitation of f at x is no greater than K^β. Let m be smallest such that $f = f_m \circ \cdots \circ f_1$ where $f_j \in \{g_i, g_i^{-1}\}_i$ and let x_i be defined inductively by $x_1 = x$ and $x_{i+1} = f_i(x_i)$. The choice of $f_m, \ldots, f_1$ implies that $x_i \in N$ for all but at most β choices of i, hence the dilitation of f_i at x_i is 1 for all but at most β choices of i. It follows that the dilitation of f at x is no greater than the product of the dilitations of f_i at x_i which is no greater than K^β. $\qquad\square$

In §5 we need the following coordinatization of S^n which generalizes the Steiner coordinatization of S^2. Another discussion of these coordinates may be read in [**TV**]. Define an equivalence relation on $B^p \times S^q$ by $(x,s) \sim (x,t)$, for all $x \in \partial B^p$ and $s,t \in S^q$. Then $B^p \times S^q / \sim$ is homeomorphic to S^n, where $n = p + q$. This gives topological coordinates to S^n; so given a homeomorphism $f : B^p \to B^p$ one may think of $(f \times \text{ identity})$ as a homeomorphism of S^n. Next we will show that in these coordinates maps of the form (conformal $\times$ identity) are conformal on S^n.

Let $\mathbb{R}^{2+n} = \mathbb{R}^{2+p} \times \mathbb{R}^q$ have form $\langle +1, -1, \ldots, -1 \rangle$ and define $S^n = \{x = (1, x_0, \ldots, x_n) \mid \langle x, x \rangle = 0\}$. Each $T \in S0(1, n+1)$ determines the an element of $\text{Möb}(S^n)$ given by $x \to T(x)/T_{-1}(x)$, where $T(x) = (T_{-1}, T_0, \ldots, T_n)(x)$. This gives an isomorphism $S0(1, n+1) \approx \text{Möb}(S^n)$. Define $B^p = \{x = (1, x_0, \ldots, x_p, 0, \ldots, 0) \mid \langle x, x \rangle = 0 \text{ and } 0 \leq x_0\}$ and $S^q = \{x = (1, x_0, 0, \ldots, 0, x_{p+1}, \ldots, x_n) \mid \langle x, x \rangle = 0\}$. So $B^p \cap S^1 = \text{point}$. Let $\text{Möb}(B^p) \subset \text{Möb}(S^n)$ and $S0(q+1) \subset \text{Möb}(S^n)$ be the obvious inclusions. Notice $S0(q+1)$ acts trivially on ∂B^p and $\text{Möb}(B^p) \times S0(q+1) \subset \text{Möb}(S^n)$. Thus we get a quotient map $\pi : B^p \times S^q \to S^n$ and the following is obvious.

1.3   LEMMA. *If $f \in \text{Möb}(B^p)$ and $\overline{f} \in \text{Möb}(S^n)$ its image, then*

$$\overline{f} \circ \pi = \pi \circ (f \times \text{ identity}).$$

□

2.   The action ς_n.

In [**FS**] we defined a topological action $\varsigma : F_2 \times S^n \to S^n$, any $n \geq 4$. Each was extension by rotation of an action $F_2 \times S^3 \to S^3$. Call these actions $\varsigma_n : F_2 \times S^n \to S^n$, $n \geq 3$. The methods of [**FS**], show that ς_3 is not conjugate to a uniformly quasiconformal action. The argument compares length and volume near the limit set. Here we explain how to generalize those arguments to show that no ς_n, $n \geq 4$ is conjugate to a uniformly quasiconformal action. Roughly we argue that for a uniformly quasiconformal action, $(n-2)$-area is comparable to $(\text{volume})^{(n-2)/n}$, but in our example area remains on the average constant while volume tends to zero.

Recall the fundamental domain for ς_n is $\mathcal{D} \approx S^n - \overset{\circ}{a}_1 - \overset{\circ}{a}_2 - \overset{\circ}{b}_1 - \overset{\circ}{b}_2$, where $a_1 \approx a_2 \approx S^{n-2} \times B^2$, $b_1 \approx b_2 \approx S^1 \times B^{n-1}$ and the a_i's are unknotted and unlinked and the S^1 factor of each b_i represents the commutator class of $\pi_1(S^n - a_1 - a_2)$. The action is generated by homeomorphisms $j_i : S^n \to S^n$ satisfying $j_i(S^n - \overset{\circ}{a}_i) = b_i$, $i = 1, 2$. See Figure 2.1.

Let A_0 be homeomorphic to $S^{n-2} \times B^2$. Define the sequence $A_0 \overset{\circ}{\supset} A_1 \overset{\circ}{\supset} A_2 \overset{\circ}{\supset} \ldots$ inductively by the following property. For any component T of A_i, $(T, T \cap A_{i+1})$ is homeomorphic to $(S^n - b_T, a_1 \cup a_2)$, where $S^1 \times B^{n-1} \approx b_T \subset S^n - a_1 - a_2$ and the S^1 factor represents the commutator class of $\pi_1(S^n - a_1 - a_2)$. Actually the topological type of $(S^n - \overset{\circ}{b_T}, a_1 \cup a_2)$ is independent of b_T. The proof is hardest in low dimensions and is the 1-handle straightening technique which carefully uses Quinn's controlled h-cobordism Theorem [Q]. This will not be used, however, in the following lemmas.

For any A homeomorphic to $S^{n-2} \times B^2$ with a Riemannian metric on $\overset{\circ}{A}$, a $(n-2)$-manifold $C \subset A$ is a *core* if C is smooth almost everywhere and it represents a generator of $H_{n-2}(A; \mathbb{Z})$. For $n \geq 6$ it is known that necessarily $\overset{\circ}{A}$ is diffeomorphic to $S^{n-2} \times B^2$, thus in these dimensions A has a core. More generally, if $f : \overset{\circ}{A} \to \overset{\circ}{B}^2$ is a smooth approximation to projection onto the second factor and is transverse to 0, then some component of $f^{-1}(0)$ is a core. The *area* of A is $\inf_{C}\{(n-2) - \text{area}(C)\}$, where C ranges over all cores of A.

2.1   LEMMA. *Let $A_0, A_1, \ldots$ as above. Also suppose A_0 is isometric to $S^{n-2} \times \overset{\circ}{B}^2$ with each $\overset{\circ}{A}_i$ inheriting the Riemannian metric. Then for each $k = 0, 1, \ldots$*

$$\text{area}(S^{n-2}) \leq \frac{1}{2^k} \sum_T \text{area}\,(T),$$

where T ranges over all components of A_k.

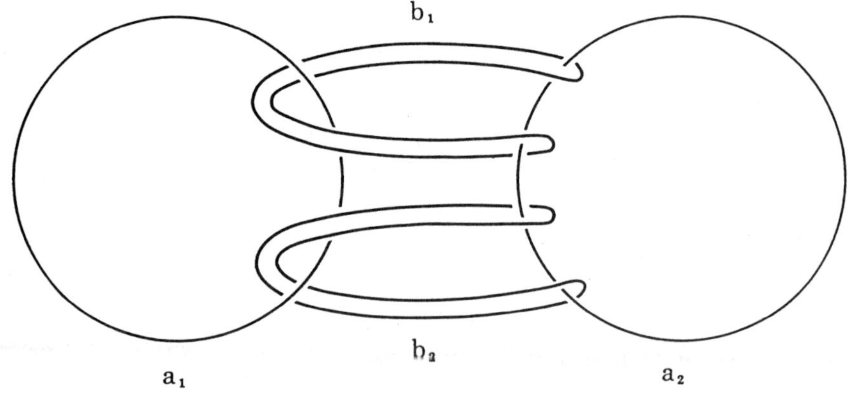

Figure 2.1

2.2   LEMMA. *Let (A, T_1, T_2) be homeomorphic to $(S^n - \overset{\circ}{b}_T, a_1, a_2)$, where $S^1 \times B^{n-1} \approx b_T \subset S^n - a_1 - a_2$ and the S^1 factor represents the commutator class, and let $(P, \partial P) \subset (A, \partial A)$ be an embedded, connected, planar surface with $[P]$ a generator of $H_2(A, \partial A; \mathbb{Z})$. If P is transverse to T_1 and T_2, then for $i = 1$ or 2, $P \cap T_i$ contains at least two planar surfaces each representing generators of $H_2(T_i, \partial T_i; \mathbb{Z})$.*

PROOF: Endow $\overset{\circ}{A}$ with a Riemannian metric and let C_i be a core of T_i, $i = 1, 2$. We will argue by contradiction. Suppose each component of $P \cap T_i$ is trivial in $H_2(T_i, \partial T_i; \mathbb{Z})$, for each i. Changing the C_i's by general positioning and by 0-surgeries, one obtains new cores C_i' of T_i, $i = 1, 2$, such that $P \cap (C_i' \cup C_2') = \phi$. Finally, by performing more 0-surgeries one obtains cores C_i'' of T, $i = 1, 2$, such that $P \cap (C_1'' \cup C_2'') = \phi$ and $\pi_1(T_1 - C_i'') \approx \mathbb{Z}$.

Let $\pi_1(T_1 - C_1'') = \langle x \rangle$, $\pi_1(T_2 - C_2'') = \langle y \rangle$ and $\pi_1(A - C_1'' - C_2'') = \pi$. Then $\pi \approx \mathbb{Z} * \mathbb{Z}$. Each component of ∂P represents a conjugate of $[x, y]$. The existence of P implies that some product of $k + 1$ conjugates of $[x, y]$ and k conjugates of $[x, y]^{-1}$ is trivial in π; which implies that $[x, y] = 0 \in \pi/[[\pi, \pi], \pi]$. This is, of course, false because $[x, y]$ generates the central $\mathbb{Z} \subset \pi/[[\pi, \pi], \pi]$.

Thus, for $i = 1$ or 2 some component of $P \cap T_i$, is a generator of $H_2(T_i, \partial T_i; \mathbb{Z})$. Homological considerations show that there are two such components. $\qquad\square$

2.3   LEMMA. *Fix k and cores $C_1, \ldots, C_{2^k}$ of the components of A_k. If $D^2 = D^2 \times pt \subset A_\circ$ and D^2 is transverse to $\cup_i C_i$, then $D^2 \cap \cup_i C_i$ contains at least 2^k points.*

PROOF: Change D^2 near ∂A_k to make D^2 transverse to A_k. Apply Lemma 2.2 k times to see that $D^2 \cap A_k$ contains at least 2^k planar surfaces each representing a generator of $H_2(A_k, \partial A_k; \mathbb{Z})$. By duality $D^2 \cap \cup_i C_i$ contains at least 2^k points. $\qquad\square$

For any function $f : X \to Y$, let $\natural_f : Y \to \mathbb{Z}^+ \cup \{\infty\}$ be defined by $\natural_f(y) = {}^\natural f^{-1}(y)$. The following is easy.

2.4   LEMMA. *If $f : M^m \to N^m$ be a smooth map between Riemannian manifolds, then $\mathrm{area}(f(M)) = \int_N \natural_f \, dA$.* $\qquad\square$

PROOF OF 2.1: Fix k and choose cores $C_1, \ldots, C_{2^k}$ for the components of A_k. Let $p : S^{n-2} \times B^2 \to S^{n-2}$ be projection. For almost all $y \in S^{n-2}$,

$p^{-1}(y)$ is transverse to $\cup_i C_i$. By Lemma 2.3

$$\text{area}(S^{n-2}) \leq \frac{1}{2^k} \sum_i \int_{S^{n-2}} \sharp_{p|C_i} \, dA.$$

By Lemma 2.4

$$\int_{S^{n-2}} \sharp_{p|C_i} \, dA \ = \ \text{area}\,(p(C_i)) \leq \int_{C_i} dA.$$

Combining inequalities and taking the infimum over all core families completes the proof. $\qquad\square$

2.5   THEOREM. *The action $\varsigma_n : F_2 \times S^n \to S^n$, $n \geq 4$, is admissible and not conjugate to a uniformly quasiconformal action.*

PROOF: Since ς_n is defined only up to topological conjugation, it suffices to show ς_n is not uniformly quasiconformal. The proof is by contradiction; suppose ς_n is K-quasiconformal.

Let $S^1 \times B^{n-1} \to S^n - a_1 - a_2$ be a smooth embedding, such that $S^1 \times \{0\}$ represents a commutator class. Then $S^1 \times B^{n-1}$ is standard and $S^n - (S^1 \times B^{n-1})^\circ$ is diffeomorphic to $S^{n-2} \times B^2$.

Let $A_0 = S^n - (S^1 \times B^{n-1})^\circ$, $A_1 = a_1 \cup a_2$, $A_2 = j_1^{-1}(a_1) \cup j_1^{-1}(a_2) \cup j_2^{-1}(a_1) \cup j_1^{-1}(a_2), \ldots$, etc.

There is a diffeomorphism $f : S^{n-2} \times B^2 \to A_0$ and $s > 0$ such that $s\|v\| \leq \|Df(v)\|$, for all $v \in T(S^{n-2} \times B^2)$. Fix k. By Lemma 2.1

$$s^2 \cdot 2^k \cdot \text{area}(S^{n-2}) \leq \sum_i \text{area}(T_i).$$

Let $C \subset \overset{\circ}{a}_1$, $C' \subset \overset{\circ}{a}_2$ be smooth cores and $C \times B^2$, $C' \times B^2$ smooth products with the product metric. Then there are smooth embeddings $C \times B^2 \to \overset{\circ}{a}_1$, $C' \times B^2 \to \overset{\circ}{a}_2$ (but which are not isometries).

For each component T_i of A_k there is a unique $f_i \in F_2$ and $C_i = C$ or C' such that $f_i(C_i \times \{z\})$ is a core of T_i, for almost every $z \in B^2$. By definition

$$\text{area}(T_i) \leq \int_{C_i \times \{z\}} |\det(Df_i|TC_i)| \, dA$$

for almost every $z \in B^2$. By Fubini's Theorem and Hölder's Theorem

$$\sum_i \text{area}(T_i) \leq M \sum_i \int_{C_i \times B^2} |\det(Df_i | TC_i)| dV$$

$$\leq M \left(\sum_i \int_{C_i \times B^2} |\det(Df_i | TC_i)|^{n/n-2} dV \right)^{\frac{n-2}{n}} \left(\sum_i \int_{C_i \times B^2} dV \right)^{\frac{2}{n}}$$

$$\leq M \cdot K^{n-2} \left(\sum_i \int_{C_i \times B^2} |\det(Df_i)| dV \right)^{\frac{n-2}{n}} M'(2^k)^{\frac{2}{n}}$$

$$\leq M \cdot K^{n-2} [\text{vol}(S^n)]^{\frac{n-2}{n}} M'(2^k)^{\frac{2}{n}},$$

where M, M' are independent of k. Since k is arbitrary, this is a contradiction. $\qquad\square$

3. The action μ.

In this section we construct an action $\mu : (\mathbb{Z}^2 \rtimes \mathbb{Z}) \times S^3 \to S^3$ which is not conjugate to a uniformly quasiconformal action. The method is quite general and gives infinitely many more examples in each dimension $n \geq 3$ and $n \neq 4$. Using manifold connected sum gives admissible actions in these dimensions which are not conjugate to uniformly quasiconformal actions.

Recall that the $(\mathbb{R}^3, \text{NIL})$ geometry is determined by the form $ds^2 = dx^2 + dy^2 + (dz - xdy)^2$ [**Sc**].

The action μ will be an extension of an action on $\mathbb{R}^3$. Start with diffeomorphisms $f, g, t : \mathbb{R}^3 \to \mathbb{R}^3$ defined by

$$f(x, y, z) = (x, y + 1, z),$$
$$g(x, y, z) = (x, y, z + 1) \text{ and}$$
$$t(x, y, z) = (x + 1, y, z + y).$$

Clearly $\langle f, g \rangle \approx \mathbb{Z} \oplus \mathbb{Z}$ and $t \circ f \circ t^{-1} = f$ and $t \circ g \circ t^{-1} = f \circ g$; hence, $\langle f, g, t \rangle$ is isomorphic to a nilpotent semi-direct product $\mathbb{Z}^2 \rtimes \mathbb{Z}$. In fact, f, g and t are isometries of $(\mathbb{R}^3, \text{NIL})$. One-point compactification determines the action $\mu : (\mathbb{Z}^2 \rtimes \mathbb{Z}) \times S^3 \to S^3$. Clearly Λ_μ is a point and $\mu|\Omega_\mu$ is a covering action. Notice each element acts quasiconformally.

3.1 THEOREM. *The action* $\mu : (\mathbb{Z}^2 \rtimes \mathbb{Z}) \times S^3 \to S^3$ *is not conjugate to a uniformly quasiconformal action.*

PROOF: Let $M = \mathbb{R}^3/\mathbb{Z}^2 \rtimes \mathbb{Z}$. By construction M has the $(\mathbb{R}^3, \text{NIL})$ structure. The intermediate quotient $\mathbb{R}^3/\langle f, g \rangle$ is homeomorphic to $\mathbb{R} \times S^1 \times S^1$.

48

Since $\langle f, g \rangle$ is normal in $\mathbb{Z}^2 \rtimes \mathbb{Z}$, t acts on $\mathbb{R} \times S^1 \times S^1$ and $\mathbb{R} \times S^1 \times S^1 / \langle t \rangle \approx M$. Clearly M is a $S^1 \times S^1$ bundle over S^1. In particular M is compact.

We now argue by contradiction that μ is not conjugate to a quasi-conformal action. Suppose for some homeomorphism $h : S^3 \to S^3$ the conjugate action $\mu^h : (\mathbb{Z}^2 \rtimes \mathbb{Z}) \times S^3 \to S^3$ is uniformly quasiconformal. Without loss of generality suppose $h(+\infty) = +\infty$. Then $\mu^h : (\mathbb{Z}^2 \rtimes \mathbb{Z}) \times \mathbb{R}^3 \to \mathbb{R}^3$ is quasiconformal. Let $(\mathbb{R}^3, \text{usual})$ be $\mathbb{R}^3$ with the conformal structure induced by the usual Euclidean structure. If $M' = \mathbb{R}^3 / \mu^h$, then M' inherits the quasiconformal structure usual/μ^h.

Clearly M and M' are homeomorphic. By uniqueness of triangulations—hence quasiconformal structures—on closed 3-manifolds [**Mo**], there is a quasiconformal map $j : (M, \text{NIL}) \to (M', \text{usual}/\mu^h)$. This lifts to a quasiconformal homeomorphism $\tilde{j} : (\mathbb{R}^3, \text{NIL}) \to (\mathbb{R}^3, \text{usual})$. But this contradicts that these two geometric structures are not quasiconformally equivalent [**P1**], [**Gr**]. $\square$

Theorem 3.1 may also be deduced from the analysis of quasiconformal affine actions in [**Ma2**].

The above example was suggested by the result of Bill Goldman that M (and any $S^1 \times S^1$ bundle over S^1 which admits the NIL structure) is not conformally flat [**Go**]. This immediately implies that μ is not conjugate to a conformal action.

We emphasize that μ is purely parabolic; in fact, each element of the action is individually conjugate to translation on $\mathbb{R}^3$. Furthermore, Ω_μ / μ is compact. The reader should compare μ with the parabolic actions of S. Kinoshita, L.S. Husch and F.W. Gehring and G.J. Martin described in [**GM**].

We now adapt μ to get an admissible action. The connected sum $M \,\sharp\, (S^1 \times S^2)$ has fundamental group $(\mathbb{Z}^2 \rtimes \mathbb{Z}) * \mathbb{Z}$ and universal covering $(M \,\sharp\, (S^1 \times S^2))^\sim \approx S^3 - \{\text{tame Cantor set}\}$. Thus compactifying the ends of $(M \,\sharp\, (S^1 \times S^2))^\sim$ one obtains an action of $(\mathbb{Z}^2 \rtimes \mathbb{Z}) * \mathbb{Z}$ on S^3.

3.2 THEOREM. *The action $(\mathbb{Z}^2 \rtimes \mathbb{Z}) * \mathbb{Z} \times S^3 \to S^3$ is admissible and not conjugate to a uniformly quasiconformal action.*

PROOF: Clearly the limit set is the ends of $(M \,\sharp\, (S^1 \times S^2))^\sim$ and the action is a covering action on the domain of discontinuity.

Since the new action restricted to $\mathbb{Z}^2 \rtimes \mathbb{Z}$ is μ of Theorem 3.1, this new action is not conjugate to a uniformly quasiconformal action. $\qquad\square$

The topology of M is important only to the extent that it admits a geometric structure on $\mathbb{R}^3$ which is not quasiconformally equivalent to $(\mathbb{R}^3, \text{usual})$. There are infinitely many NIL closed 3-manifolds. Furthermore, there are closed manifolds admitting each of the other structures on $\mathbb{R}^3$, namely H^3, $\mathsf{H}^2 \times \mathbb{R}$, $SL_2\mathbb{R}$, and SOLV [**Sc**]. These structures are also quasiconformally inequivalent to $(\mathbb{R}^3, \text{usual})$ [**P1**], [**P2**], [**Gr**].

Finally, there are closed n-manifolds admitting structures on $\mathbb{R}^n$ quasiconformally inequivalent to $(\mathbb{R}^n, \text{usual})$ for all $n \geq 4$. The quasiconformal hauptvermutung for closed manifolds, however, is known in all dimensions $n \neq 4$ [**Mo**], [**Su**]. Thus Theorems 3.1 and 3.2 have analogs is every dimension $n \geq 5$.

4. The action η.

In this section we construct an action $\eta : F_r \times S^3 \to S^3$ for sufficiently large r, which is uniformly quasiconformal. This example is interesting because the Cantor set limit set Λ_η is "more wildly" embedded than the limit set Λ_ψ [**FS**]. This is made precise below. We do not know whether or not η is conjugate to a conformal action.

Let $a_1, \ldots, a_r, b_1, \ldots, b_r$ be solid tori in S^3 as in Figure 4.1. Fix the conformal type of each torus such that all the tori are conformally equivalent (for r large this is independent of r). There are diffeomorphisms $h_i : S^3 \to S^3$ such that $h_i(S^3 - \overset{\circ}{a}_i) = b_i$, $i = 1, 2, \ldots, r$.

The image of $h_1|S^3 - \overset{\circ}{a}_1$ is shown in Figure 4.2. From the figure it is clear that (by increasing r if necessary) one may suppose $h_i|a_1 \cup \cdots \cup \hat{a}_i \cup \cdots \cup a_r \cup b_1 \cup \cdots \cup b_r$ is conformal. The image of $h_1^{-1}|S^3 - \overset{\circ}{b}_1$ is shown in Figure 4.3. Unlike $h_1|S^3 - \overset{\circ}{a}_1$, it is not immediately clear where h_1^{-1} is conformal. The tori b_2 and b_r seem to be distorted. But observe that b_2 and b_r contain the images of the tori under h_2 and h_r respectively. These tori in b_2 and b_r are loosely linked so that (by increasing r again if necessary) one may suppose

$$h_1^{-1}|(a_1 \cup \cdots \cup a_r \cup b_3 \cup \cdots \cup b_{r-1})$$
$$\cup\, h_2(a_1 \cup a_3 \cup \cdots \cup a_r \cup b_1 \cup \cdots \cup b_r)$$
$$\cup\, h_r(a_1 \cup \cdots \cup a_{r-1} \cup b_1 \cup \cdots \cup b_r)$$

is conformal.

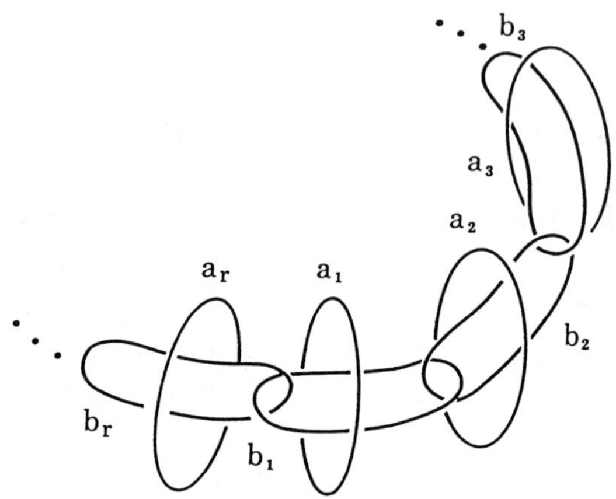

Figure 4.1

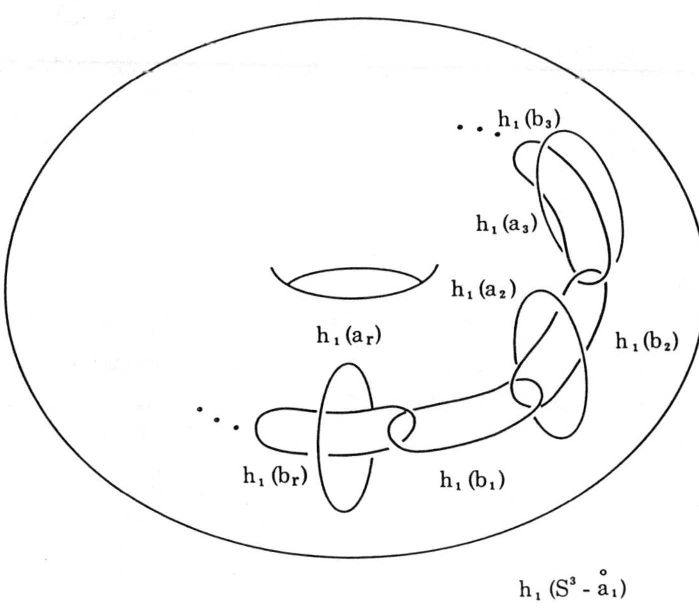

Figure 4.2

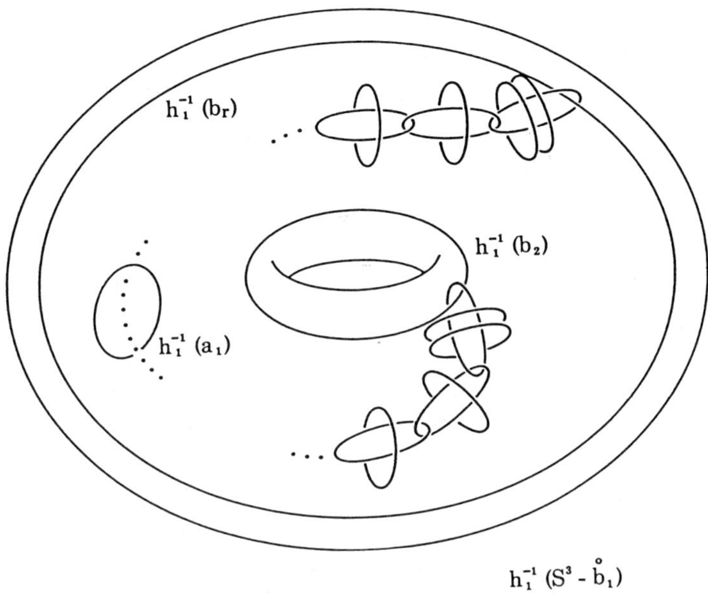

Figure 4.3

A similar statement is true of each h_i^{-1}.

Combining the above statements for h_1 and h_1^{-1} one has

$$h_1|(a_2 \cup \cdots \cup a_r \cup b_1 \cup \cdots \cup b_r)$$
$$\cup\, h_1^{-1}(a_1 \cup \cdots \cup a_r \cup b_3 \cup \cdots \cup b_{r-1})$$
$$\cup\, h_1^{-1} \circ h_2(a_1 \cup a_3 \cup \cdots \cup a_r \cup b_1 \cup \cdots \cup b_r)$$
$$\cup\, h_1^{-1} \circ h_r(a_1 \cup \cdots \cup a_{r-1} \cup b_1 \cup \cdots \cup b_r)$$

is conformal. Again a similar statement is true of each h_i.

This defines the action $\eta : F_r \times S^3 \to S^3$. The limit set is a Cantor set. Clearly η is admissible. By construction there is a neighborhood N of Λ_η such that each h_i is conformal on N. Lemma 1.2 gives the following.

4.1 **THEOREM.** *For r sufficiently large the admissible action $\eta : F_r \times S^3 \to S^3$ is smooth and uniformly quasiconformal.* $\qquad\square$

The techniques of [**FS**] may be used to show η extends to an action of the 4-ball.

Recall that the limit set of example ψ of [**FS**] has the property that every proper sub-Cantor set of Λ_ψ has simply connected complement. This is a consequence of the "loose" linking of tori in the construction of ψ. The linking is crucial to proving ψ is uniformly quasiconformal.

Though the example η is also uniformly quasiconformal, its limit set does not have the above property. The linking of the tori in the construction of η is unsymmetrical.

It is unknown whether there is a uniformly quasiconformal action $F_r \times S^3 \to S^3$ which is not conjugate to a conformal action. Both $\psi|F_r \times S^3$ and η are candidates. We expect that neither $\psi|F_r \times S^3$ nor η are conjugate to conformal actions and that a proof of this would require finding obstructions to conformal imbedding which go beyond the obvious restrictions arising from the moduli of included curve families.

5.   The actions ψ_n.

In this section we construct for each $n \geq 4$ and r sufficiently large an admissible action $\psi_n : (F_r \rtimes \mathbb{Z}_{2r}) \times S^n \to S^n$ which is smooth and uniformly quasiconformal but is not conjugate to a conformal action. As in [**FS**], our argument actually shows that $\psi_n|F_r \rtimes \mathbb{Z}_2$ is not conjugate to a conformal action. Let $S^3 \to S^n$ be standard. Then $\psi_n|(F_r \rtimes \mathbb{Z}_{2r}) \times S^3$ is the action ψ of [**FS**].

Recall that the action ψ used tori linked as in Figure 5.1. An important property of this link of tori is that each torus is unknotted. Thus there are homeomorphisms $h_i : S^3 \to S^3$ such that $h_i(S^3 - \overset{\circ}{a}_i) = b_i$. The symmetry of the link allowed a homeomorphism $g : S^3 \to S^3$ with fixed point set a circle and of order $2r$ which permutes the tori $a_1 \to a_2$, $a_2 \to a_3, \ldots$, etc. These homeomorphisms were also carefully chosen to satisfy all plausible relations. Thus they generated an action $\psi : (F_r \rtimes \mathbb{Z}_{2r}) \times S^3 \to S^3$. Furthermore, the loose linking permitted ψ to be uniformly quasiconformal.

It was proved in [**FS**] that ψ extends to an admissible action on S^4. We now show how ψ extends to all dimensions and that the extension may be chosen to be uniformly quasiconformal. There are several ingredients in this extension—we begin with the topological ingredients.

Let B^3 be the 3-ball of Figure 5.1. Notice ∂B^3 is symmetric about the link of tori and that $g|B^3$ permutes the components of the link inside B^3. Recall from §1 that $S^n \approx B^3 \times S^{n-3} / \sim$ (ignore the conformal structure for the moment). Let $P_i = (a_i \cap B^3) \times S^{n-3} / \sim$ and $Q_i = (b_i \cap B^3) \times S^{n-3} / \sim$. Thus each P_i and Q_i is homeomorphic to $S^{n-2} \times B^2$, and $a_i = P_i \cap S^3$,

$b_i = Q_i \cap S^3$. Unfortunately, $S^n - \overset{\circ}{P}_i \neq Q_i$. This is remedied as follows.

Recall that the *join* between two spaces X and Y, denoted $X * Y$, is the space $X \times Y \times [-1,1]/ \sim$, where $(x,y,-1) \sim (x',y,-1)$ and $(x,y',1) \sim (x,y,1)$ for all $x, x' \in X$ and $y, y' \in Y$. In particular, $S^{p-1} * S^q \approx S^{p+q}$.

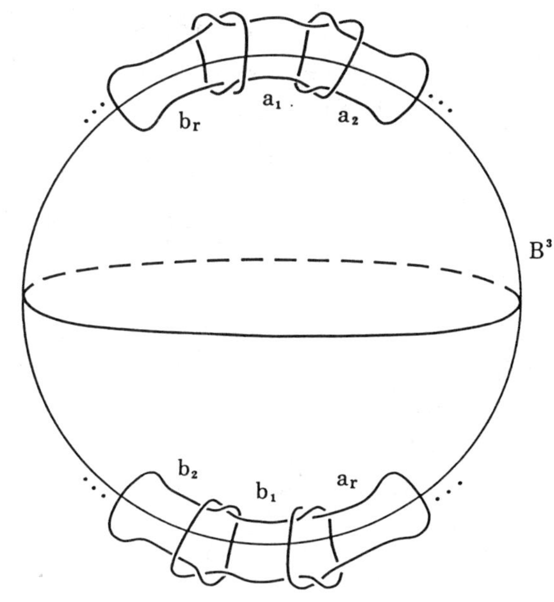

Figure 5.1

Let $\tau \subset S^3$ be a solid Clifford torus. Let $A = \tau * S^{n-4}$, $B = \left(S^3 - \overset{\circ}{\tau}\right) * S^{n-4}$. Then $A \approx B$, $A \cup B = S^3 * S^{n-4} = S^n$ and $S^n - \overset{\circ}{A} \approx B$.

If P_i, Q_i are as above then define $A_i \subset P_i$ and $B_i \subset Q_i$ such that

$$\left(S^n, A_i, a_i\right) \approx \left(S^n, A, \tau\right) \quad \text{and} \quad \left(S^n, B_i, b_i\right) \approx \left(S^n, B, S^3 - \overset{\circ}{\tau}\right).$$

Furthermore, choose this collection invariant under the extension of g to S^n.

It should now be clear how to (topologically) extend ψ to S^n; f_i should be extended to map $S^n - \overset{\circ}{A}_i$ to B_i. We now show how to insure the extension is quasiconformal.

Let $f : B^3 \to B^3$ be the diffeomorphism described by Figure 5.2. Notice that f greatly expands $a_1 \cap B^3$ and $f|(a_2 \cup \cdots \cup a_r \cup b_1 \cup \cdots \cup b_r) \cap B^3$

is conformal. This assertion requires r large and some care; compare with
[**FS**]. We do not claim $f(B^3 - \overset{\circ}{a}_1) = b_1 \cap B^3$.

Again using the identification $B^3 \times S^{n-3}/ \sim \, \approx S^n$, define the diffeomor-
phism $F : S^n \to S^n$ by $F = (f, 1)$. It will be shown that $F|P_2 \cup \cdots \cup P_r \cup$
$Q_1 \cup \cdots \cup Q_r$ is conformal. Concentrate on $F|P_2$ and recall that $f|b_2 \cap B^3$
is conformal. By construction there is a conformal map $f' : B^3 \to B^3$ such
that $f'|b_2 \cap B^3 = f|b_2 \cap B^3$ (this crucially uses the fact that $b_2 \cap \partial B^3 \neq \phi$).
By Lemma 1.3 $(f', 1)$ is conformal; hence, $(f', 1)|P_2 = F|P_2$. Arguing on
each component the claim follows.

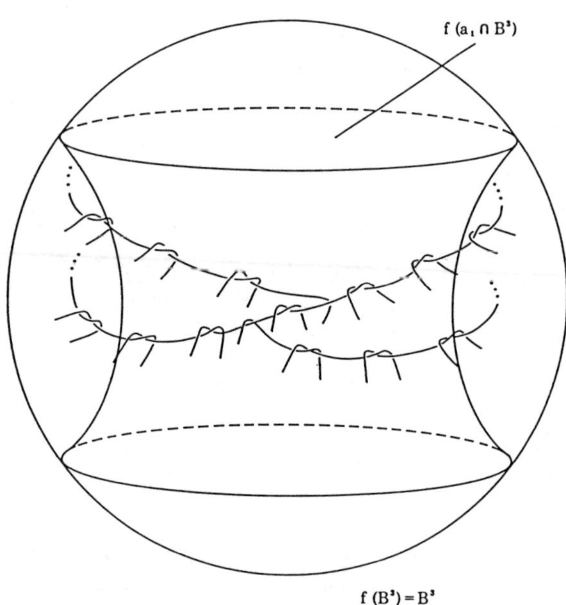

Figure 5.2

Unfortunately, $F(S^n - \overset{\circ}{A}_1) \neq B_1$, so we need to compose F with another
diffeomorphism. Recall from [**FS**] the map $h : S^3 \to S^3$ with $h(S^3 - \overset{\circ}{a}_1) =$
b_1. Notice that $F|S^3$ is the double of f. Define $j : S^3 \to S^3$ to satisfy
$h = j \circ (F|S^3)$. Clearly $j|F(a_2 \cup \cdots \cup a_r \cup b_1 \cup \cdots \cup b_r)$ is conformal.

This, in particular, achieves $j \circ f(S^n - \overset{\circ}{a}_1) = b_1$. Just as in the three
dimensional case, the components of $F(A_2 \cup \cdots \cup A_r \cup B_1 \cup \cdots \cup B_r)$ make

up a quite flexible chain which is increasingly flexible as r grows. Thus for r sufficiently large, j has an extension $J : S^n \to S^n$ which is conformal near this chain. Furthermore, the components of $J \circ F(A_2 \cup \cdots \cup A_r \cup B_1 \cup \cdots \cup B_r)$ have small diameter, so we may also insist that $J(F(S^n - \overset{\circ}{A}_1)) = B_1$ without causing distorsion of the sets $A_2, \ldots, A_r, B_1, \ldots, B_r$.

Define $H = J \circ F$; it clearly has the desired properties. Let G extend g.

As in [**FS**] $G^r \circ H | \partial A_1$ is isotopic to an involution. By carefully isotoping H we may assume $G^r \circ H | \partial A_1$ is an involution. Now define $H_1 : S^n \to S^n$ by

$$H_1(x) = \begin{cases} H(x), & X \in S^n - \overset{\circ}{A}_1 \\ G^r \circ H \circ G^r(x), & x \in \overset{\circ}{A}_1. \end{cases}$$

Clearly H_1 is continuous and may be arranged to be a diffeomorphism. We omit the details.

Let $H_2 = G \circ H_1 \circ G^{-1}$, $H_3 = G^2 \circ H_1 \circ G^{-2}, \ldots$ etc. Then $\langle H_i \rangle \approx F_r$. By construction $G^r \circ H_i \circ G^{-r} = H_i^{-1}$, so $\langle H_i, G \rangle \approx F_r \rtimes \mathbb{Z}_{2r}$. Call this action ψ_n.

5.1   THEOREM. *For r sufficiently large, the admissible action ψ_n : $(F_r \rtimes \mathbb{Z}_{2r}) \times S^n \to S^n$ is smooth and uniformly quasiconformal, but it is not conjugate to a conformal action.*

PROOF: By Lemma 1.2 ψ_n is uniformly quasiconformal.

To prove ψ_n is not conjugate to a conformal action we argue as in [**FS**]. The element $G^{r+k} \circ H_1 \circ G^{-k}$, $k = 0, 1, \ldots, 2r - 1$, is an involution with fixed point set an unknotted $(n-2)$-sphere. Collectively the fixed point sets form a link $L = \{\ell_1, \ldots, \ell_{2r}\}$ whose components are pairwise unlinked. However, L is a non-trivial link because $\pi_1(S^n - L) \approx \pi_1(B^3 - L)$ [**Ro**] and a simple computation shows $\pi_1(B^3 - L)$ is not free. If ψ_n were conjugate to a conformal action then L would be equivalent to a link of round spheres. But this contradicts the following lemma. $\qquad\square$

5.2   LEMMA. *Let $L = \{\ell_1, \ldots, \ell_k\} \subset S^n$ be a link of round spheres (not necessarily all of the same dimension). Then L is equivalent to the unlink if and only if $\langle \ell_i, \ell_j \rangle = 0$ for all pairs $i \neq j$, where $\dim \ell_i = \dim \ell_j + 1 = n$.*

PROOF: Only one direction needs proof. Suppose L satisfies the algebraic linking hypothesis. Consider $S^n \subset S^{n+1}$ as an equator. Let $\overline{L} = \{\overline{\ell}_1, \ldots, \overline{\ell}_k\} \subset S^{n+1}$ be the collection of round spheres determined by $\overline{\ell}_i \cap S^n = \ell_i$ and $\overline{\ell}_i$ is perpendicular to S^n.

In general, round spheres intersect in round spheres or a point. Since $\ell_i \cap \ell_j = \phi$ for $i \neq j$, $\bar{\ell}_i \cap \bar{\ell}_j = \phi$ or two points ($\approx S^\circ$). The linking hypothesis is now equivalent to $\bar{\ell}_i \cap \bar{\ell}_j = \phi$, for all $i \neq j$.

Let B^{n+1} be a hemisphere of S^{n+1} with $\partial B^{n+1} = S^n$. By construction $\bar{L} \cap B^{n+1}$ is a slice of the link L. The components of $\bar{L} \cap B^{n+1}$ are convex with respect to the radial coordinates from the origin of B^{n+1}, so they may be isotoped into disjoint imbedded balls in S^n, showing L is a trivial link. $\qquad\square$

The action ψ_n has several interesting properties. Firstly, the examples ψ_n nullify the hope expressed by Gehring and Palka that every uniformly quasiconformal action on S^n is quasiconformally conjugate to a conformal action [**GP**]. Previous examples were constructed by C.H. Giffen [**Gi**], P. Tukia [**Tu**], G. Martin [**Ma1**], J. McKemie [**Mc**], Martin and Gehring [**MG**], and Freedman and Skora [**FS**].

Secondly, each element of the action is standard, i.e., conjugate to either a hyperbolic or elliptic element of Möb(S^n).

Also ψ_n extends ψ, so $\Lambda_{\psi_n}|S^3 = \Lambda_\psi$. Though Λ_ψ is a wild Cantor set in S^3, when included into S^n it becomes tame (this may be visualized directly or it follows by a general result [**K1**]). In fact $\psi_n|F_r \times S^n$ is conjugate to the Schottky action: for $n \geq 5$ this is a consequence of the Cantor set being tame [**Fr**]. But for all $n \geq 4$ one may explicitly visualize disjoint n-balls $C_1, \ldots, C_r, D_1, \ldots, D_r \subset S^n$ such that $H_i(S^n - \overset{\circ}{C}_i) = D_i$ and then appeal to Lemma 1.4 of [**FS**] which says it is conjugate to the Schottky action.

Finally, the above implies that $\Omega_{\psi_n}/F_r \approx \underset{r}{\natural} \, S^1 \times S^{n-1}$. Since F_r is normal in $F_r \rtimes \mathbb{Z}_{2r}$, g induces a cyclic rotation of order $2r$ on $\underset{r}{\natural} \, S^1 \times S^{n-1}$. Theorem 4.1 implies this rotation is non-standard. In particular, the involution g^r on $\underset{r}{\natural} \, S^1 \times S^{n-1}$ is non-standard.

Michael H. Freedman, Department of Mathematics, University of California, San Diego, La Jola, CA 92093

Richard Skora, Department of Mathematics, Indiana University, Bloomington, IN 47405
Current address: Department of Mathematics, SUNY, Stony Brook, NY 11794

REFERENCES

[**Fr**] Freedman, M.H., *A geometric reformulation of 4-dimensional surgery*, Top. Appl. **24** (1986), 133–141.

[**FS**] Freedman, M.H. and Skora, R., *Strange actions of Groups on Spheres*, J. Diff. Geo. **25** (1987), 75–98.

[**GP**] Gehring, F.W. and Palka, B.P., *Quasiconformally homogeneous domains*, J. d'Analyse **30** (1976), 172–199.

[**GM**] Gehring, F.W. and Martin, G.J., *Discrete Quasiconformal Groups, I*, (to appear Proc. Lond. Math. Soc.).

[**Gi**] Giffen, C.H., *The generalized Smith Conjecture*, Amer. J. Math. **88** (1966), 187–198.

[**Go**] Goldman, W.M., *Conformally flat manifolds with nilpotent holonomy and the uniformization problem for 3-manifolds*, Trans. A.M.S. **278** (1983), 573–583.

[**Gr**] Gromov, M., "Structures métriques pour les variétés riemanniennes," notes de cours rédigées par J. Lafontaine et P. Pansu, CEDIC-Fernand-Nathan, Paris, 1981.

[**Kl**] Klee, Jr., Y.L., *Some topological properties of convex sets*, Trans. A.M.S. **78** (1955), 30–45.

[**Ma1**] Martin, G.J., *Discrete Quasiconformal groups that are not the quasiconformal conjugates of Möbius groups*, Ann. Acad. Sci. Fenn. Ser. A.I. Math. **11** (1986), 179–202.

[**Ma2**] Martin, G.J., *On Quasiconformal and Affine Groups*, (preprint).

[**MG**] Martin, G.J. and Gehring, F.W., *Generalizations of Kleinian Groups*, (preprint).

[**Mc**] McKemie, J., "Quasiconformal Groups and Quasisymmetric Embeddings," Ph.D. Thesis, University of Texas, Austin, 1985.

[**Mo**] Moise, E.E., *Affine structures in 3-manifolds, I-V*, Ann. Math., **54** (1951), 506–553; **55** (1952), 172–176, 203–214, 215–222; **56** (1952), 96–114.

[**P1**] Pansu, P., *An isoperimetric inequality in the Heisenberg group*, Rend. Sem. Math. Torino., Fas Spe **7** (1983), 159–174.

[**P2**] Pansu, P., (private communication).

[**Q**] Quinn, F., *Ends of Maps, III Dimensions 4 and 5*, J. Diff. Geo. **17** (1982), 503–521.

[**Ro**] Rolfsen, D., "Knots and Links," Publish or Parish, Inc., Berkeley, 1976.

[**Sc**] Scott, P., *The geometries of 3-manifolds*, Bull. London Math. Soc. **15** (1983), 401–487.

[**Su**] Sullivan, P., *Hyperbolic Geometry and Homeomorphisms*, Geometric Topology ed. James Cantrell, Academic Press, Inc., New York (1979), 543–555.

[**TV**] Tukia, P. and Väisälä, J., *Lipshitz and quasiconformal approximation and extension*, Ann. Acad. Sci. Fenn. Ser. A.I. Math. **6** (1981), 303–342.

[**Tu**] Tukia, P., *A quasiconformal group not isomorphic to a Möbius group*, Ann. Acad. Sci. Fenn. Ser. A.I. Math. **6** (1981), 149–160.

Quasiconformal groups and the conical limit set

BY J.B. GARNETT, F.W. GEHRING AND P.W. JONES

INTRODUCTION

1. Quasiconformal groups. For $n \geq 2$ we let $\mathbb{R}^n$ denote euclidean n-space with the standard orthonormal basis $e_1, \ldots, e_n$ and $\overline{\mathbb{R}}^n = \mathbb{R}^n \cup \{\infty\}$ its one point compactification equipped with the chordal metric.

Suppose that G is a family of self homeomorphisms of $\overline{\mathbb{R}}^n$ which forms a group under composition. Then G is *discontinuous* at a point $x \in \overline{\mathbb{R}}^n$ if there exists a neighborhood U of x such that $g(U) \cap U \neq \emptyset$ for at most finitely many $g \in G$; G is *discontinuous* if it is discontinuous at some $x \in \overline{\mathbb{R}}^n$. We let $O(G)$ denote the set of all x at which G is discontinuous and call $L(G) = \overline{\mathbb{R}}^n \backslash O(G)$ the *limit set* of G. Next G is a *K-quasiconformal* group if each $g \in G$ is K-quasiconformal. Hence G is a 1-quasiconformal group if and only if it is a Möbius group, i.e. a subgroup of $\mathrm{M\ddot{o}b}(\overline{\mathbb{R}}^n)$, the group of all Möbius transformations of $\overline{\mathbb{R}}^n$ generated by reflections in $(n-1)$-spheres and hyperplanes.

If H is a Möbius group and f a quasiconformal self mapping of $\overline{\mathbb{R}}^n$, then

$$(1.1) \qquad\qquad G = f \circ H \circ f^{-1}$$

is easily seen to be a quasiconformal group. Moreover when $n = 2$, each quasiconformal group G can be written in this form [S], [T1]. Examples exist to show this is not the case when $n \geq 3$ [FS], [M], [T2].

Möbius groups have been studied for many years and it now appears that much of the highly developed theory for this class also holds for the larger family of quasiconformal groups. The purpose of this paper is to point out quasiconformal analogues of some well known results on the conical limit set of a Fuchsian group.

2. Notation. We assume from now on that G is a discontinuous K-quasiconformal group acting on $\overline{\mathbb{R}}^n$ each of whose elements g maps the unit ball $\mathsf{B} = \mathsf{B}^n$ onto itself. Then $\partial\mathsf{B}$ is closed and invariant under G and $L(G) \subset \partial\mathsf{B}$ by Corollary 3.8 in [GM1]. Next if h is a Möbius transformation which maps $\partial\mathsf{B}$ onto $\overline{\mathbb{R}}^{n-1}$, then $H = h^{-1} \circ (G|\partial\mathsf{B}) \circ h$ is a quasiconformal group acting on $\overline{\mathbb{R}}^{n-1}$ and hence $L(G) = \partial\mathsf{B}$ or $L(G)$ is nowhere dense in

$\partial \mathsf{B}$ by Theorem 4.9 in [**GM1**]. We say that G is a *Fuchsian group* of the *first kind* if $L(G) = \partial \mathsf{B}$ and of the *second kind* if $\partial \mathsf{B} \backslash L(G) \neq \emptyset$.

Since G is discontinuous in B, G is countable and we call

$$(2.1) \qquad \rho(x, \alpha; G) = \sum_{g \in G} (1 - |g(x)|^2)^\alpha$$

the *Poincaré series* for G; when $K = 1$, each $g \in G$ is a Möbius transformation and

$$(2.2) \qquad \rho(x, \alpha; G) = (1 - |x|^2)^\alpha \sum_{g \in G} |g'(x)|^\alpha.$$

From [**GM2**] it follows that for each $\alpha > 0$, $\rho(x, \alpha; G) < \infty$ for every $x \in \mathsf{B}$ if and only if $\rho(O, \alpha; G) < \infty$, and that

$$(2.3) \qquad \rho(O, \alpha; G) < \infty$$

for $\alpha > n - 1$. We say that G is of *convergence type* if (2.3) holds for $\alpha = n - 1$ and of *divergence type* otherwise.

For each $x \in \mathsf{B}$ let $L(x, G)$ denote the set of $y \in \partial \mathsf{B}$ for which there exists a sequence of elements $g_j \in G$ such that $g_j(x) \to y$. Next let denote $L_c(x, G)$ the subset of such y for which we can choose $\{g_j\}$ so that, in addition, $g_j(x) \to y$ in some Stolz cone $Cone(y, a)$, where

$$(2.4) \qquad Cone(y, a) = \{x \in \mathsf{B}: |x - y| < a(1 - |x|)\}.$$

for $y \in \partial \mathsf{B}$ and $1 < a < \infty$. Since G is discontinuous in B, $L(G) = L(x, G)$ for all $x \in \mathsf{B}$ by Lemma 4.4 of [**GM1**]. By Theorem 11 in [**G1**] or Theorem 18.1 in [**V1**],

$$(2.5) \qquad \frac{|g(x_1) - g(x_2)|}{1 - |g(x_1)|} \leq \Theta_K \left(\frac{|x_1 - x_2|}{1 - |x_1|} \right)$$

for each $g \in G$ and $x_1, x_2 \in \mathsf{B}$ with $|x_1 - x_2| < 1 - |x_1|$; here $\Theta_K : (0, 1) \to (0, \infty)$ is an increasing function which depends only on K and n. If $x_1, x_2 \in \mathsf{B}$ with $|x_1 - x_2| \leq (1 - |x_1|)/3$ and if $g_j(x_1) \to y$ in $Cone(y, a)$, then $|x_1 - x_2| \leq (1 - |x_2|)/2$ and (2.5) implies that $g_j(x_2) \to y$ in $Cone(y, b)$ where $b = b(a, K, n)$. It follows that $L_c(x, G) = L_c(O, G)$ for all $x \in \mathsf{B}$ and we call

$$(2.6) \qquad L_c(G) = L_c(O, G)$$

the *conical limit set* for G. Clearly

$$(2.7) \qquad g(L_c(G)) = L_c(G) \quad \text{for} \quad g \in G.$$

For $x \in \mathbb{B}$ and $1 < a < \infty$ let

$$(2.8) \qquad Cap(x, a) = \{y \in \partial\mathbb{B} : |y - x| < a(1 - |x|)\}.$$

Then $Cap(x, a)$ is either a spherical cap or the whole of $\partial\mathbb{B}$, $y \in Cap(x, a)$ if and only if $x \in Cone(y, a)$, and

$$(2.9) \qquad L_c(G) = \bigcup_{\ell=2}^{\infty} (\bigcap_{k=1}^{\infty} (\bigcup_{j=k}^{\infty} Cap(g_j(0), \ell))),$$

where $\{g_j\}$ is any enumeration of G.

3.   Main results. We shall establish the following relations between the conical limit set $L_c(G)$ and the exponent of convergence for the Poincaré series $\rho(x, \alpha; G)$. Here and in what follows we let m denote $(n-1)$-dimensional Lebesgue measure.

3.1   THEOREM.   *If $n \geq 3$, then $m(L_c(G)) = 0$ or $m(L_c(G)) = m(\partial\mathbb{B})$.*

3.2   THEOREM.   *If G is of convergence type, then $m(L_c(G)) = 0$.*

3.3   THEOREM.   *If $n \geq 3$ and if G is of divergence type, then $m(L_c(G)) = m(\partial\mathbb{B})$.*

Theorem 3.1 was also established by Tukia [**T3**] by a different argument. This result is an analogue for the conical limit set of the Ahlfors measure zero problem for the topological limit set [**A1**].

Theorem 3.3 with $K = 1$ is due to Sullivan [**S**]; our proof is a modification of an argument for this case due to Thurston [**A2**]. (See also [**Ga**].) Theorem 3.3 implies that each Fuchsian group of the second kind is of convergence type [**GM2**].

PRELIMINARY RESULTS

4.   Spherical caps. We begin with a result on the distortion of spherical caps.

4.1   LEMMA. *If $x_1, x_2 \in \mathsf{B}$ with $|x_1| \leq |x_2|$ and if $g \in G$ with $g(x_1) = 0$,
then*

$$(4.2) \qquad g(Cap(x_1, a) \cap Cap(x_2, a)) \subset Cap(g(x_2), b)$$

where $b = b(a, K, n)$.

PROOF: Choose $y \in Cap(x_1, a) \cap Cap(x_2, a)$ and $z \in \partial \mathsf{B}$ so that $|g(z) - g(x_2)| = 1 - |g(x_2)|$. Then

$$(4.3) \qquad \frac{|g(y) - g(x_2)|}{1 - |g(x_2)|} = \frac{|g(y) - g(x_2)|}{|g(z) - g(x_2)|} \frac{|g(z) - g(x_1)|}{|g(y) - g(x_1)|} \leq \Phi_K \left(\frac{|y - x_2||z - x_1|}{|z - x_2||y - x_1|} \right),$$

where $\Phi_K \colon (0, \infty) \to (0, \infty)$ is an increasing function which depends only
on K and n [**V2**]. Next

$$\frac{|z - x_1|}{|y - x_1|} \leq \frac{|z - x_2| + |y - x_1| + |y - x_2|}{1 - |x_1|} < \frac{|z - x_2|}{1 - |x_2|} + 2a \leq (2a + 1) \frac{|z - x_2|}{1 - |x_2|}$$

and

$$(4.4) \qquad \frac{|y - x_2||z - x_1|}{|z - x_2||y - x_1|} < (2a + 1) \frac{|y - x_2|}{1 - |x_2|} < (2a + 1)a.$$

Thus $y \in Cap(g(x_2), b)$ by (4.3) and (4.4) where $b = \Phi_K((2a + 1)a)$.

5.   Area estimates. We require next estimates for the distortion of area
under elements of G.

5.1   LEMMA. *If $n \geq 3$ and if $g \colon \overline{\mathsf{R}}^n \to \overline{\mathsf{R}}^n$ is K-quasiconformal with
$g(\mathsf{B}) = \mathsf{B}$ and $g(0) = 0$, then*

$$(5.2) \qquad m(g(E)) \leq c_1 m(E)^\alpha$$

for each measurable $E \subset \partial \mathsf{B}$, where $c_1 = c_1(K, n)$ and $\alpha = \alpha(K, n)$.

PROOF: Let H denote the upper half space $\{x \in \mathsf{R}^n \colon x \circ e_n > 0\}$; by
symmetry it suffices to establish (5.2) under the assumption that E lies in
$\partial \mathsf{B} \cap \overline{\mathsf{H}}$. Next by composing g with a rotation of B, we may also assume
that $g(-e_n) = -e_n$. Now

$$(5.3) \qquad h(x) = 2 \frac{x + e_n}{|x + e_n|^2} - e_n$$

is a Möbius transformation which maps H onto B and E into the unit
$(n - 1)$-cube $Q = \{x \in \mathsf{R}^{n-1} \colon |x \circ e_j| \leq 1, j = 1, \ldots, n - 1\}$ in R^{n-1}. Then

$$63$$

$f = h \circ g \circ h^{-1}$ is K-quasiconformal in $\overline{\mathsf{R}}^n$ with $f(\mathsf{H}) = \mathsf{H}$, $f(\infty) = \infty$ and $f(e_n) = e_n$.

If $x \in Q$ and $y = f^{-1}(0)$, then

$$(5.4) \qquad |f(x) - e_n| = \frac{|f(x) - f(e_n)|}{|f(y) - f(e_n)|} \leq \Phi_K\left(\frac{|x - e_n|}{|y - e_n|}\right) \leq \Phi_K(n^{1/2}),$$

where Φ_K is as in (4.3); thus $m(f(Q)) \leq c$ where $c = c(K, n)$. Next $|h'(x)| \leq 1$ in E and $m(h(E)) \leq m(E)$. Because $f|\mathsf{R}^{n-1}$ is a K-quasiconformal self mapping of R^{n-1}, Hölder's inequality and Section 5 in $[\mathbf{G2}]$ imply that

$$(5.5) \qquad m(h \circ g(E)) \leq c\frac{m(f(h(E)))}{m(f(Q))} \leq c'\left(\frac{m(h(E))}{m(Q)}\right)^\alpha \leq c''m(E)^\alpha,$$

where $c'' = c''(K, n)$ and $\alpha = \alpha(K, n)$. (Cf. Lemma 5 in $[\mathbf{CF}]$.) Finally $|h'(x)| \geq 1/2$ in $g(E) \subset \partial\mathsf{B}$, $m(g(E)) \leq 2^{n-1}m(h \circ g(E))$ and (5.2) follows from (5.5).

5.6 COROLLARY. *If $n \geq 3$ and $x_1 \in \mathsf{B}$ and if $g \in G$ with $g(x_1) = 0$, then*

$$(5.7) \qquad m(g(E)) \leq c_2\left(\frac{m(E)}{m(C)}\right)^\alpha, \qquad \frac{m(E)}{m(C)} \leq c_2 m(g(E))^\alpha$$

for each measurable $E \subset C = Cap(x_1, a)$, where $c_2 = c_2(a, K, n)$ and α is as in (5.2).

PROOF: Choose a Möbius transformation f such that $f(\mathsf{B}) = \mathsf{B}$ and $f(x_1) = 0$. Then

$$(5.8) \qquad |f'(y)| = \frac{1 - |x_1|^2}{|y - x_1|^2} \geq \frac{1}{a^2(1 - |x_1|)}$$

for $y \in E \subset C$ $[\mathbf{A2}]$ and

$$(5.9) \qquad \frac{m(E)}{m(C)} \leq cm(f(E)) \leq c_2 m(g(E))^\alpha$$

by Lemma 5.1 applied to $h = f \circ g^{-1}$. The first part of (5.7) follows similarly.

5.10   COROLLARY. *If $n \geq 3$ and $x_1 \in B$ and if $g \in G$ with $g(x_1) = 0$, then*

$$(5.11) \qquad m(\partial B \setminus g(C)) \leq c_3 a^{(1-n)\alpha},$$

where $C = Cap(x_1, a)$, $c_3 = c_3(K, n)$ and α is as in (5.2).

PROOF: Let U be the open ball with center x_1 and radius $r = a(1 - |x_1|)$. Then $C = \partial B \cap U$ and we may assume that

$$(5.12) \qquad a(1 - |x_1|) \leq 1 + |x_1|$$

since otherwise $C = \partial B$ and $\partial B \setminus g(C) = \emptyset$.

Let $f \colon B \to B$ be a Möbius transformation which maps the line through x_1 and 0 onto itself and x_1 onto 0. Then $z_1 = x_1(1 - r/|x_1|)$ and $z_2 = x_2(1 + r/|x_1|)$ are diametral points of ∂U and an elementary calculation shows that

$$(5.13) \qquad dia(f(\partial U)) = |f(z_1) - f(z_2)| = \left| \frac{2a(1 + |x_1|)}{a^2|x_1|^2 - (1 + |x_1|)^2} \right|.$$

If $a \geq 5$, then (5.12) implies that $2(1 + |x_1|) \leq a|x_1|$, that $\infty \in f(U)$, and hence with (5.13) that

$$(5.14) \qquad dia(\partial B \setminus f(U)) \leq 12a^{-1}.$$

Since (5.14) holds trivially when $a < 5$, we can apply Lemma 5.1 to $h = g \circ f^{-1}$ to obtain

$$(5.15) \qquad m(\partial B \setminus g(C)) = m(h(\partial B \setminus f(U))) \leq c_3 a^{(1-n)\alpha}.$$

PROOF OF MAIN RESULTS

6. Proof of Theorem 3.1. Suppose that $m(L_c(G)) > 0$, let y be a point of density for $L_c = L_c(G)$ and fix $\varepsilon > 0$. By (2.9) we can choose $\ell \geq 2$ and a sequence $\{g_j\}$ in G so that

$$(6.1) \qquad c_3 \ell^{(1-n)\alpha} < \frac{\varepsilon}{2},$$

where c_3 and α are as in (5.11), and so that

$$(6.2) \qquad y \in Cap(g_j(0), \ell) = C_j$$

for all j. Next because y is a point of density for L_c, we can fix j so that

$$(6.3) \qquad c_2 \left(\frac{m(C_j \backslash L_c)}{m(C_j)} \right)^\alpha < \frac{\varepsilon}{2},$$

where c_2 and α are as in (5.7). Then (6.1), (6.3) and Corollaries 5.6 and 5.10 with $g = g_j^{-1}$ imply that

$$(6.4) \quad m(\partial \mathbb{B} \backslash L_c) = m(\partial \mathbb{B} \backslash g(L_c)) \leq m(\partial \mathbb{B} \backslash m(g(C_j))) + m(g(C_j \backslash L_c)) < \varepsilon.$$

7. Proof of Theorem 3.2. By (2.9),

$$(7.1) \qquad L_c = \bigcup_{\ell=2}^{\infty} E_\ell \quad \text{where} \quad E_\ell = \bigcap_{k=1}^{\infty} \left(\bigcup_{j=k}^{\infty} Cap(g_j(0), \ell) \right).$$

Hence

$$(7.2) \qquad m(E_\ell) \leq \sum_{j=k}^{\infty} m(Cap(g_j(0), \ell)) \leq c \sum_{j=k}^{\infty} (1 - |g_j(0)|)^{n-1} = c\varepsilon_k,$$

where $c = c(\ell, n)$. Since G is of convergence type, $\varepsilon_k \to 0$ as $k \to \infty$, $m(E_\ell) = 0$ and $m(L_c) = 0$.

8. Proof of Theorem 3.3. Suppose $m(L_c) < m(\partial \mathbb{B})$. Fix $a > 1$, let b and c_2, α be as in (4.2) and (5.7), respectively, and choose $\varepsilon > 0$ so that

$$(8.1) \qquad c_2 \varepsilon^\alpha \leq \frac{1}{2}.$$

Then $m(L_c) = 0$ by Theorem 3.1, and with (2.9), we can choose a finite subset G_0 of G containing the identity g_0 such that

$$(8.2) \qquad m(E_0) < \varepsilon \quad \text{where} \quad E_0 = \bigcup_{g \in G \backslash G_0} Cap(g(0), b).$$

Next by induction we can choose distinct elements $g_j \in G \backslash G_0$ such that

$$(8.3) \qquad |g_j(0)| = \min \left\{ |g(0)| : g \in G \backslash \left(\bigcup_{k=0}^{j-1} g_k \circ G_0 \right) \right\}$$

for $j = 1, 2, \ldots$.

Let $x_j = g_j(0)$, $C_j = Cap(x_j, a)$ and suppose that $k > j$. Then $|x_j| \leq |x_k|$ and $g = g_j^{-1} \circ g_k \notin G_0$ by (8.3). Hence

$$(8.4) \qquad g_j^{-1}(C_j \cap C_k) \subset Cap(g_j^{-1}(x_k), b) = Cap(g(0), b) \subset E_0$$

by Lemma 4.1 and

$$(8.5) \qquad g_j^{-1}(C_j \cap E_j) \subset E_0 \quad \text{where} \quad E_j = \bigcup_{k>j} C_k.$$

Corollary 5.6 with $g = g_j^{-1}$ implies that

$$(8.6) \qquad \begin{aligned} m(C_j \backslash E_j) &= m(C_j) - m(C_j \cap E_j) \\ &\geq m(C_j) - c_2 m(E_0)^\alpha m(C_j) \\ &\geq m(C_j)/2 \end{aligned}$$

for $j = 1, 2, \ldots$, and since the $C_j \backslash E_j$ are disjoint, we obtain

$$(8.7) \qquad \sum_{j=0}^{\infty} (1 - |g_j(0)|)^{n-1} \leq c \sum_{j=1}^{\infty} m(C_j \backslash E_j) \leq c'$$

where $c' = c'(a, n)$.

For each $g \in G$ there exists a g_j such that $g = g_j \circ h$ and $h \in G_0$. Then (2.5) with $x_1 = 0$ and $x_2 = h(0)$ implies that

$$(8.8) \qquad \frac{|g_j(0) - g(0)|}{1 - |g_j(0)|} \leq \Phi_K(|h(0)|)$$

and hence that

$$(8.9) \qquad \sum_{g \in G} (1 - |g(0)|)^{n-1} \leq c'' N \sum_{j=0}^{\infty} (1 - |g_j(0)|)^{n-1} \leq c' c'' N,$$

where $c'' = c''(K, n)$ and $N = card(G_0)$, contradicting the hypothesis that G is of divergence type.

J.B. Garnett, Department of Mathematics, University of California, Los Angeles, California

F.W. Gehring, Department of Mathematics, University of Michigan, Ann Arbor, Michigan

P.W. Jones, Department of Mathematics, Yale University, New Haven, Connecticut

References

[**A1**] Ahlfors, L.V., *Finitely generated Kleinian groups*, Amer. J. Math. **84** (1964), 413–429.

[**A2**] Ahlfors, L.V., "Möbius transformations in several dimensions," Univ. of Minnesota, 1981.

[**CF**] Coifman, R.R., and Fefferman, C., *Weighted norm inequalities for maximal functions and singular integrals*, Studia Math. **51** (1974), 241–250.

[**FS**] Freedman, M.H., and Skora, R., *Strange actions of groups on spheres*, J. Diff. Geo. **25** (1987), 75–98.

[**Ga**] Garnett, J.B., "Applications of harmonic measure," University of Arkansas Lecture Notes in Math. Sciences 8, Wiley & Sons, 1986.

[**G1**] Gehring, F.W., *Rings and quasiconformal mappings in space*, Trans. Amer. Math. Soc. **103** (1962), 353–393.

[**G2**] Gehring, F.W., *The L^p-integrability of the partial derivatives of a quasiconformal mapping*, Acta Math. **130** (1973), 265–277.

[**GM1**] Gehring, F.W., and Martin, G.J., *Discrete quasiconformal groups I*, Proc. London Math. Soc. **55** (1987), 331–358.

[**GM2**] Gehring, F.W., and Martin, G.J., *Discrete quasiconformal groups II*, (to appear).

[**M**] Martin, G.J., *Discrete quasiconformal groups that are not the quasiconformal conjugates of Möbius groups*, Ann. Acad. Sci. Fenn. **11** (1986), 179–202.

[**S**] Sullivan, D., *On the ergodic theory at infinity of an arbitrary discrete group of hyperbolic motions*, in "Riemann surfaces and related topics: Proceedings of the 1978 Stony Brook Conference," Annals of Math. Studies **97**, Princeton Univ. Press 1981, 465–496.

[**T1**] Tukia, P., *On two-dimensional quasiconformal groups*, Ann. Acad. Sci. Fenn. **5** (1980), 73–78.

[**T2**] Tukia, P., *A quasiconformal group not isomorphic to a Möbius group*, Ann. Acad. Sci. Fenn. **6** (1981), 149–160.

[**T3**] Tukia, P., *On quasiconformal groups*, J. d'Analyse Math. **46** (1986), 318–346.

[**V1**] Väisälä, J., "Lectures on n-dimensional quasiconformal mappings," Lecture Notes in Mathematics **229**, Springer-Verlag, 1971.

[**V2**] Väisälä, J., *Quasimöbius maps*, J. d'Analyse Math. **44** (1984/85), 218–234.

Generic fundamental polyhedra for kleinian groups

BY T. JORGENSEN AND A. MARDEN

1. Introduction.

For aficionados of fundamental polyhedra in the study of the kleinian groups, it is helpful to be able to choose in each particular case a polyhedron with the simplest possible local structure about its edges and vertices. For example, in the study of small deformations as in [4], a fundamental polyhedron for one group is compared to those of nearby groups; if the one polyhedron is as simple as possible, the nearby ones will tend to be as well. It is the purpose of the present note to find such polyhedra. Indeed, we will show that the "generic" fundamental polyhedra for a given group are as simple as the algebraic/geometric structure of the group allows. When the group has no elliptic transformations, the features of the generic polyhedra will be precisely identified. In groups with torsion, on the other hand, certain configurations of elliptic transformations, for example three elliptics whose axes are pairwise coplanar, involve additional difficulties and we have decided to leave these cases aside.

The polyhedra we deal with are Dirichlet regions $P_0(y)$ centered at a point y of hyperbolic 3-space H^3 or at an ordinary point on the sphere at infinity $\partial \mathsf{H}^3$. The most satisfying situation occurs when there are no parabolic transformations in the group G. Then for a dense set of points $y \in \mathsf{H}^3 \cup \Omega(G)$, in the G-orbit of $P_0(y)$ every edge of $P_0(y)$ is surrounded by exactly three polyhedra and every vertex is shared by exactly four polyhedra. This statement is not quite true in general when there are parabolics and the best one can obtain is described later.

We have organized this investigation as follows. In Chapter 2 we have collected the formulas for various hyperbolic planes and listed associated properties. This enables us in Chapter 3 to describe the equations representing such unwanted situations as three bisecting planes intersecting in a line and four intersecting in a point. We show that the equations are not identically satisfied (except in some special situations when they are!) by a separate geometric argument. This knowledge is drawn together in Chapter 4. We define "generic" polyhedra in terms of the desired properties in §4.3 and prove that for a dense set of points $y \in \mathsf{H}^3 \cup \Omega(G)$, $P_0(y)$ indeed has these properties (Theorem 4.6). When the group G is geometrically finite, the center of y can be chosen from a dense open set (Corollary 4.8).

2. Formulas.

2.1 We will denote the upper half space model of hyperbolic space by H^3, that is

$$\mathsf{H}^3 = \{(z,t) \mid z \in \mathbb{C},\ t > 0\}.$$

Its boundary, $\partial\mathsf{H}^3 = \mathbb{C} \cup \{\infty\}$, is the sphere at infinity.

If $P \subset \mathsf{H}^3$ is a (hyperbolic) plane, its supporting circle (which is a euclidean circle or line) on $\partial\mathsf{H}^3$ will be denoted by ∂P.

The action on $\mathsf{H}^3 \cup \partial\mathsf{H}^3$ of a Möbius transformation T,

$$T \sim \begin{pmatrix} a & b \\ c & d \end{pmatrix}, \quad ad - bc = 1,$$

is given by the formula

$$(z,t) \mapsto \left(-\frac{1}{c}\left\{ \frac{\overline{(cz+d)}}{|cz+d|^2 + |c|^2 t^2} - a \right\},\ \frac{t}{|cz+d|^2 + |c|^2 t^2} \right),$$

if $c \neq 0$, otherwise,

$$(z,t) \mapsto (a^2 z + ab,\ |a|^2 t).$$

Letting $t = 0$ one regains the usual action of T on the extended complex plane, $T(z) = (az + b)(cz + d)^{-1}$.

2.2 Given a Möbius transformation T and a point $x \in \mathsf{H}^3$ not fixed by T, the plane which is the perpendicular bisector of the line segment $[x, T^{-1}x]$ is denoted by $B(x;T)$; that is, if $d(\cdot,\cdot)$ denotes hyperbolic distance, then

$$B(x;T) = \{y \in \mathsf{H}^3 \mid d(y,x) = d(y,T^{-1}x)\}.$$

Note that $y \in B(x;T)$ if and only if $x \in B(y;T^{-1})$. The transformation T sends $B(x;T)$ onto $B(x;T^{-1})$.

LEMMA. *Suppose T is a Möbius transformation and $p \in \partial\mathsf{H}^3$ is a given point, not a fixed point of T. Then $p \in \partial B(x;T^{-1})$ if and only if $x \in I(p;T)$ where $I(p;T)$ is the plane*

$$\left| z - \left\{ p + \frac{(p^2 c + pd - pa - b)(\overline{a - pc})}{|a - pc|^2 - 1} \right\} \right|^2 + t^2 = \frac{|p^2 c + pd - pa - b|^2}{(|a - pc|^2 - 1)^2},$$

if $|a - pc| \neq 1$. If $p = \infty$ this reduces to the isometric plane $I(\infty;T)$ for T,

$$|z + d/c|^2 + t^2 = |c|^{-2}.$$

If $|a - pc| = 1$, then $I(x; T)$ is the vertical plane

$$2\,\mathrm{Re}\,\{e^{i(\theta - \varphi)}(z - p)\} = |p^2 c + pd - pa - b|,$$

where

$$e^{i\theta} = a - pc, \qquad e^{i\varphi} = \frac{p^2 c + pd - pa - b}{|p^2 c + pd - pa - b|}.$$

PROOF: The case $p = \infty$ follows immediately from the formula for the action of T on H^3. For $p \neq \infty$, conjugate T by $z \mapsto -1/(z - p)$. This gives the general result. Note that p is a fixed point of T if and only if $cp^2 + dp - ap - b = 0$.

The plane $I(p; T)$ is vertical in the euclidean sense in H^3 if and only if p lies on the isometric circle of T^{-1}. (The term "isometric" plane in referring to $I(\infty; T)$ is justified because $I(\infty; T) = \{y \in \mathsf{H}^3 \mid |T'(y)| = 1\}$).

2.3 Given a point $p \in \partial \mathsf{H}^3$ not fixed by T we will define the bisecting plane $B(p; T)$ between p and $T^{-1}(p)$ to be

$$B(p; T) = I(p; T).$$

This definition is natural because of the following fact.

LEMMA. *If the point $x \in \mathsf{H}^3$ converges to $p \in \partial \mathsf{H}^3$, and $Tp \neq p$, then $B(x; T)$ converges to $I(p; T)$.*

PROOF: We may assume $p = \infty$. The center of $\partial I(\infty; T)$ is $T^{-1}(\infty)$. Let H denote the horosphere at ∞ which is tangent to $I(\infty; T)$; it meets $I(\infty; T)$ at the point $x_0 = (-d/c, 1/|c|) \in \mathsf{H}^3$ (in the notation of §2.1). The horosphere $T^{-1}(H)$ at $T^{-1}(\infty)$ is also tangent to $I(\infty; T)$ at x_0. Thus in the sense of horospherical distance, $I(\infty; T)$ is the perpendicular bisector of the line in H^3 between $T^{-1}(\infty)$ and ∞: the reflection in $I(\infty; T)$ interchanges the points ∞ and $T^{-1}(\infty)$. Thus $B(x; T) \to I(p; T)$.

Geometrically then, the identification $B(p; T) = I(p; T)$ is justified as follows: There is a unique horosphere H at p which is tangent to the horosphere $T^{-1}(H)$ at $T^{-1}(p)$. The bisecting plane $B(p; T) \equiv I(p; T)$ is the plane tangent to both horospheres H and $T^{-1}(H)$.

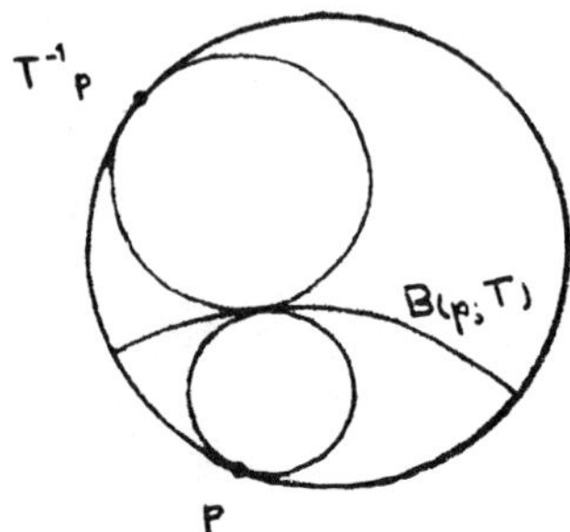

2.4 LEMMA. *If T is elliptic or parabolic then $\partial B(p;T)$ passes through the fixed points of T. If T is loxodromic then $\partial B(p;T)$ does not pass through any fixed point of T; in fact as p approaches a fixed point ς of T, the plane $B(p;T)$ converges to ς.*

PROOF: Assume that T is parabolic, say, $Tz = z + b$, $b > 0$. In this case,

$$\partial B(p;T) = \{z \mid \mathrm{Re}(p - z) = b/2\}.$$

For $-\infty < \mathrm{Re}\, p < \infty$ this gives the set of all vertical lines in $\mathbb{C}$. Note that if $\mathrm{Re}\, p \to \infty$ then $B(p;T) \to \infty$.

If T is elliptic we may assume that $Tz = e^{2i\theta}z$, where $\theta \not\equiv 0 (\mathrm{mod}\ \pi)$. In this case,

$$\partial B(p;T) = \{z :\ \mathrm{Im}\ e^{i(\theta - \varphi)}(p - z) = |p|\ \sin\ \theta\},$$

where we have set $e^{i\varphi} = p/|p|$. As φ ranges over values $0 \leq \varphi \leq 2\pi$, $\partial B(p;T)$ runs through the set of all straight lines through the origin.

If $|a - \varsigma c| \neq 1$, we see by inspection of the formulas of Lemma 2.1 that $B(p;T) \to \varsigma$ as $p \to \varsigma$, where ς is a fixed point of T.

If $|a - \varsigma c| = 1$ and $c \neq 0$, then ς lies on the isometric circle of T^{-1} so that T is elliptic or parabolic. If $c = 0$ and $|a| = 1$, then too we see directly that T is elliptic or parabolic.

If T is loxodromic and we take T to be of the form $Tz = k^2 z$, $|k| > 1$, then the formulas of Lemma 2.2 show that $\partial B(p;T)$ goes through neither 0 nor ∞; it actually separates the two points. Its equation is

$$\left| z - \frac{p}{|k|^4 - 1}\left(\frac{\bar{k}^2}{k^2} - 1\right) \right| = \frac{|p||k^4 - 1|}{|k|^2(||k|^4 - 1|)}.$$

COROLLARY. *If T is loxodromic, then for no $x \in \mathsf{H}^3$ does the boundary of the perpendicular bisecting plane of the segment $[x, Tx]$ pass through a fixed point of T.*

2.5 To make further calculations we have to make use of the ball model B^3 of hyperbolic space. The basic facts we need are as follows.

First $\partial \mathsf{H}^3 = \mathbb{C} \cup \{\infty\}$ is sent to $\partial \mathsf{B}^3$ by stereographic projection,

$$x_1 = \frac{z + \bar{z}}{|z|^2 + 1}, \quad x_2 = \frac{z - \bar{z}}{i(|z|^2 + 1)}, \quad x_3 = \frac{|z|^2 - 1}{|z|^2 + 1},$$

$$\partial \mathsf{B}^3 = \{(x_1, x_2, x_3) \mid x_1^2 + x_2^2 + x_3^2 = 1\}.$$

A circle in $\partial \mathsf{H}^3$,

$$\{x \in \mathbb{C} \mid |z - c| = r\},$$

is sent to the circle on $\partial \mathsf{B}^3$,

$$(*) \qquad \begin{cases} 2(\operatorname{Re} c)x_1 + 2(\operatorname{Im} c)x_2 + (|c|^2 - r^2 - 1)x_3 = |c|^2 - r^2 + 1, \\ x_1^2 + x_2^2 + x_3^2 = 1. \end{cases}$$

The euclidean line,

$$2 \operatorname{Re} \bar{c}z = m,$$

is sent to the circle on $\partial \mathsf{B}^3$,

$$(**) \qquad \begin{cases} 2(\operatorname{Re} c)x_1 + 2(\operatorname{Im} c)x_2 + mx_3 = m \\ x_1^2 + x_2^2 + x_3^2 = 1. \end{cases}$$

2.6 A hyperbolic plane P in B^3 is the intersection with B^3 of a sphere or plane P^* orthogonal to $\partial \mathsf{B}^3$. We will refer to the euclidean center of P^* as the center of P, where the center is ∞ if P^* is flat. A circle in $\partial \mathsf{B}^3$ supports a uniquely determined hyperbolic plane. The centers C of the planes supported by the circles $(*)$ and $(**)$ are, respectively,

$$C = \frac{(2\operatorname{Re} c, 2\operatorname{Im} c, |c|^2 - r^2 - 1)}{|c|^2 - r^2 + 1},$$

$$C = \frac{(2\operatorname{Re} c, 2\operatorname{Im} c, m)}{m}.$$

If a denominator is zero, we take $C = \infty$.

2.7 LEMMA. *In* $\mathbb{B}^3$, *three hyperbolic planes* P_1, P_2, P_3 *have a common line of intersection, or* ∂P_1, ∂P_2, ∂P_3 *have a common point of mutual tangency, if and only if their centers lie on a euclidean straight line in* $\mathbb{R}^3$ *exterior to* $\mathbb{B}^3$, *tangent to* $\partial \mathbb{B}^3$. *Four planes* P_1, P_2, P_3, P_4 *have a common point of intersection in* $\mathbb{B}^3$, *or* ∂P_1, ∂P_2, ∂P_3, ∂P_4 *have a common point of intersection on* $\partial \mathbb{B}^3$, *if and only if their centers lie on a euclidean plane in* $\mathbb{R}^3$ *exterior to* $\mathbb{B}^3$, *tangent to* $\partial \mathbb{B}^3$.

These facts are easily verified.

2.8 LEMMA. *In* $\mathbb{B}^3$, *two distinct hyperbolic planes* P_1, P_2 *are parallel, that is,* ∂P_1 *is tangent to* ∂P_2, *if and only if their euclidean centers lie on a line tangent to* $\partial \mathbb{B}^3$. *Three distinct hyperbolic planes* P_1, P_2, P_3 *are such that* $\partial P_1 \cap \partial P_2 \cap \partial P_3 \neq \emptyset$ *if and only if their centers lie on a plane tangent to* $\partial \mathbb{B}^3$.

2.9 Given $p \in \partial \mathbb{H}^3$ and the Möbius transformation $T \neq$ id. consider the plane $B(p; T)$. Via stereographic projection we may regard $B(p; T)$ as located in $\mathbb{B}^3$. Viewed in this way, $B(p; T)$ has a center $C(p; T) \in \mathbb{R}^3$. The purpose of this section is to record the formula for $C(p; T)$. Referring back to §2.6 we introduce the additional notation,

$$A = a - pc,$$
$$B = b - pc.$$

Then the formula for $C(p; T)$ is

$$C(p; T) = \frac{-(2\mathrm{Re}(p + \overline{A}B), 2\mathrm{Im}(p + \overline{A}B), |A|^2 - |B|^2 + |p|^2 - 1)}{|A|^2 + |B|^2 - |p|^2 - 1}.$$

When $p = \infty$ this reduces to

$$C(\infty; T) = \frac{-(2\mathrm{Re}\,\overline{c}d, 2\mathrm{Im}\,\overline{c}d, |c|^2 - |d|^2 + 1)}{|c|^2 + |d|^2 - 1}.$$

When the denominator is zero, the center is ∞.

If p is a fixed point of T and $|a - pc| \neq 1$,

$$C(p; T) = \frac{(2\mathrm{Re}\,p, 2\mathrm{Im}\,p, |p|^2 - 1)}{|p|^2 + 1},$$

which is the image of p under stereographic projection. If however $|a - pc| = 1$, the formula is indeterminate.

LEMMA. *If $C(p; T_1) = C(p; T_2)$ for three distinct points $p \in \mathbb{C} \cup \{\infty\}$, then $T_1 \equiv T_2$.*

PROOF: In ∂H^3, the circles $\partial B(p; T_2)$ have the same centers if and only if $T_1^{-1}(p) = T_2^{-1}(p)$. (The condition $C(p; T_1) = C(p; T_2)$ implies more strongly that $\partial B(p; T_1) \equiv \partial B(p; T_2)$.)

2.10 In contrast to the planes $B(p; T)$, the formula for the center of a bisecting plane $B(x, T)$ is much more complicated. For T acting in $\mathbb{B}^3$, given $x \in \mathbb{B}^3$, the formula for the center $C(x; T^{-1})$ of $B(x; T^{-1})$ is,

$$C(x; T^{-1}) = [x + (1 - |x|^2)(b + x)^*]^*,$$

where,

$$b = a^* - (|a^*|^2 - 1)(x^* + a^*)^*,$$
$$a = -x + (1 - |x|^2)((Tx)^* - x)^*,$$

and $y^* = y/|y|$ is the notation for reflection in $\partial \mathbb{B}^3$ (see [1]).

3. The equations.

3.1 LEMMA. *Suppose T_1, T_2, T_3 are distinct Möbius transformations not equal to the identity nor all elliptic, which do not have a common fixed point. Then*

$$B(p; T_1) \cap B(p; T_2) \cap B(p; T_3)$$

contains a line, or the circles $\partial B(p; T_1)$, $\partial B(p; T_2)$, $\partial B(p; T_3)$ are mutually tangent at a common point, if and only if $p \in \mathbb{C}$ satisfies the two real algebraic equations arising after elimination of k from the relation

(1) $$C(p; T_1) - C(p; T_2) = k(C(p; T_1) - C(p; T_2))$$

for some $k \in \mathbb{R}$, $k \neq 0$. This pair of equations is proper: In other words, it is not satisfied by every $p \in \mathbb{C}$. Therefore its solution set is nowhere dense in $\mathbb{C}$.

COROLLARY. *Under the same hypothesis on T_1, T_2, T_3, or if T_1, T_2, T_3 are distinct elements of a cyclic loxodromic group,*

$$B(x; T_1) \cap B(x; T_2) \cap B(x; T_3)$$

contains a line, or the circles $\partial B(x; T_1)$, $\partial B(x; T_2)$, $\partial B(x; T_3)$ are mutually tangent at a common point, if and only if $x \in \mathsf{H}^3$ satisfies the two real algebraic equations determined by the relation

$$\text{(2)} \qquad C(x; T_1) - C(x; T_2) = k(C(x; T_1) - C(x; T_2))$$

for some $k \in \mathsf{R}$, $k \neq 0$. This pair of equations is proper: it is not satisfied by every $x \in \mathsf{H}^3$ and therefore its solution set is nowhere dense.

PROOF: Because the transformations are distinct, $C(p; T_i) = C(p; T_j)$, $i \neq j$, for at most three values of $p \in \mathsf{C} \cup \{\infty\}$ (Lemma 2.9). By Lemma 2.7, the planes $B(p; T_j)$ intersect as described if and only if (1) is satisfied, and (1) includes the sporadic cases that two or three of the planes are identical. The problem here is to show that (1) is not satisfied for all $p \in \mathsf{C}$.

Under the hypothesis, there is a loxodromic or parabolic fixed point ς, of T_3 say, which is not fixed by T_1, say. As $p \to \varsigma$, $B(p; T_1) \to B(\varsigma; T_1)$. On the other hand, if p approaches ς suitably (if T_3 is parabolic), $B(p; T_3) \to \varsigma$. Thus if (1) were satisfied for all p (not a fixed point), then $\varsigma \in \partial B(\varsigma; T_1)$, which is impossible by Lemma 2.2.

To prove that (2) cannot be identically satisfied in H^3, we need only allow $x \in \mathsf{H}^3$ to approach a point $p \in \partial \mathsf{H}^3$ which does not satisfy (1). If instead, T_1, T_2, T_3 are distinct elements of a cyclic loxodromic group, take x to be a point on their common axis. The three planes $B(x; T_i)$ as well as the three circles $\partial B(x; T_i)$ are then mutually disjoint.

3.2 LEMMA. *Suppose T_1, T_2, T_3, T_4 are distinct Möbius transformations every three of which satisfies the hypothesis of Lemma 3.1. Then,*

$$B(p; T_1)^- \cap B(p; T_2)^- \cap B(p; T_3)^- \cap B(p; T_4)^- \neq \emptyset \text{ in } \mathsf{H}^3 \cup \partial \mathsf{H}^3,$$

if and only if $p \in \mathsf{C}$ satisfies the real algebraic equation determined by the relation,

$$\text{(3)} \quad C(p; T_1) - C(p; T_2) = k_1(C(p; T_1) - C(p; T_3)) + k_2(C(p; T_1) - C(p; T_4)),$$

for some $k_1 \neq 0$, $k_2 \neq 0$ in R. This equation is proper: it is not satisfied by every $p \in \mathsf{C}$ and therefore its solution set is nowhere dense.

COROLLARY. *Suppose T_1, T_2, T_3, T_4 are distinct Möbius transformations every three of which satisfy the hypothesis of Corollary 3.1. Then*

$$B(x; T_1)^- \cap B(x; T_2)^- \cap B(x; T_3)^- \cap B(x; T_4)^- \neq \emptyset \text{ in } \mathsf{H}^3 \cup \partial \mathsf{H}^3,$$

if and only if $x \in \mathsf{H}^3$ satisfies the real algebraic equation determined by the relation

$$(4) \quad C(x;T_1) - C(x;T_2) = k_1(C(x;T_1) - C(x;T_3)) + k_2(C(x;T_1) - C(x;T_4))$$

for some $k_1 \neq 0$, $k_2 \neq 0$ in R. The equation is proper: it is not satisfied by every $x \in \mathsf{H}^3$ and therefore its solution set is nowhere dense.

PROOF: The equation resulting from (3) would be identically satisfied in p if for three of the elements T_i, T_j, T_k of T_1, T_2, T_3, T_4, the centers $C(p;T_1)$, $C(p;T_j)$, $C(p;T_k)$ lay along a line for all p. This possibility is excluded by hypothesis and Lemma 3.1. For a dense set of values $p \in \mathsf{C}$, the vectors $C(p;T_1) - C(p;T_3)$ and $C(p;T_1) - C(p;T_4)$ determine a proper plane in R^3. That $C(p;T_1) - C(p;T_2)$ cannot lie in this plane for the dense set of values of p is seen by letting p approach a loxodromic or parabolic fixed point which is not fixed by all T_i.

As before the Corollary is proved by allowing x to approach $p \in \partial \mathsf{H}^3$ which does not satisfy (3), or to lie on the common axis of the three loxodromic elements.

3.3   LEMMA. *Suppose T_1, T_2, T_3 are distinct Möbius transformations satisfying the hypothesis of Lemma 3.1. Then*

$$\partial B(p;T_1) \cap \partial B(p;T_2) \cap \partial B(p;T_3) \neq \emptyset$$

if and only if there is a vector $\vec{\rho}$ which satisfies

$$(5) \quad \begin{aligned} \vec{\rho} \cdot C(p;T_i) &= 1, \quad i = 1,2,3, \\ |\vec{\rho}| &= 1. \end{aligned}$$

The real algebraic equation for $p \in \mathsf{C}$ that arises from (5) is proper.

COROLLARY. *Suppose T_1, T_2, T_3 are distinct Möbius transformations satisfying the hypothesis of Corollary 3.1. Then*

$$\partial B(x;T_1) \cap \partial B(x;T_2) \cap \partial B(x;T_3) \neq \emptyset$$

if and only if there is a vector $\vec{\rho}$ which satisfies

$$(6) \quad \begin{aligned} \vec{\rho} \cdot C(x;T_i) &= 1, \quad i = 1,2,3, \\ |\vec{\rho}| &= 1. \end{aligned}$$

The real algebraic equation for $x \in \mathsf{H}^3$ that arises from (6) is proper.

PROOF: For a dense set of values p, the vectors $C(p; T_1) - C(p; T_2)$, $C(p; T_1) - C(p; T_3)$ are linearly independent and span a proper plane in R^3. Equation (5) says that this plane is tangent to $\partial \mathsf{B}^3$ at the point $\vec{p}$. Equivalently, (5) says that the circles $\partial B(p; T_i)$ on $\partial \mathsf{B}^3$ all pass through $\vec{p}$. Once again the device of letting p approach a loxodromic or parabolic fixed point shows that (5) is proper.

That (6) is also proper is shown by allowing x to approach a suitable point $p \in \partial \mathsf{B}^3$, or taking x on the common axis of the T_i.

3.4 In the same way one obtains

LEMMA. *Suppose $T_1, T_2 \neq$ id. are distinct and not both are parabolic with the same fixed point. Then $\partial B(p; T_1)$ and $\partial B(p; T_2)$ are tangent if and only if there is $k \in \mathsf{R}$, $k \neq 0, 1$ such that*

$$(7) \qquad \begin{cases} C(p; T_i) \cdot \vec{p} = 1, & i = 1, 2 \\ \vec{p} = (1 - k)C(p; T_1) + kC(p; T_2) \\ |\vec{p}| = 1. \end{cases}$$

The system of two real algebraic equations for $p \in \mathsf{C}$ that arises from (7) is proper.

COROLLARY. *With T_1, T_2 as above, $\partial B(x; T_1)$ is tangent to $\partial B(x; T_2)$ if and only if there is $k \in \mathsf{R}$, $k \neq 0, 1$ such that*

$$(8) \qquad \begin{cases} C(x; T_i) \cdot \vec{p} = 1, & i = 1, 2 \\ \vec{p} = (1 - k)C(x; T_1) + kC(x; T_2) \\ |\vec{p}| = 1. \end{cases}$$

The system of two real algebraic equations for $x \in \mathsf{H}^3$ that arises from (8) is proper.

4. Generic polyhedra.

4.1 We are now nearly ready to apply our results to a kleinian group G without elliptic elements. The transformations considered are all assumed different from the identity.

For each triple T_1, T_2, T_3 of distinct elements without a common fixed point define

$$\mathcal{E}^*(T_1, T_2, T_3) = \{p \in \mathsf{C} \cup \{\infty\} \mid p \text{ satisfies (1) or (5)}\}.$$

For each quadruple T_1, T_2, T_3, T_4 of distinct elements of G no three of which have a common fixed point define

$$\mathcal{V}^*(T_1, T_2, T_3, T_4) = \{p \in \mathbb{C} \cup \{\infty\} \mid p \text{ satisfies (3)}\}.$$

For each triple T_1, T_2, T_3 of distinct elements not all parabolic with a common fixed point, define

$$\mathcal{E}(T_1, T_2, T_3) = \{x \in \mathsf{H}^3 \mid x \text{ satisfies (2) or (6)}\}.$$

For each quadruple T_1, T_2, T_3, T_4 of distinct elements no three of which are parabolic with a common fixed point, define

$$\mathcal{V}(T_1, T_2, T_3, T_4) = \{x \in \mathsf{H}^3 \mid x \text{ satisfies (4)}\}.$$

For each pair of distinct elements T_1, T_2 which are not parabolic with a common fixed point, define

$$\mathcal{T}^*(T_1, T_2) = \{p \in \mathbb{C} \cup \{\infty\} \mid p \text{ satisfies (7)}\},$$
$$\mathcal{T}(T_1, T_2) = \{x \in \mathsf{H}^3 \mid x \text{ satisfies (8)}\}.$$

Let $\mathcal{E}^*$, $\mathcal{V}^*$, $\mathcal{T}^*$ denote the subsets of $\mathbb{C} \cup \{\infty\}$ which are the unions of $C^*(T_1, T_2, T_3)$, $\mathcal{V}^*(T_1, T_2, T_3, T_4)$, $\mathcal{T}^*(T_1, T_2)$ over all admissible triples, quadruples and pairs respectively of G. Let $\mathcal{E}$, $\mathcal{V}$, $\mathcal{T}$ denote the corresponding subsets of H^3. Finally define

$$\mathcal{G}^* = \mathbb{C} \cup \{\infty\} \backslash (\mathcal{E}^* \cup \mathcal{V}^* \cup \mathcal{T}^*),$$
$$\mathcal{G} = \mathsf{H}^3 \backslash (\mathcal{E} \cup \mathcal{V} \cup \mathcal{T}).$$

By our previous results (and the Baire category theorem), $\mathcal{G}^*$ is dense in $\mathbb{C} \cup \{\infty\}$ and $\mathcal{G}$ is dense in H^3.

4.2 We will consider at once all the Dirichlet fundamental polyhedra $\{P_0(y)\}$ for G where, for $y \in \mathsf{H}^3$,

$$P_0(y) = \{x \in \mathsf{H}^3 \mid d(x, y) \le d(y, Tx), \forall T \in G\},$$

and for $y \in \partial \mathsf{H}^3$,

$$P_0(y) = \{x \in \mathsf{H}^3 \mid x \text{ lies in the closure of the component of } \mathsf{H}^3 \backslash B(y; T)$$
$$\text{that is adjacent to } y, \forall T \in G, T \ne \text{ id.}\}.$$

80

However when $y \in \partial H^3$, $P_0(y)$ is well defined and is a fundamental polyhedron only when y has special properties, for example, when $y \in \Omega(G)$.

The point y is called the *center* of $P_0(y)$. Set

$$P(y) = P_0(y)^- \cap (H^3 \cup \Omega(G)).$$

Then $P(y) \cap \Omega(G)$ is a fundamental set for the action of G on $\Omega(G)$. It is shown in [**3**] as a consequence of the Ahlfors Finiteness Theorem, that if G is finitely generated then $P(G) \cap \Omega(G)$ has a finite number of components, each of which is either a point, or a finite sided circular polygon.

Associated with each edge e of $P_0(y)$ is an edge cycle of length k, $(T_1 = \text{id.}, T_2, \ldots, T_k, T_{k+1} = T_1)$, where $P(y)$, $T_2 P(y), \ldots, T_k(P(y))$ is the cyclic arrangement of polyhedra about e in the G-orbit of $P(y)$. Equivalently, k is the number of disjoint edges of $P_0(y)$ that are equivalent to e under G.

The order k of a vertex of $P_0(y)$ is the number of distinct vertices of $P_0(y)$ that are equivalent to v under G. Equivalently, k is the number of polyhedra, $P(y)$, $T_2 P, \ldots, T_k(P)$ in the G-orbit of $P(y)$ that share the vertex v. The transformations $T_i \in G$, are said to be associated with v.

A *cusp* of $P(y)$ is a parabolic fixed point that lies in the euclidean closure of $P(y)$. It is of rank one or two according to the rank of the parabolic subgroup that fixes it. Boundary vertices and edges of $P(y)$ are those that lie in $\Omega(G)$. Associated with each boundary vertex is a vertex cycle analogous to the edge cycles of $P_0(y)$.

The full line containing an edge e of $P_0(y)$ is denoted by $\ell(e)$.

4.3 The polyhedron $P(y)$ is called *generic* if it has the following properties.

(i)   Each edge e of $P_0(y)$ for which $\ell(e)$ does not end at a parabolic fixed point has an edge cycle of length three. If $\ell(e)$ ends at a parabolic fixed point ς, then e has an edge cycle of length three or four, and every transformation entering into the cycle fixes ς.

(ii)   Three edges emanate from each vertex v of $P_0(y)$. For at most one of them e, $\ell(e)$ ends at a parabolic fixed point ς. The order of v is either four or five. In the latter case, three of the four transformations $\neq$ id. associated with v are parabolic and fix the end point ς of $\ell(e)$ for an edge e emanating from v.

(iii)   Every boundary vertex v^* is an end point of exactly one edge e of $P_0(y)$. The vertex cycle at v^* has length three or four. In the latter case, either all the transformations are parabolic and fix the other end of e, or they all lie in a cyclic loxodromic subgroup and $y \in \partial H^3$.

(iv) No edges of $P_0(y)$ end at a rank one cusp ς of $P(y)$ but two faces of $P(y)$ are tangent to ς with a face pairing transformation that fixes ς. Every rank two cusp ς is the end point of four or six edges of $P_0(y)$.

REMARK: We shall see that if an edge e in (i) or (ii) has order four, then $\mathrm{Stab}(\varsigma)$ is a rank two parabolic group associated with a rectangular torus.

4.4 There are still two special cases that must be dealt with.

LEMMA. *Suppose H is a cyclic loxodromic group and that $y \in \partial\mathsf{H}^3$ is not a fixed point of H. Then $P_0(y) = P(y) \cap \mathsf{H}^3$ is generic and $P(y) \cap \Omega(H)$ is connected. Each boundary vertex v^* is the end point of a single edge of $P_0(y)$. The vertex cycle about v^* has length three or four.*

These properties were proved in [3]. One can show that they remain valid when the base point y is chosen in H^3, but we shall not need this fact.

4.5 LEMMA. *Suppose H is a rank two parabolic group with common fixed point ∞. For all $y \in \mathsf{H}^3 \cup \mathsf{C}$, $P(y) \cap \mathsf{C}$ is either a hexagon or it is a rectangle and $P(y)$ is the vertical chimney arising from this. In the former case the cycle about each of the six edges of $P_0(y)$ has length three, and in the latter case, the cycle about each of the four edges of $P_0(y)$ has length four.*

PROOF: We may restrict our attention to the action of H in C, and we will assume that $y \in \mathsf{C}$. $P(y) \cap \mathsf{C}$ is convex and therefore connected. By considering the Euler characteristic on the torus C/H we deduce that $P(y) \cap \mathsf{C}$ has either four or six sides. As usual, the sides are arranged in pairs. Corresponding to each pair (s, s') is a side pairing transformation $S \in H$, $S(s) = s'$. The sides s, s' are parallel since they are orthogonal to the straight line L through y, $S(y)$ (and $S^{-1}(y)$). Thus if $P(y) \cap \mathsf{C}$ has four sides it must be a rectangle: if it were not then the translation S along L could not send s onto s'.

4.6 THEOREM. *Let G be a kleinian group without elliptic elements. There are dense sets of points $\mathcal{G}^* \subset \partial\mathsf{H}^3$, $\mathcal{G} \subset \mathsf{H}^3$ such that for any $y \in \mathcal{G}$, or for any $y \in \mathcal{G}^* \cap \Omega(G)$, $P(y)$ is a generic fundamental polyhedron for G.*

PROOF: Assume that $y \in \mathcal{G}$. The proof for $y \in \mathcal{G}^*$ is very similar.

Consider first an edge e of $P_0(y) = P(y) \cap \mathsf{H}^3$, and the corresponding cycle of distinct transformations $(\mathrm{id.}, T_2, T_3, \dots)$. If this cycle has length ≥ 4 then the bisecting planes $B(y; T_i^{-1})$, $i = 2, 3, 4$ intersect in a line or plane containing e. If this happens, by Corollary 3.1, T_2, T_3, T_4 belong to a rank two parabolic subgroup G_0. For this case to occur, according to Lemma 4.5, G_0 represents a rectangular torus and the cycle about e has exactly length four. Thus edges are generic in the sense of (i).

Consider next a vertex v of $P_0(y)$ and the associated distinct transformations (id., $T_2, T_3, \ldots$). Suppose that there were four of these $\neq$ id., T_2, T_3, T_4, T_5. By Corollary 3.2, three of them, say T_2, T_3, T_4 would be parabolic with a common fixed point and therefore lie in a parabolic subgroup G_0, necessarily of rank two. For this to happen, G_0 would have to be a rectangular torus group and T_2, T_3, T_4 associated with an edge e' of the polyhedron $P(y)$ for G_0. No other transformation associated with v can lie in G_0. Therefore there can be only one other transformation T_5 associated with v for otherwise the quadruple (T_2, T_3, T_5, T_6) would violate our choice of $y \in \mathcal{G}$. Now T_2, T_3, T_4 are associated with an edge e' for G_0 and e' is cut by $B(y; T_5^{-1})$ and no additional bisecting plane at v. Therefore there is an edge of $P(y)$ emanating from v that is contained in e'. We conclude that the order of v is four, unless the exceptional case as described above arises, in which case it is five. In any case, there are exactly three edges of $P(y)$ that emanate from v.

Two of the edges e_1, e_2 from v cannot both be associated with parabolic transformations. For the face of $P_0(y)$ that contains both e_1 and e_2 is contained in $B(y; T_i^{-1})$ for one of the parabolic transformations, say T_1, associated with v and with the edge cycle for e_1. But then $\partial B(y; T_1^{-1})$ not only goes through the fixed point ς_1 of T_1, but it goes through the fixed point $\varsigma_2 \neq \varsigma_1$ of a different parabolic transformation, say T_2, which is also associated with v. Thus $\partial B(y; T_1^{-1})$, $\partial B(y; T_2^{-1})$, $\partial B(y; T_2^{-2})$ have the point ς_2 in common in violation of our choice of $y \in \mathcal{G}$. Thus vertices of $P_0(y)$ are generic in the sense of (ii).

Now let $v^* \in P(y) \cap \Omega(G)$ be a boundary vertex. There is at least one edge e of $P_0(y)$ that ends at v^* because our choice of $y \in \mathcal{G}$ does not allow the possibility that two faces of $P_0(y)$ are tangent at v^*. Consider the vertex cycle at v^*, (id., $T_2, T_3, \ldots$). Again our choice of $y \in \mathcal{G}$ does not allow the possibility that there are more than two transformations besides the identity in this cycle, unless the transformations generate a rectangular torus and fix the other end of e. (Here there is a difference between $y \in \mathsf{H}^3$ and $y \in \partial \mathsf{H}^3$—see Lemma 4.4.) Thus there can be only one edge ending at v^*. Furthermore, the vertex cycle for each boundary vertex v^* has length three, with the one exception for $y \in \mathsf{H}^3$ when $P(y)$ contains the entire edge ending at v^* of a rectangular chimney for a rank two parabolic subgroup.

Finally, suppose that ς is a parabolic fixed point in the closure of $P(y)$ and $G_0 = \mathrm{Stab}(\varsigma)$. Our choice of $y \in \mathcal{G}$ implies that $\varsigma \in \partial B(y; T)$ if and only if $T \in G_0$. Furthermore, no plane $B(y; T)$, $T \notin G_0$, separates ς from y.

We may assume that $\varsigma = \infty$. If G_0 has rank two, its fundamental polyhedron $P(y; G_0)$ is a vertical chimney based on a hexagon or rectangle in $\mathbb{C}$. For some horoball $H = \{(z, t) \in \mathsf{H}^3 \mid t > h\}$, sufficiently large h, $P(y) \cap H = P(y; G_0) \cap H$. If G_0 has rank one, its fundamental polyhedron $P(y; G_0)$ is a vertical slab in H^3, based on a parallel strip S in $\mathbb{C}$. Take any bounded segment S_0 of S and the vertical chimney $S_0^* \subset P(p; G_0)$ lying over S_0. For sufficiently large h, $P(y) \cap S_0^* \cap H = S_0^* \cap H$.

4.7 COROLLARY. *Suppose G is a finitely generated kleinian group without elliptic transformations. Then for y in a dense subset $\mathcal{G} \subset \mathsf{H}^3$, $F(y) = P(y) \cap \Omega(G)$ is a finite union of finite sided circular polygons. Each vertex cycle of $F(y)$ has length three and if ς is a parabolic fixed point in $F(y)^-$, then two sides of some component of $F(y)$ end at ς and are paired by a parabolic element fixing ς; no other sides of $F(y)$ end at ς.*

PROOF: The only new ingredient is the result from [2].

4.8 COROLLARY. *Suppose G is a geometrically finite kleinian group without elliptic transformations. Then there is a dense open set $\mathcal{G}_0 \subset \mathsf{H}^3$ for which the fundamental polyhedron $P(y)$, $y \in \mathcal{G}_0$, is generic.*

PROOF: Consider a generic polyhedron $P(y_0)$ for G, $y_0 \in \mathcal{G}$. It has a finite number of faces. As y moves a small amount in H^3 near y_0, the structure of $P(y)$ with respect to cusps and edges of cycle length four is unchanged. Indeed for all y sufficiently close to y_0, every face, edge, or vertex pairing transformation of $P(y)$ is at the same time one of the same type for $P(y_0)$. That is, $P(y)$ is generic as well. Now choose $\mathcal{G}_0$ as the union of such neighborhoods for all $y_0 \in \mathcal{G}$.

4.9 REMARK: Let G be a arbitrary kleinian group. There exists a dense set of points $\mathcal{G} \subset \mathsf{H}^3$ such that for any $y \in \mathcal{G}$, the polyhedron $P(y)$ is as generic as the algebraic/geometric structure of G allows.

PROOF: A particular algebraic equation which characterizes an unwanted geometric situation is either identically satisfied or not. If it is identically satisfied, then the situation will persist no matter what the choice is for the center of the polyhedron. Remove from our countable set of equations all such identically satisfied equations. The meaning of a generic polyhedron $P(x)$ is that x does not satisfy any of the remaining equations.

T. Jorgensen, Department of Mathematics, Columbia University, New York, NY 10027

A. Marden, Department of Mathematics, University of Minnesota, Minneapolis, MN 55455

85

REFERENCES

1. Ahlfors, L.V., "Möbius Transformations in Several Dimensions," University of Minnesota Lecture Notes, Minneapolis, 1981.
2. Beardon, A.F. and Jorgensen, T., *Fundamental domains for finitely generated Kleinian groups*, Math. Scand. **35** (1975), 21–26.
3. Jorgensen, T., *On cyclic groups of Möbius transformations*, Math. Scand. **33** (1974), 250–260.
4. Jorgensen, T. and Marden, A., *Algebraic and geometric convergence of Kleinian groups.*

Quasiconformal Actions on Domains in Space

BY GAVEN J. MARTIN

§1. Introduction.

The purpose of this paper is to investigate the topological and analytical restrictions on a domain D in euclidean n-space $\mathbf{R}^n$ on which an infinite discrete quasiconformal group can act. We will see that the restrictions are indeed severe, unlike the case of a discrete group of topological or differentiable homeomorphisms.

We have several goals in mind. Firstly we would like to investigate the consequences of the universal cover of an n-manifold quasiconformally embedding in n-space. Theorems due to Gromov [Gr.], Sullivan [Su.] and Tukia [Tu.1], [Tu.2] imply that if the universal cover of a closed n-manifold is quasiconformally equivalent to the unit ball, then that manifold has a hyperbolic structure. That is, there is a discrete and faithful embedding of the fundamental group into the group of hyperbolic isometries of the unit ball $\mathbf{B}^n$. Actually, in the case $n = 3$ we do not need to assume that the manifold is closed. More recently we have shown, see [M.G.], that there is a semi-conjugacy by a pseudo-isometry between the fundamental group of the manifold and a discrete subgroup of hyperbolic isometries which yields a homotopy equivalence between the manifold and a manifold of constant negative curvature, sometimes this will imply that the underlying manifold is homeomorphic to a hyperbolic space form, for instance if the manifold is Haken. Thus we seek natural conditions to imply that if a domain admits an infinite discrete quasiconformal group action, then it is quasiconformally equivalent to the unit ball. We will see that if the boundary of the domain is sufficiently regular at a single conical limit point of the group, then the domain is quasiconformally equivalent to the unit ball. The results and proofs in this direction are similar to those obtained by Gehring and Palka [G.P.] in their study of quasiconformally homogeneous domains, although there are some added complications in our situation and we will obtain some new results concerning quasiconformally homogeneous domains.

Secondly, we would like to begin an investigation into quasiconformal actions on manifolds. The natural place to start is with manifolds that are reasonably well understood, namely domains in space. We will find that there are both topological and analytical restrictions on a domain in space to admit an infinite quasiconformal group action. For example we will find domains D_1 and D_2 which are diffeomorphic, but D_1 admits an infinite discrete quasiconformal action while D_2 does not. We also exhibit a domain D for which the only discrete quasiconformal action is the trivial action $\{Id.\}$. We will also show that if D admits an infinite quasiconformal group G and if a limit point of G is a manifold

Research supported in part by the A.P. Sloan Foundation and NSF Grant 8120790.

point of D, then D is contractible. Many of the results we obtain in this direction will also be true for so called convergence groups, see [G.M.]. These are groups of topological homeomorphisms with the compactness properties of quasiconformal mappings. However for simplicity we will restrict our attention to the quasiconformal case.

1.1 NOTATION AND DEFINITIONS. Throughout this paper D   will denote a subdomain of $\mathbf{R}^n$. We denote the usual basis vectors of $\mathbf{R}^n$ by $e_1, e_2,..., e_n$. A group of self homeomorphisms of D is called a <u>quasiconformal group</u> if there is a finite K   such that each $g \in G$ is K-quasiconformal. A group G will be called <u>discrete</u> if G is a discrete subgroup of the group of self homeomorphisms of D with the compact open topolgy. We will see later that the choice of topology on the group of self homeomorphisms of D will be of little consequence for discrete quasiconformal groups. A <u>Möbius (or conformal) transformation</u> of cl $(\mathbf{R}^n)$ is the finite composition of reflections in spheres and hyperplanes of cl $(\mathbf{R}^n)$. We denote the group of all Möbius transformations of cl $(\mathbf{R}^n)$ by <u>Möb(n)</u>. We remark here that a one-quasiconformal mapping of D is conformal and so by the generalized version of Liouvilles theorem (see [Ge.1]) is the restriction of a Möbius transformation when n > 2. We say that a group of homeomorphisms of D acts <u>properly discontinuously</u> in D if for each compact subset F of D there are only finitely many $g \in G$ for which

$$g(F) \cap F \neq \emptyset.$$

Let f:D$\rightarrow$ D be a homeomorphism. We set

$$< f > = \{ f^n : n \in \mathbf{Z} \}.$$

We say that f is a <u>(discrete) quasiconformal automorphism</u> of D if <f> is a (discrete) quasiconformal group. We note that the elements of a (discrete) quasiconformal group are always (discrete) quasiconformal automorphisms and that a conformal self map of D is a conformal automorphism.

One may easily construct a quasiconformal group acting on a domain D by conjugating a conformal group acting on D by a quasiconformal self homeomorphism of D. Gehring and Palka [G.P.] first asked in their study of quasi-conformally homogeneous domains, if this were the only way to construct such quasiconformal groups. It was subsequently shown by Sullivan [Su.] and Tukia [Tu.1] that this is indeed the case for two dimensional quasiconformal groups. For higher dimensions this is not the case, see [Tu.2], [Ma], [F.S.] and [G.M.] for a variety of examples in all dimensions greater than two.

Quasiconformal groups also arise naturally as groups of isometries of certain metrics on domains in space, for example those locally Lipschitz equivalent to the hyperbolic metric of the unit ball in $\mathbf{R}^n$. For more details and applications see [M.G.].

One of the main tools that we will use in our study is the following version of the Caratheodory convergence theorem for quasiconformal mappings of space domains, see [Vä. Cor.19.3 and 37.4].

1.2 THEOREM. *Let D be a domain with at least two boundary points and let $f_j:D \to D$ be a family of K-quasiconformal homeomorphisms. Then there is a subsequence, which we again relabel $\{f_j\}$, converging uniformly on compact subsets of D, which we denote c-uniformly in D, to a mapping f and either*

(1) f is constant and this value lies in ∂D or

(2) $f:D \to D$ is a K-quasiconformal homeomorphism.

It is clear that if the sequence $\{f_j\}$ lay in a discrete (in the compact open topology) quasiconformal group then the conclusion (2) could not occur. In fact it is not difficult to see that discreteness in any reasonable topology will imply that (2) could not occur. This is what we meant when we said that the topology of the group of self homeomorphisms of D will be of little consequence to us. Henceforth we will assume that D has at least two points in the boundary. The case that D has one or no boundary points is extensively covered in [G.M.] and in these cases it it is easy to see that there are infinite discrete conformal groups acting. The following result is an easy consequence of Theorem 1.2.

1.3 COROLLARY. *Let G be a discrete quasiconformal group acting on D. Then G is properly discontinuous in D.*

We conclude this section by providing the following easy example.

1.4 EXAMPLE. *The domain $D = \mathbf{R}^n\text{-}\{0,1,\infty\}$ admits no infinite discrete quasiconformal group.*

PROOF: Let $\{f_j\}$ be an infinite sequence of K-quasiconformal self homeomorphisms of D. By the removable singularity theorem for quasiconformal mappings, see [Vä. Thm 17.3], we see that each f_j extends to a quasiconformal mapping of cl $(\mathbf{R}^n)$ which must permute the points $0, 1$ and ∞ amongst themselves. Thus, by [Vä 20.5], the family $\{f_j\}$ is normal and (since no subsequence can converge to a constant map) so contains a subsequence converging to a K-quasiconformal self mapping of D. Thus every discrete quasiconformal group acting on D is finite.

It is easy to see that there are many finite quasiconformal groups acting on D, namely the rotations about the line through 0 and 1.

§2.1 A topological condition.

We now introduce a simple topological condition which we will use to show that certain domains do not admit infinite discrete quasiconformal actions.

2.2 DEFINITION. We refer to [Ru.] for the definition of an i-complex. All that is important to us is that the boundary of an i+1-simplex is an i-complex (it is homeomorphic to S^i). We will say that a domain D is <u>LBC(i) at $x \in \partial D$</u>, for i-locally boundary connected at x, if there is a neighbouhood of x in the closure of D containing no essential i-complex of D. That is, there is a neighborhood U_i of x in the closure of D such that for any i-complex Δ_i and any continuous map $h:\Delta_i \to U_i \cap D$, $h(\Delta_i)$ is homotopically trivial in $U_i \cap D$.

We say that D is i-locally connected at the boundary, or more simply D is LBC(i), if D is LBC(i) at each of its boundary points. Let G be a discrete quasiconformal group acting on D. We say x is a <u>limit point</u> of G if there is an infinite sequence of elements of G converging c-uniformly in D to x. It follows from Theorem 1.2 that $x \in \partial D$. Notice that if G is finite there can be no limit points and from the same theorem if G is infinite there is at least one limit point. For more details on the limit set of a discrete quasiconformal group see [G.M.].

The condition LBC(i) does not, of course, preclude the possibility that there is an essential i-complex in D, it merely asserts that locally, near the boundary there is no essential complex. As an example we see that the domain $D = \mathbf{R}^n - \{B^n(j,1/4) : j \in \mathbf{Z} \}$ is LBC(i) for every i at each boundary point other than ∞. At the point ∞ we see D is not LBC(n-1). The translation $x \to x+1$ generates an infinite discrete quasiconformal group acting on D. Our interest in the condition follows from

2.3 THEOREM. *Let G be a discrete quasiconformal group acting on D. Suppose that D is LBC(i) at a limit point x of G. Then there is no essential i-complex in D.*

PROOF: Let U be a neighborhood of x in the closure of D such that $U \cap D$ contains no essential i-complex, and let $\{f_j\}$ be a sequence of elements of G converging c-uniformly in D to x, both given by the hypotheses. Let Δ_i be any i-complex in

D. Since Δ_i is compact in D and since $f_j \to x$, we see for sufficiently large j, $f_j(\Delta_i)$ is a complex in U and hence trivial. Since f_j is a homeomorphism of D we must have that Δ_i is also trivial in D. The theorem is proved.

2.4 COROLLARY. *Let D be LBC(i) and suppose that $\pi_i(D) \neq 0$. Then every infinite sequence $\{f_j\}$ of K-quasiconformal self homeomorphisms of D contains a subsequence which converges c-uniformly in D to a K-quasiconformal self homeomorphism of D.*

PROOF: The hypothesis that $\pi_i(D) \neq 0$ implies that there is an essential i-complex in D, while we see as in the proof of Theorem 2.3 that this implies that no subsequence of $\{f_j\}$ can converge to a point of the boundary. The result then follows.

2.5 COROLLARY. *Let D be LBC(i) and $\pi_i(D) \neq 0$. Then every discrete quasiconformal group acting on D is finite.*

2.6 COROLLARY. *Let D be LBC(i) for all $0 < i < n$ and suppose that D admits an infinite discrete quasiconformal group. Then D is contractible.*

PROOF: It follows from the above corollaries, together with the hypothesis there is a discrete group acting, that every homotopy group must vanish for $0 < i < n$, while $\pi_n(D) = 0$ since D has at least two boundary points. Thus D has the homotopy type of a point.

We remark here that if D is a n-submanifold of $\mathbf{R}^n$ with boundary, then the boundary of D is collared in D and so ∂D is LBC(i) for all i. Thus if D admits an infinite quasiconformal group then D is contractible. There can be at most one component of the boundary of D, since the boundary could not be collared at a limit point which must lie on the boundary (as some components of the boundary must accumulate at such points). When $n = 3$, the only 2-manifold that bounds a contractible subset of $\mathbf{R}^3$ is the two-sphere S^2, which must then bound a ball by the Schönfliess theorem (recall that it is collared in D). Thus the only 3-submanifold of $\mathbf{R}^3$ with boundary that admits an infinite discrete quasiconformal group is $\mathbf{B}^3$.

2.7 EXAMPLES. The solid torus $S^1 \times \mathbf{B}^{n-1}$ in $\mathbf{R}^n$ cannot admit an infinite discrete quasiconformal group. Neither does the annulus $A = \{x \in \mathbf{R}^n: 1 < |x| < 2\}$. The annulus A does admit an infinite discrete group of diffeomorphisms acting properly discontinuously and without fixed points, namely the group generated by the homeomorphism $f(r,\zeta) = (2-(2-r)^2,\zeta)$, where we have used the coordinates $A = \{(r,\zeta) : r \in (1,2), \ \zeta \in S^{n-1}\}$. The quotient $A/\langle f \rangle = S^1 \times S^{n-1}$. If we pull back any metric of $S^1 \times S^{n-1}$ to A so that f acts as isometries, then this metric cannot be quasiconformally equivalent to the euclidean metric of A. If we define $g: \mathbf{R}^n\text{-}\{0\} \to A$ by $g(r,\zeta) = (2-e^{-r},\zeta)$, then $g^{-1} \circ f \circ g(r,\zeta) = (2r,\zeta)$ and so $\langle f \rangle$ is conjugate to a quasiconformal (actually conformal) group of $\mathbf{R}^n\text{-}\{0\}$. The importance of this example follows from the results of §4 where we will see that if the universal cover of a compact n-manifold quasiconformally embedds in $\mathbf{R}^n$ with "nice" boundary, then that universal cover is the ball. In the above example, we have two embeddings of $S^{n-1} \times \mathbf{R}$ (the universal cover of $S^{n-1} \times S^1$) in $\mathbf{R}^n$, one with nice boundary (two codimension one spheres) which cannot be quasiconformal and one which is quasiconformal but whose boundary is not nice (it is not codimension one, for instance). Notice that A and $\mathbf{R}^n\text{-}\{0\}$ are diffeomorphic, and that A does not admit an infinite discrete quasiconformal group while $\mathbf{R}^n\text{-}\{0\}$ does. Thus the diffeomorphism type of a domain does not determine whether it admits an infinite discrete quasiconformal group.

We now begin to show that a domain with finitely many, but at least two, boundary components cannot admit an infinite discrete quasiconformal group. The main tool is the following theorem whose proof can be found in [Vä. Thm. 21.11].

2.8 THEOREM. *Let $\{f_j\}$ be a sequence of K-quasiconformal self homeomorphisms of a domain D and suppose that the boundary of D has at least three points and finitely many, but at least two, components. Then there is a subsequence of the $\{f_j\}$ converging c-uniformly in D to a K-quasiconformal self homeomorphism of D.*

The following definition is merely to circumvent the difficulty in considering a quasiconformal action of a domain D which does not extend to the boundary of D.

2.9 DEFINITION. Let F be a family of self homeomorphisms of a domain D. We will say that the orbit of a component E of the boundary of D is *finite under* F if there is a finite collection $\{E_k\}$ of components of the boundary of D such that if $\{x_i\}$ is a sequence in D converging to a point of E, then for all $f \in F$, there is a k such that $\text{dist}(f(x_i),E_k) \to 0$ as $i \to \infty$.

Notice that if ∂D has finitely many components, then the orbit of any component of ∂D is finite. The motivation for the definition is that often a quasiconformal homeomorphism of a domain will extend to some of the boundary components. For instance if D is a uniform domain, see [G.Mo.]. In this case the orbit of some component will always be finite if there is only finitely many other components of D with the same topological type.

2.10 THEOREM. *Let $\{f_j\}$ be a sequence of K-quasiconformal self homeomorphisms of a domain D. Suppose that the boundary of D has at least two components and that there is an isolated component E of ∂D with a finite orbit under $\{f_j\}$. Then there is a subsequence of the $\{f_j\}$ converging c-uniformly in D to a K-quasiconformal self homeomorphism of D.*

PROOF: From Theorem 2.8 we may suppose that ∂D has infinitely many components. We need only show that conclusion (1) of Theorem 1.2 cannot occur. Suppose for contradiction that it did. Then there is a subsequence which we relabel as $\{f_j\}$ converging c-uniformly in D to a constant value $x \in \partial D$. Since E is isolated we may find a compact set F of D separating E and $\partial D\text{-}E$. We might find such an F as a component of the boundary of the union of all the simplicies meeting E of a sufficiently fine triangulation of $\mathbf{R}^n$.

Let $E_1, E_2,...., E_k$ be the smallest (by inclusion) finite orbit of E under the sequence $\{f_j\}$. We denote by $q(A)$ the spherical diameter of a subset A of $\mathbf{R}^n$. Let $r = q(\partial D\text{-}UE_j)$ and let $s = \min\{q(E_i):j=1,2,...,k\}$. Since ∂D has infinitely many components we see $r > 0$. We claim that in this situation E must be a point. If E were not a point then neither could E_i, $i = 1,2,...,k$, by the removable singularity theorem. Thus $s > 0$. We also find that for all j

$$q(f_j(F)) \geq \min\{r,s\} > 0.$$

This is because $f_j(F)$ separates some E_i from $\partial D\text{-}E_i$ and these have spherical diameter at least s and r respectively. This conclusion, however, is incompatible with the assumption that the sequence $\{f_j\}$ converges c-uniformly in D to $x \in \partial D$, for then $q(f_j(F)) \to 0$ as $j \to \infty$. Thus we must conclude that E, and hence all the E_i, are points. Since E is an isolated point it is removable, so then each E_i is an isolated point. Thus we may extend each f_j to a K-quasiconformal self homeomorphism of $D' = D \cup \{E_1, E_2,..., E_k\}$. From Theorem 1.2 and our above assumptions, we see that $\{f_j\}$ must converge c-uniformly in D' to a point of $\partial D'$. This is clearly impossible, as each f_j permutes the set $\{E_1,E_2,...,E_k\}$ which lies in the interior of D'. We have reached the desired contradiction and the result follows.

2.11 COROLLARY. *Let D be a domain in R^n and suppose that the boundary of D has at least three points and two components. Let G be a discrete quasiconformal group acting on D. If either*

> *(1) The boundary of D has finitely many components or*
> *(2) There is an isolated component of the boundary of D with a finite orbit under G,*

then G is finite.

2.12 COROLLARY. *Suppose that D is a domain in R^n and that the boundary of D has at least two points. Then for any discrete quasiconformal group G acting on D the stabilizer of a point $x \in D$, $G_x = \{g \in G : g(x) = x\}$, is a finite subgroup of G.*

Thus in order for a domain to admit an infinite discrete quasiconformal group, it is necessary that D have either one or infinitely many boundary components. It is clear that every contractible planar domain admits an infinite discrete quasiconformal group. This is since either the domain is the whole plane, in which case the fact is trivial, or the Riemann mapping theorem implies the existence of such a conformal group. This is not the case in higher dimension as the following example implies.

2.13 EXAMPLE. *There is a subdomain D of R^n, $n \geq 3$, such that D is homeomorphic to B^n and such that every discrete quasiconformal group acting on D is finite.*

PROOF: Let $D = B^n - [0, e_1]$. Then it easy to see that D is a uniform domain. That is, there are constants a and b such that for each pair of points x, y in D there is an arc ß joining x to y in D such that length(ß) $\leq$ a| x-y| and for each $z \in$ ß, min{length(ß(x,z)), length(ß(y,z))} $\leq$ b dist(z,∂D). If f is any quasiconformal self homeomorphism of D, then both D and f(D) = D are uniform domains and so by a theorem of Gehring and Martio, f has a homeomorphic extension to its closure, which is B^n. Since $[0, e_1]$ has finite n-1 dimensional measure, we find from [Vä. Thm. 35.1] that $[0, e_1]$ is an exceptional set. That is each such f extends to a quasiconformal self homeomorphism of B^n. It is clear from topological reasons, that each such f must fix the origin. This easily implies from our earlier results that every discrete quasiconformal group acting on D is finite. It remains only to observe that D is homeomorphic to B^n.

Actually, in dimension three we can slightly modify D to obtain a domain homeomorphic to B^3 such that the only orientation preserving discrete quasiconformal group acting on D is {Identity}. We illustrate the domain D below. It is homeomorphic to

$\mathbf{B}^3$ - {$[e_1/2,e_1]$,$[-e_1/2,-e_1]$,$[e_2/2,e_2]$,$e_1/2+[-e_2/4,e_2/4]$,$-3e_1/4+[-e_2/4,e_2/4]$}.

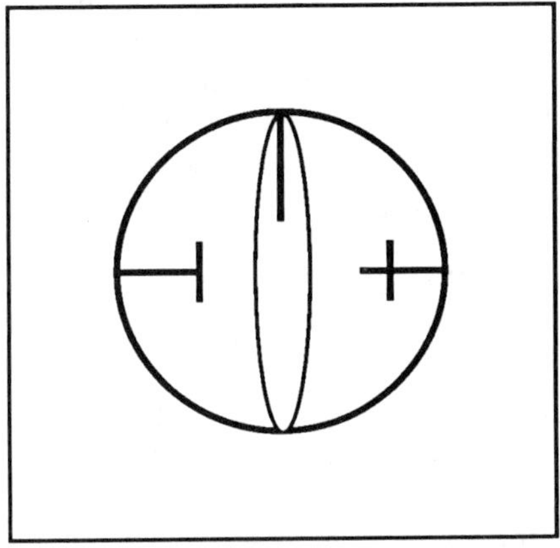

As before the domain is uniform and so every quasiconformal self mapping of D extends to a quasiconformal self homeomorphism of the closure of $\mathbf{B}^3$. Thus every discrete quasiconformal group acting on D is easily seen to be finite. Moreover, each such map must preserve each spike and fix the topologically distinguished points on them, namely the endpoints of two, the cross point of the third and the intersection of each with S^2. If f is a periodic self homeomorphism of D, then f must fix the line segment of the boundary of D connecting each pair of distinguished points. This is since the map must preserve this interval and a periodic map of an interval which fixes its endpoints must fix the entire interval. If in addition f is orientation preserving, we can extend the extension of f to $\mathbf{B}^3$, by reflection in S^2, to a periodic orientation preserving self homeomorphism of S^3. From Smith Theory [M.B.], we know that the fixed point set of f is a topological circle, and from our construction, this set must contain the three segments mentioned above, and be symmetric in the two sphere S^2. It is not difficult to see that this is impossible (any arc in $\mathbf{B}^3$ containing these three segments will, upon adding its reflection in S^2, yield a curve containing a wedge of circles). It is interesting to observe that reflection in the plane generated by e_1 and e_2 is a quasiconformal periodic self homeomorphism of D which is orientation reversing.

In the above example we used some topology which was special to dimension three, for instance the Smith conjecture fails in higher dimensions. Next, in section three, we will construct a domain in $\mathbf{R}^n$ for all $n \geq 2$ (and which will be simply connected if $n \geq 3$) for which the only discrete quasiconformal group acting is the identity. The construction is a little complicated. We outline the proof of the following proposition to show that (at least in dimension two) the domain must be infinitely connected and so in some sense the complicated construction is necessary.

2.14 PROPOSITION. *Let D be a finitely connected planar domain. Then D admits a finite nontrivial discrete quasiconformal action.*

PROOF: It is a standard fact that such a domain as D is conformally equivalent to the complex plane punctured at finitely many points or to the unit disk with n punctures and m subarcs of circles centered at the origin removed. We consider the latter case. The former being similar. Let p be the greatest common divisor of m and n. If $p = 1$ we can find a Z_2 action by choosing a diffeomorphism of the disk which is the identity near the boundary of the disk and moves each of the n punctures and m subarcs to subsets of the real line, and then using this diffeomorphism to conjugate the reflection in the real axis. If $p > 1$, we can find a Z_p action by choosing a diffeomorphism which is the identity near the boundary of the disk and which maps the n subarcs of circles to n equally spaced and equal length subarcs of the circle of radius $1/2$ as well as mapping the m punctures to equally spaced points on the circle of radius $1/4$. The desired action is then obtained by conjugating the rotation through the angle π/p by this diffeomorphism. In both cases the group will be uniformly quasiconformal since each diffeomorphism of the disk will be quasiconformal and conjugating a conformal group by a quasiconformal mapping will yield a quasiconformal group.

§3. A domain which admits no nontrivial discrete quasiconformal group.

We need some basic facts from the theory of quasiconformal mappings. We let $\mu(x,y)$ denote the hyperbolic distance between a pair of points x and y in the unit ball B^n of R^n. From [G.O.] Theorem 3 and (2.19) there is a constant c depending only on n and K such that if f is a K-quasiconformal self homeomorphism of B^n, then

$$(3.1) \qquad \mu(f(x),f(y)) \leq c \max\{\mu(x,y),\mu(x,y)^a\}, \qquad a = K^{1/(n-1)}.$$

Notice too that f^{-1} must also satisfy this inequality. The quasihyperbolic metric of a domain D with at least two boundary points is defined from the conformal metric tensor $ds^2 = \text{dist}(x,\partial D)^{-2}|dx|^2$. The quasihyperbolic distance between two points in D is denoted $k_D(x,y)$. Later we shall need the fact that the quasihyperbolic metric is a complete metric in a domain D and defines the usual topology there and that the inequality (3.1) is valid for μ replaced by k_D and f a K-quasiconformal self map of D, see [G.O.].

3.2 DEFINITION. We inductively define a sequence of real valued functions with rapid growth as follows. Set

$$(3.3) \qquad \exp_0(t) = t \quad \text{and} \quad \exp_m(t) = \exp(\exp_{m-1}(t)).$$

3.4 LEMMA. *Let x and y be points of B^n. Let $\{x_n\}$ and $\{y_n\}$ be disjoint sequences converging to x and y respectively such that for two distinct positive integers p and q we have*

(1) $\mu(x_n,x) = [\exp_p(n)]^{-1}$ and

(2) $\mu(y_n,y) = [\exp_q(n)]^{-1}$.

Let $D = B^n - \{\{x_n\},\{y_n\},x,y\}$. Then any quasiconformal self homeomorphism of D extends to a quasiconformal self homeomorphism of B^n which fixes both x and y.

PROOF: We may assume without loss of generality that $p > q$. We first show that any K-quasiconformal self homeomorphism of D extends to a K-quasi-conformal self homeomorphism of B^n. This will follow from repeated application of the removable singularity theorem. Since both x_1 and y_1 are isolated (by (1) and (2)) we can find a K-quasiconformal self mapping f_1 of $D \cup \{x_1,y_1\}$ agreeing with f on D. We inductively proceed to find f_m defined on $D_m = D \cup \{x_1,...,x_m,y_1,...,y_m\}$ and agreeing with f_{m-1} on D_{m-1}. From the sequence $\{f_m\}$ we can extract a subsequence converging to a K-quasiconformal self homeomorphism f_0 of $B^n - \{x,y\}$ and agreeing with f on D, see [Vä. Thm. 21.9] (clearly the sequence cannot converge c-uniformly to a constant value). Finally, we again appeal to the removable singularity theorem to extend f_0 to a K-quasiconformal self homeomorphism of B^n agreeing with f on D. We will denote this extension also by f. Since f must map the boundary of D to itself and since it extends homeomorphically over this set we see $f(\{x_n\})$ lies in the set $\{\{x_n\},\{y_n\}\}$ Furthermore, since x and y are the only limit points of these sequences we find that either $f(x) = x$ (and hence $f(y) = y$ and we are done) or else $f(x) = y$. We seek to show that this latter conclusion is impossible. This is essentially because f is almost bilipschitz in the hyperbolic metric and since the sequences $\{x_n\}$ and $\{y_n\}$ approach their limits at exponentially different rates. Thus suppose that $f(x) = y$. We may assume that x and y are distinct. Since f is a homeomorphism of B^n we see that for all sufficiently large integers j

$$(3.5) \qquad f(x_j) \in \{y_n\}.$$

Thus we set $n(j)$ as the integer such that $f(x_j) = y_{n(j)}$. We have

$$[\exp_p(j)]^{-1} = \mu(x,x_j),$$
$$[\exp_q(n(j))]^{-1} = \mu(y,y_{n(j)}), \text{ and from } 3.1$$
$$\mu(y,y_{n(j)}) = \mu(f(x),f(x_j)) \leq c\,\mu(x,x_j)^a.$$

This last inequality follows since $\mu(x,x_j) \leq 1$. Since $p > q$, these three equations yield a constant M, depending only on the dimension n and the coefficient of quasiconformality K such that for all sufficiently large j

$$n(j) \geq e^j - M.$$

This last inequality implies that there is an infinite sequence of values y_k not attained by any $f(x_j)$ which is clearly impossible. The lemma is then proved.

The preceeding lemma indicates how we should construct the domain we are seeking.

3.6 THEOREM. *There is a subset F of B^n such that $\Omega = B^n - F$ is a domain with the following properties.*

(1)   Every K-quasiconformal self homeomorphism f of Ω extends to a K-quasiconformal self homeomorphism of R^n which is the identity on $R^n\text{-}B^n$.

(2) Ω is simply connected if $n \geq 3$.

(3)   The only discrete quasiconformal automorphism of Ω is the identity.

PROOF: Let F_m be the vertices of a triangulation of $S^{n-1}(0,1-1/m)$ with mesh $1/m$ (that is, no simplex of the triangulation has diameter greater than $1/m$). It is clear that $S^{n-1}(0,1)$ is contained in the closure of $F' = \cup\, F_m$. Next order the points of F' as $x_1,x_2,...,x_j,...$ and let $d_j = \text{dist}(x_j,F'\text{-}\{x_j\})/4 > 0$. For each j let $\{y^j_k\}$ be a sequence of points of $B^n(x_j,d_j)$ converging to x_j and such that

$$(3.7) \quad \mu(x_j,y^j_k) = [\exp_j(k)]^{-1}.$$

Set $F = F' \cup \{y^j_k\}$. Since F is closed it is clear that $\Omega = B^n - F$ is a domain and it is not difficult to see that Ω is simply connected if $n \geq 3$.

Let f be a K-quasiconformal self homeomorphism of Ω. It is clear (as in the proof of the previous lemma) that the set F is removable. That is each K-quasiconformal self homeomorphism f of Ω extends to a K-quasiconformal self homeomorphism f' of $\mathbf{B}^n$. The map f' also extends by reflection to a K-quasiconformal self homeomorphism f" of $\mathbf{R}^n$. This is not the extension we desire, however we will show that f" | $\mathbf{S}^{n-1}$ = identity. To see this it is only necessary to show that f' must fix each point of F'. Since f and f' agree on Ω and since the only limit points of the set F in $\mathbf{B}^n$ are the points of F' which are ordered, we see that for each integer j there is an integer k(j) such that

$$f(x_j) = x_{k(j)}.$$

As in the proof of Lemma 3.4 we find that the condition (3.7) will imply that

$$j = k(j).$$

Thus f' is the identity on $\mathbf{S}^{n-1}$ and we may define an extension h as follows.

$$h(x) = \begin{cases} f'(x) & \text{if } x \in \mathbf{B}^n \\ x & \text{if } x \in \mathbf{R}^n - \text{int}(\mathbf{B}^n). \end{cases}$$

Then h is the desired extension to satisfy conclusion (1) of our theorem. To establish (3) we observe that every quasiconformal group acting on Ω extends to a quasiconformal group acting on $\mathbf{B}^n$ and which is the identity on $\mathbf{S}^{n-1}$. Thus, from our earlier results we see that each such discrete group is finite. Hence every element of such a group is periodic and has an extension to $\mathbf{R}^n$ which is the identity on $\mathbf{R}^n - \mathbf{B}^n$. The following theorem of Newman [Ne.] asserts that no periodic map can pointwise fix an open set. This implies that the only possible discrete quasiconformal action on Ω is the identity. This last fact then completes the proof of the theorem.

THEOREM. (Newman Thm.1.) *Let O be an open subset of* cl $(\mathbf{R}^n)$. *For each p > 1 there is a positive constant δ depending on n, p and O such that no periodic homeomorphism of* cl $(\mathbf{R}^n)$ *of period p moves every point of O a distance less than δ.*

3.8 REMARKS. It would be very interesting to know if the assumption of discreteness in Theorem 3.6 (3) is really necessary. If f is a quasiconformal automorphism of Ω which is not discrete (and so of infinite order), then there is a sequence of integers n(j) such that

$$f^{n(j)} \to \text{Identity} \quad \text{c-uniformly in } \Omega \text{ as } j \to \infty.$$

Actually more is true. Namely that each $f^{n(j)}$ extends to $\mathbf{R}^n$ and $f^{n(j)}$ converges uniformly to the identity. To see this, we find from Theorem 1.2 that there is a quasiconformal homeomorphism $h:\Omega \to \Omega$ and a subsequence $\{f^{m(j)}\}$ of the sequence $\{f^j\}$ such that $f^{m(j)} \to h$ c-uniformly in Ω (also $f^{-m(j)} \to h^{-1}$ c-uniformly in Ω).

Setting $n(j) = m(j+1)-m(j)$ we see $f^{n(j)} \to \text{Identity}$ c-uniformly in Ω. If the difference $n(j+1)-n(j)$ is bounded (or indeed some finite linear combination of m successive $n(j)$'s has a finite lim inf), then f is the identity. This is because

$$f^{n(j+1)-n(j)} \to \text{Identity} \quad \text{c-uniformly in } \Omega$$

and since $n(j+1)-n(j)$ must take on some finite value infinitely often we see f is periodic. Since it extends to a homeomorphism which is again periodic and which is the identity on an open set we see that $f = \text{Identity}$, by Newmans' theorem. Thus if f is the extension to $\mathbf{R}^n$ of a quasiconformal automorphism of Ω which is not discrete, then there is a sequence of integers $n(j)$ such that $n(j+1)-n(j) \to \infty$ as $j \to \infty$ and such that

$$f^{n(j)} \to \text{Identity} \quad \text{uniformly in } \mathbf{R}^n.$$

Such homeomorphisms are naturally analogous to the irrational rotations of $\mathbf{R}^n$. It seems reasonable to conjecture that a version of Newmans' theorem should be true for such generalized irrational rotations. That is, they cannot be the identity on an open set of $\mathbf{R}^n$, without being the identity throughout $\mathbf{R}^n$. Unfortunately the methods of proof in [Ne.] do not seem to generalize to this case.

§4. Domains which admit an infinite quasiconformal group.

In previous sections we have investigated conditions which imply that there is no infinite discrete quasiconformal group acting. In this section we consider a domain D on which an infinite quasiconformal group acts and then try to find some conditions, such as boundary regularity, which imply that the domain is quasiconformally equivalent to the unit ball. We need some terminology and we again recall the (equivalent) definition of a limit point.

4.1 DEFINITIONS. Given a sequence of domains $\{D_j\}$, the <u>kernel of the sequence D_j</u>, denoted $\operatorname{Ker} D_j$, is the set of all points of $\mathbf{R}^n$ which have a neighborhood which is contained in all but a finite number of the sets D_j.

Let G be an infinite quasiconformal group acting on a domain D. A point $x \in \partial D$ is a <u>limit point for G</u> if there is a point $y \in D$ such that x is a limit point of the orbit $G(y) = \{g(y) : g \in G\}$ of y. If G is discrete, it is not difficult to see from the compactness Theorem 1.3, that for any two point z and w of D, the limit points of the sets $G(z)$ and $G(w)$ are identical. A limit point x of G will be called a <u>conical limit point</u> is there is a point $y \in D$ and a subsequence $\{x_n\}$ of $G(y)$ converging to x, and a sequence $\{t_n\}$ of positive real numbers such that some compact set F of D contains infinitely many of the points $t_n(x_n-y)+y$ and such that F lies in the kernel $\operatorname{Ker}(t_n(D-y)+y)$.

It is clear that $\lim t_n = \infty$ and that this definition agrees with the usual definition of a conical limit point in the case of group actions on $\mathbf{B}^n$ or in case there is a Stolz cone W lying in D whose vertex lies at x and which contains infinitely many points of the orbit $G(y)$ of some point y.

A domain D is <u>K-quasiconformally homogeneous</u> if there is an infinite family of K-quasiconformal self homeomorphisms of D acting transitively on D. That is, for each pair of points $x,y \in D$, there is a K-quasiconformal self homeomorphism f of D such that $f(x) = y$.

A boundary point x of a domain D has a <u>Q.C.-bicollared neighborhood</u> if there is a neighborhood O of x in $\mathbf{R}^n$ and a quasiconformal homeomorphism of the pair $(O, O \cap \partial D)$ onto the pair $(\mathbf{B}^n, \mathbf{B}^n \cap \mathbf{R}^{n-1})$. We say that $x \in \partial D$ is a <u>manifold point</u> if there is a homeomorphism of the pair $(O \cap D, O \cap \partial D)$ onto the pair $(\mathbf{B}^n_+, \mathbf{B}^n \cap \mathbf{R}^{n-1})$, where $\mathbf{B}^n_+$ is the intersection of $\mathbf{B}^n$ with the closed upper half space (whose boundary is identified as $\mathbf{R}^{n-1}$). A point $x \in D$ is a <u>Q.C.-flat point of D</u> if there is a quasiconformal embedding $g:D \to \mathbf{R}^n$ such that $g(x) = 0$ and there is a neighborhood U of 0 such that

$$\partial g(D) \cap U = U \cap \mathbf{R}^{n-1}.$$

Being quasiconformally bicollared at a boundary point x is of course a much stronger condition than x being a manifold point. However a Q.C.-flat point need not be a manifold point and a manifold point need not be a Q.C.-flat point. As an example, it is not difficult to see that in the domain $D = \mathbf{B}^n - \mathbf{B}^{n-1}_+$, the origin is a Q.C.-flat point (the domain is quasiconformally equivalent to the upper half space) but not a manifold point.

We say that a point $x \in \partial D$ *admits a tangent plane* if for any sequence of reals numbers $\{t_j\}$ with $\lim t_j = \infty$ we have $\mathrm{Ker}\ D_j = \mathbf{H}^n$, where $\mathbf{H}^n$ is some half space, $D_j = t_j(D\text{-}x)$ and $D\text{-}x = \{y\text{-}x : y \in D\}$.

The condition that D admits a tangent plane at x means that if we blow up D by homotheties based at x then we obtain a half space in the limit. It is necessary that near x the component of the boundary of D containing x be isolated from other boundary components. For example the domain $\mathbf{H}^n\text{-}\{e_n/m : m=1,2,...\}$ does not admit a tangent plane at the origin.

4.2 LEMMA. *Suppose that D is a domain whose boundary near x is an open subset of a smooth hypersurface. Then D is quasiconformally bicollared and admits a tangent plane at x.*

PROOF: A small normal neighborhood of the boundary near x will provide the quasiconformal bicollar upon restricting to a slightly smaller subset. The desired tangent plane will be the tangent hyperplane to the boundary at x.

4.3 LEMMA. *Suppose that D is quasiconformally bicollared at $x \in \partial D$. Then D is flat at x.*

PROOF: Let U be a neighborhood of x and let

$$h:(U,U \cap \partial D) \to (\mathbf{B}^n, \mathbf{B}^n \cap \mathbf{R}^{n\text{-}1})$$

be a quasiconformal map given by the fact that ∂D is quasiconformally bicollared at x. By the quasiconformal Schönfliess theorem, see [G.V.], we can restrict h to a slightly smaller neighborhood U' containing x and then extend $h \mid U'$ to a quasiconformal homeomorphism g of $\mathbf{R}^n$ such that $h \mid U' = g \mid U'$. The map $g \mid D$ is then the desired quasiconformal embedding to establish that D is Q.C.-flat at x.

In view of the fact that D is homotopically trivial near a manifold point or a Q.C.-flat point of the boundary, from our previous results, for example Corollary 2.6, we find

4.4 PROPOSITION. *Suppose that D is a domain which admits an infinite discrete quasiconformal group. If a limit point of G is either a manifold point or a Q.C.-flat point, then D is contractible.*

If D is a quasiconformally homogeneous domain, then every point of the boundary of D is a limit point (of a sequence of quasiconformal mappings converging c-uniformly in D to that point) and so we find as above

4.5 COROLLARY. *Let D be a quasiconformally homogeneous domain. If ∂D contains a single manifold point, or a single Q.C.-flat point, then D is contractible.*

4.6 REMARKS. Tukia [Tu.2] has found an example of a quasiconformally homogeneous domain which is not quasiconformally equivalent to the ball. It is however, topologically equivalent to the ball (and so contractible). At present there appears to be no known examples of quasiconformally homogeneous domains which are contractible, but not homeomorphic to $\mathbf{B}^n$. In fact it is not easy to construct contractible domains which are not homeomorphic to $\mathbf{B}^n$. The Whitehead spaces, see [Wh.], are examples of such domains. It is not difficult to see that these particular spaces are not quasiconformally homogeneous from, for instance, Theorem 6.3 of [G.P.] (the boundary of these spaces are connected sets which are locally like a Cantor set crossed with an interval). We raise these points in order to consider whether Corollaries 4.4 and 4.5 are in some sense best possible (for instance is the conclusion that D is topologically a ball valid in these circumstances? We will see soon that actually a single Q.C.-flat point will imply that a quasiconformally homogeneous domain is quasiconformally equivalent to a ball).

The following Theorem is a modification of Theorem 6.3 in [G.P.] to our situation of discrete quasiconformal actions. It will prove quite useful in later applications.

4.7 THEOREM. *Let D be a domain in $\mathbf{R}^n$ which admits a discrete quasiconformal group G. If D admits a tangent plane at a conical limit point of G, then D is quasiconformally homeomorphic to $\mathbf{B}^n$.*

PROOF: We may suppose that the origin is the conical limit point in question and that the tangent plane is $\mathbf{R}^{n-1}$. Then by the hypothesis that the origin is a conical limit point, there is a compact subset F of D, a point $y \in D$, an infinite subsequence $\{g_j\}$ of G and a sequence $\{t_j\}$ of positive real numbers tending to infinity, such that

(1) $g_j(y) \to 0$ and $j \to \infty$.

(2) For all j, $t_j \cdot g_j(y) \in F$.

Now the sequence $f_j = t_j \cdot g_j : D \to D_j = t_j \cdot D$, is easily seen to be normal (see, for instance [Vä.Thm. 20.5]) and so has a subsequence which converges c-uniformly in D to either a K-quasiconformal homeomorphism f of D to a component of

Ker D_j or to a point of $\mathbf{R}^n$-{Ker D_j U Ker($\mathbf{R}^n$-D_j)}. Again, the hypothesis that the origin is a conical limit point will imply that $\hat{F}$ lies in Ker D_j, so that the latter conclusion cannot apply, while the assumption that D admits a tangent plane at the origin imply that Ker D_j = $\mathbf{H}^n$, the upper half space. Since the upper half space is conformally equivalent to the unit ball we are done.

It is not difficult to see that the rather complicated definition of a conical limit point is necessary for the proof of the theorem to work. For example, the domain illustrated below admits a discrete quasiconformal group G which has the origin as a limit point, which admits a tangent plane at the origin and the origin satisfies all the conditions to be a conical limit point, except the condition that F lies in Ker D_j. The group G is obtained by conjugating the parabolic translation $z \rightarrow z+1$, with the conformal mapping η of D onto $\mathbf{H}^2$ (which has a quasiconformal extension to the whole plane). The orbit of the point $\eta(i)$ is illustrated as is the compact set F. It is not difficult to see (since the orbit of any point is tangential to the origin) that for no sequence of real numbers $\{t_j\}$ and subsequence $\{g_j\}$ of G, does the sequence $t_j.g_j$ converge to a homeomorphism.

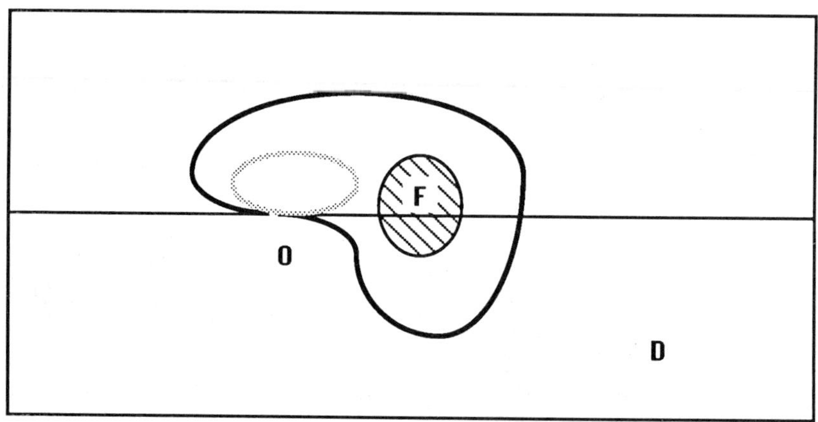

Another difficulty with a suitable definition for conical limit points is the problem that the boundary of D may be highly irregular, for instance near cusps and nonmanifold points. Fortunately we will be able to show in an important case that concerns us, that there are plenty of conical limit points. This is in the case of compact quotients or groups of the first kind.

4.8 COROLLARY. *Let D be a domain which is either smooth in a neighborhood of, or quasiconformally bicollared at, a conical limit point of an infinite discrete quasiconformal group G. Then D is quasiconformally equivalent to the unit ball.*

4.9 DEFINITION. We say a discrete quasiconformal group G acting on a domain
D is cocompact, if there is a compact subset F of D such that

$$D = \cup\{g(F) : g \in G\}.$$

That is, there is a compact set whose orbit under the group G covers every point of
D.

4.10 THEOREM. *Let D be a domain which admits a cocompact quasiconformal
group action G. If D is Q.C.-flat at a single point, then D is quasiconformally
equivalent to the unit ball.*

PROOF: Let f be the quasiconformal embedding which flattens D at some point
z. Then the group $H = f \circ G \circ f^{-1}$ acting on the domain f(D) is a cocompact
quasiconformal group. The following lemma asserts that the flattened point is a
conical limit point of H, while the definition of a flat point asserts that each point of
$\partial f(D)$ in a sufficiently small neighborhood of the flattened point admits a tangent
plane. The result then follows from Theorem 4.8.

4.11 LEMMA. *Let D be a domain which admits a cocompact quasiconformal
group. Suppose that there is some point $x \in \partial D$ and a neighborhood U of x in
R^n such that $B^{n-1}(x,r) = \partial D \cap U$. Then x is a conical limit point.*

PROOF: We may assume that x = 0, and by homothetically expanding, we may
further assume that B^n lies in U. such an expansion will clearly not alter the fact
that x is a conical limit point. Let E be the compact set whose orbit covers D. We
recall the definition of the quasihyperbolic metric and some of its properties from the
beginning of section three. Let c be the constant of (3.1) for quasiconformal self
homeomorphisms of D (with μ replaced by k_D). Set $M = c(k_D(E)+1)$. Now
since B^n lies in U, and since $\partial D \cap U \cap B^n = B^{n-1}$, the quasihyperbolic metric
of D and the hyperbolic metric of the upper half space agree in $B^n(1/4) \cap D$ (they
have the same metric tensor there). Thus let W be the Stolz cone of hyperbolic
radius M based at the origin and let F be the hyperbolic ball of radius M centered
at $e_n/4$. Now the orbit of E covers D so there is a $g_j \in G$ such that $e_n/j \in g_j(E)$.
Let y be any point of E. Since each g_j is K-quasiconformal,

$$k_D(g_j(E)) \leq c(k_D(E)+1) \leq M.$$

So that $g_j(E)$ lies in W and in particular $g_j(y) \in$ W. Then there is a real number t_j such that $t_j.g_j(y) \in$ F. Clearly F lies in Ker $t_j.D$ and so it follows that the origin is a conical limit point.

4.12 REMARK. Actually it follows from Corollary 3.5 of [G.P.] (and the proof of Lemma 4.3), that if D admits a cocompact quasiconformal action, then D is quasiconformally homogeneous. However, the fact that the domain is quasiconformally homogeneous does not imply that that the original group (from which the desired quasiconformal family is constructed) has a conical limit point.

4.13 COROLLARY. *If D is quasiconformally homogeneous and if ∂D has a single Q.C.-flat point, then D is quasiconformally equivalent to the unit ball.*

PROOF: Since D is quasiconformally homogeneous, any point is a compact set whose orbit covers D. The proof of Theorem 4.10 now implies the result.

And as a Corollary to Theorem 6.3 of [G.P.] we find (where the definition of a k-tangent is as defined there)

4.14 COROLLARY. *Let D be a domain in R^n which admits a cocompact quasiconformal action. If D admits a k-dimensional tangent, then D is quasiconformally equivalent to*

$$(1) \quad B^n \qquad if \ k = n\text{-}1$$
$$(2) \quad R^n - R^k \quad if \ 0 < k < n\text{-}1.$$

We now obtain the following corollary, which is a generalization of a result of Gromov [Gr.] and partly follows from the proof of Gromov's result by Tukia [Tu.3].

4.15 THEOREM. *Let M be a closed smooth Riemannian n-manifold. Suppose that the universal cover M' of M admits a quasiconformal embedding in R^n which is Q.C.-flat at a single boundary point (or admits an (n-1)-dimensional tangent in the sense of [G.P.]). Then M' is quasiconformally equivalent to the unit ball and hence M has the homotopy type of a hyperbolic manifold. In fact there is a discrete and faithful embedding of $\pi_1(M)$ in $M\ddot{o}b(n\text{-}1)$, the group of hyperbolic isometries of B^n.*

PROOF: Since M is smooth and Riemannian, the fundamental group, $G = \pi_1(M)$, of M acts as isometries on M'. Let f be the quasiconformal embedding given by the hypotheses and set $D = f(M')$. Then D is a domain in $\mathbf{R}^n$ which is Q.C.-flat at some point, again by hypothesis. Also the group $f \circ G \circ f^{-1}$, is a cocompact quasiconformal group acting on D and so the group is infinite. Hence by Theorem 4.11, D is quasiconformally equivalent to the unit ball, $\mathbf{B}^n$. Thus, after again conjugating by a quasiconformal mapping, we have a quasiconformal group H acting on $\mathbf{B}^n$ and whose quotient, $\mathbf{B}^n/H$, is a closed n-manifold M. The group H naturally extends by reflection to a quasiconformal group (again denoted H) acting on $\mathbf{R}^n$. By a Theorem of Tukia [Tu.3], the group $H \mid S^{n-1}$ is quasiconformally conjugate to a Möbius group. These conjugations, and then the natural extension to $\mathbf{B}^n$ (which is unique) provide the discrete and faithful embedding into Möb(n-1) that we seek.

4.16 REMARKS. (1) The hypothesis that M be smooth is unnecessary if $n \neq 4$ as Sullivan, see [T.V.], has shown that every closed n-manifold, $n \neq 4$, admits a quasiconformal structure. This structure will provide a quasiconformal structure on the universal cover, in which the fundamental group of M will act quasiconformally (actually conformally). The rest of the proof remains the same.

(2) In the case that $n=2$, the result is trivial as the hypotheses imply that the surface is of hyperbolic type. When $n = 3$, the conclusion implies that the three manifold M has a hyperbolic structure in the sense of Thurston. It is not difficult to show that a cocompact quasiconformal group acting on $\mathbf{B}^3$ must have a quotient which is geometrically atoroidal (in fact there can be no subgroups of G isomorphic to $\mathbf{Z} \times \mathbf{Z}$). Thus if M is Haken, one can assert via Thurston theory, that M is quasiconformally homeomorphic to a hyperbolic space form (a manifold of constant negative curvature).

(3) Recall that a Q.C.- flat point need not even be a manifold point. Also from Lemma's 4.2 and 4.3, we see that the conclusion of Theorem 4.13 would still be valid if we assumed that the universal cover of M admits a quasiconformal embedding into $\mathbf{R}^n$ which is either smooth in a neighborhood of a single point, or is quasiconformally bicollared at a single point.

(4) From a similar line of reasoning one can prove via Proposition 4.4 that if the universal cover of a closed n-manifold, quasiconformally embedds in $\mathbf{R}^n$ and is has a single manifold point in its boundary, then the universal cover is contractible, and hence M is a $K(\pi(M),1)$.

(5) Finally we recall Example 2.7, which shows that in both Theorem 4.15 and Remark 4.16 (4), some hypothesis about the boundary of the embedding of the universal cover is necessary.

We now investigate what can be said in the case that D admits a discrete quasiconformal group with a fundamental domain of finite quasihyperbolic volume. In order to avoid proving hard estimates for the distortion of quasihyperbolic volume under quasiconformal mappings we will consider actions which are locally Lipschitz in the quasihyperbolic metric (notice that from the inequality (3.1) quasiconformal mappings are not necessarily locally Lipschitz, however they will be for suitably distant points. Such maps are called pseudo isometries). Furthermore, in view of Tukia's example (a uniform domain homeomorphic to the ball which is homogeneous with respect to a family which is bilipschitz in the quasihyperbolic metric, and hence quasiconformally homogeneous, but the domain is not quasiconformally equivalent to the ball) we are particularly interested in what can be said in the case that D is uniform.

4.17 THEOREM. *Let D be a uniform domain in R^n, n $\neq$ 4, which admits a discrete group G which is locally L-Lipschitz in the quasihyperbolic metric (and so uniformly quasiconformal). Suppose that there is a set F whose orbit under G covers D and the quasihyperbolic volume of F is finite. If D has a single quasiconformally bicollared boundary point, then D is quasiconformally equivalent to the unit ball.*

OUTLINE OF PROOF: From the proof of Lemma 4.3, there is a quasiconformal map of R^n which flattens D at some point which we may assume to be the origin. Furthermore, since D has a quasiconformally bicollared point, R^n - D contains an open set. Since D is uniform and since the image of a uniform domain under a quasiconformal mapping of R^n is again uniform (see [G.O.]), the flattened domain (which we will again call D) is uniform and its complement contains an open set. By composing with an auxillary Möbius transformation we may suppose that this open set contains ∞ as an interior point, that is D is bounded (this may mean that the flattened portion of ∂D lies on an (n-1)-sphere which we may then locally flatten to a hyperplane). From the approximation theorem of [T.V.], if n $\neq$ 4, we may assume that the flattening map is bilipschitz in the quasihyperbolic metric. Thus, after conjugation by the flattening map we are in the following situation.

(*)   D is a bounded uniform domain, part of whose boundary coincides with
        R^{n-1}

and D admits a group G which is locally M-Lipschitz in the quasihyperbolic metric, and has a finite quasihyperbolic volume fundamental set F.

In the situation (*) the restriction to $n \neq 4$ is unnecessary. We wish to establish the existence of a conical limit point on the part of the boundary of D which coincides with $\mathbf{R}^{n-1}$, so that the result will then follow from Theorem 4.10. We do this via a series of claims.

Claim 1. *For each $g \in G$, the quasihyperbolic volume of g(F) is finite.*

To see this we observe that locally $k_D(x,y) \approx |x\text{-}y|/d(x,\partial D)$. If $g \in G$, then both g and g^{-1} are locally lipschitz in the quasihyperbolic metric, and so locally

$$^1/_M \, |x\text{-}y|/d(x,\partial D) \; \leq \; |g(x)\text{-}g(y)|/d(g(x),\partial D) \; \leq \; M \, |x\text{-}y|/d(x,\partial D).$$

Also we see that g is quasiconformal and so differentiable almost everywhere with respect to n-dimensional Lebesgue measure |dx|. Since the quasihyperbolic volume measure is $d(x,\partial D)^{-n} \, |dx|$, the result follows from the above inequality at points of differentiability of g where we find

$$|g'(x)|/d(g(x),\partial D) \; \leq \; M \, d(x,\partial D)$$

so that

$$Vol_D(g(F)) \; \leq \; M^n \, Vol_D(F).$$

Where $Vol_D(E)$ is the quasihyperbolic volume of E.

Claim 2. *If E is an open subset of the boundary of D which lies in $\mathbf{R}^{n-1}$, then every point of E is a limit point of G.*

To see this we observe that in a small neighborhood of E (in D) the quasihyperbolic metric of D and the hyperbolic metric of $\mathbf{H}^n$ agree, and so any small Stolz cone has infinite quasihyperbolic volume and so cannot be filled by any finite number of images of F under G. The claim now follows easily from the local uniform convergence of some subsequence of any sequence of elements in G.

Claim 3. E contains a conical limit point.

Here we use the hypothesis that D is uniform and so each quasiconformal self homeomorphism of D extends to a homeomorphism of the boundary of D, see [G.Mo.]. The boundary action will be uniformly quasisymmetric since D is a bounded uniform domain. For the definition and basic properties of quasisymmetric mappings see [T.V.2]. Since G is infinite and has more than one limit point, there is an element $g \in G$ of infinite order. The group $\langle g \rangle$ is infinite and so has a limit point in ∂D, which must be a fixed point of g (as g extends to ∂D). We can now proceed as in [G.M.], using the fact that every point of E is a limit point and that the group extends to ∂D (which will be quasisymmetric in a neighborhood of E in D), to find an element of infinite order f (which we call a loxodromic element of G) for which

$$f^j \to x \quad \text{and} \quad f^{-j} \to y, \quad x,y \in E \text{ and } x \neq y.$$

It is not difficult to see that such loxodromic fixed points are conical limit points using standard distortion arguments.

4.18 REMARKS. It is not difficult to see that Theorem 4.17, is true in more generality. In particular the proof really only depends on the validity of the claims 2 and 3 where we did not use the full strength of the hypotheses. In particular, the following corollary, suggested by P. Tukia, follows from the above proof since we need only establish the existence of a loxodromic fixed point in the Q.C. bicollared neighborhood as we did in claim (3).

4.19 COROLLARY. *Let D be a uniform domain in R^n which admits a discrete quasiconformal group of the first kind. That is, every point of ∂D is a limit point of G. If D has a single quasiconformally bicollared boundary point, then D is quasiconformally equivalent to the unit ball.*

4.20 EXAMPLE. As a further example, the infinite cylinder

$$C = \{(x,y,z) \in \mathbf{R}^3 : x^2 + y^2 < 1\}$$

is quasiconformally equivalent to the ball, see [Vä]. Also every point of the boundary is quasiconformally bicollared, except ∞. C admits the infinite conformal group generated by $(x,y,z) \to (x,y,z+1)$. Any fundamental domain is not of finite quasihyperbolic volume so the above results do not apply. However it seems quite difficult to prove that C is quasiconformally equivalent to the ball from the information at the single limit point ∞ (which is a manifold cusp point of C).

Yale University, New Haven, CT 06520

REFERENCES

[FS] M. Freedman and R. Skora. *Strange actions of groups on spheres*, J. Differential Geometry **25** (1987).

[GV] D. Gauld and M. K. Vamanamurthy. *Quasiconformal extensions of mappings in n-space*, Ann. Acad. Sci. Fenn. Ser. A.I. **3** (1977).

[Ge1] F. W. Gehring. *Rings and quasiconformal mappings in space*, Trans. A.M.S. **103** (1962).

[GM] F. W. Gehring and G. J. Martin. *Discrete quasiconformal groups I, II*, J. London Math. Soc. (3) **55** (1987)

[GMo] F. W. Gehring and O. Martio. *Quasiextremal distance domains and extension of quasiconformal mappings*, J. D'Analyse Math. **45** (1985).

[GO] F. W. Gehring and B. G. Osgood. *Uniform domains and the quasi-hyperbolic metric*, J. D'Analyse Math. **36** (1979).

[GP] F. W. Gehring and B. Palka. *Quasiconformally homogeneous domains*, J. D'Analyse Math. **30** (1976).

[Gr] M. Gromov. *Hyperbolic manifolds, groups and actions*, in "Annals of Math. Studies", No. 97, Princeton University Press.

[Ma] G. J. Martin. *Discrete quasiconformal groups that are not the quasiconformal conjugates of Möbius groups*, Ann. Acad. Sci. Fenn. Ser. A.I. **11** (1986).

[MG] G. J. Martin and F. W. Gehring. *Generalizations of Kleinian groups*, M.S.R.I. Preprint series.

[Ne] M. H. A. Newman. *A theorem on periodic transformations of spaces*, Quarterly J. Math. **2** (1931).

[Ru] T. B. Rushing. "Topological embeddings", Academic Press, 1973.

[Su] D. Sullivan. *The ergodic theory at infinity of an arbitrary discrete group of hyperbolic motions*, in "Annals of Math. Studies", No. 97, Princeton University Press.

[Tu1] P. Tukia. *On two dimensional quasiconformal groups*, Ann. Acad. Sci. Fenn. Ser. A.I. **5** (1980).

[Tu2] P. Tukia. *A quasiconformal group not isomorphic to a Möbius group*, Ann. Acad. Sci. Fenn. Ser. A.I. **6** (1981).

[Tu3] P. Tukia. *On quasiconformal groups*, to appear, J. d'Analyse.

[TV] P. Tukia and J. Väisälä. *Quasiconformal approximation and extension*, Ann. Acad. Sci. Fenn. Ser. A.I. **6** (1981)

[TV2] P. Tukia and J. Väisälä. *Quasisymmetric embeddings of metric spaces*, Ann. Acad. Sci. Fenn. Ser. A.I. **5** (1980).

[Vä] J. Väisälä. "Lectures on n-dimensional quasiconformal mappings", Lecture notes in Math. 229, Springer Verlag, 1971.

[Wh] J. H. C. Whitehead. *A certain open manifold whose group is unity*, Quarterly J. Math. Ser. (2) **6** (1935).

Convergence and Möbius Groups

BY GAVEN J. MARTIN AND PEKKA TUKIA

§1. Introduction.

In this paper we collect together a few results on the topological conjugacy of convergence groups to conformal or Möbius groups. Convergence groups were first introduced in [G.M. I,II] and their basic properties were established as well as the classification of the structure of elementary convergence groups. Convergence groups have been found to be a natural generalization of Kleinian or Möbius groups to the topological category. In particular if one wishes to topologically characterize Möbius groups amongst groups of homeomorphisms of S^n, the unit sphere of R^{n+1}, then one is led to the necessary condition that such a group is a convergence group. This is not in general a sufficient condition when $n \geq 2$ as there are many nonstandard convergence groups, see [G.M. I,II], [F.S.] and [M.G.] for a variety of examples. We will see however that under certain reasonable restrictions the condition of being a convergence group will suffice in dimension two and three.

This paper proceeds in three parts. Firstly, we suppose that there is an algebraic isomorphism between a convergence group and a Möbius group and find a necessary and sufficient condition for this isomorphism to induce a continuous mapping of the limit sets. We then seek a sufficient condition for a continuous map of the limit sets induced by an isomorphism, to be a homeomorphism. This will be in terms of points of approximation.

Next we show that every discrete convergence group of the closed unit disk cl (B^2) is topologically conjugate to a Fuchsian group. This will imply that discrete convergence groups of S^2 which have a closed invariant topological disk are the topological conjugates of Kleinian groups.

Finally, we consider the case of a convergence group acting on cl (B^3) and show that if it is torsion free and if the associated quotient manifold M is compact and Haken, then M can be endowed with a hyperbolic structure which will yield a conjugacy to a Möbius group on B^3. We will do this by showing that there is no incompressible torus in M.

1.1 DEFINITIONS. A convergence group of S^n is a group G of homeomorphisms of S^n with the following compactness property. If F is an infinite subset of G, then either

(1.2)   there are points x_0 and y_0 of S^n (possibly equal) and a sequence $\{f_j\}$ of elements of F such that

$$f_j \to x_0 \quad \text{locally uniformly in } S^n - \{y_0\}$$

and

$$f_j^{-1} \to y_0 \quad \text{locally uniformly in } S^n - \{x_0\}$$

or

(1.3)   there is a self homeomorphism f of S^n and a sequence $\{f_j\}$ of elements of F such that

$$f_j \to f \quad \text{and} \quad f_j^{-1} \to f^{-1} \quad \text{uniformly in } S^n \text{ as } j \to \infty.$$

Here "locally uniformly in $S^n - \{x_0\}$" means uniformly on compact subsets of $S^n - \{x_0\}$. If G is a discrete subgroup of the group of homeomorphisms of S^n (with the usual compact open topology), then case (1.3) cannot occur. Thus by a discrete convergence group G we mean a convergence group for which condition (1.3) holds for no infinite subfamily F of G.

We note here that every uniformly quasiconformal group G acting on S^n, and hence every Möbius or Kleinian group, is actually a convergence group, see [G.M. Cor. 3.5]. Furthermore, it is clear that any topological conjugate of a convergence group is again a convergence group.

We say that G acts discontinuously at a point $x \in S^n$, if there is a neighborhood U of x such that for all but finitely many $g \in G$

$$g(U) \cap U = \emptyset.$$

We denote by $O(G)$ the set of points of S^n at which G acts discontinuously and define the limit set of G as

$$L(G) = S^n - O(G).$$

We say that G acts properly discontinuously in a region D, if for each compact subset K of D there are only finitely many $g \in G$ for which

$$g(K) \cap K \neq \emptyset.$$

We recall the following facts whose proofs can be found in [G.M. I] and which we will use frequently.

1.4 FACTS.

(1)   L(G) is the smallest, nonempty, closed G-invariant set and contains no more than two points or is a perfect set. If $\#L(G) \leq 2$, we call G an *elementary convergence group.* (see [G.M. Thms. 4.5 & 6.13])

(2)   Either $L(G) = S^n$ or L(G) is nowhere dense in S^n, and G acts properly discontinuously in O(G). (see [G.M. Thm. 4.9 & Cor. 3.8])

(3)   If G is an abelian convergence group, then G is elementary. (see [G.M. Thm. 5.15])

(4)   Each element g of a discrete convergence group G is one of the following three types

(a) <u>Elliptic:</u> there is a positive integer j such that

$$g^j = \text{Identity}.$$

(b) <u>Parabolic:</u> there is a point $x \in S^n$ such that

$$g^{\pm j} \to x \quad \text{locally uniformly in } S^n - \{x\}$$

as $j \to \infty$.

(c) <u>Loxodromic:</u> there are distinct points x and y of S^n such that

$$g^j \to x \quad \text{locally uniformly in } S^n - \{y\}, \text{ and}$$
$$g^{-j} \to y \quad \text{locally uniformly in } S^n - \{x\}$$

as $j \to \infty$ (see [G.M. Thm 5.18]).

In case (b), x is the unique fixed point of g and we call such an x a parabolic fixed point of g, while in case (c) x and y are the only fixed points of g, we call x the attractive loxodromic fixed point of g and y the repulsive loxodromic fixed point of g.

(5)   If G is nonelementary, then the loxodromic fixed points are pairwise dense in $L(G) \times L(G)$ (see [G.M. Thm. 6.17]).

(6)   If G is nonelementary and if x and y are points of L(G), then there is a sequence $\{f_j\}$ in G such that

$$f_j \;\to\; x \quad \text{locally uniformly in } S^n - \{y\}$$

$$f_j^{-1} \to y \quad \text{locally uniformly in } S^n - \{x\}.$$

(this follows from (5), see [M.G.]).

§2. Isomorphisms of convergence groups.

Since most of what we establish in this and succeeding sections will be quite trivial in the case of elementary convergence groups, we will assume henceforth that G is a nonelementary discrete convergence group. A topological conjugacy between discrete converence groups G and H is a self homeomorphism f of S^n such that

$$f \circ G \circ f^{-1} = H.$$

Such a topological conjugacy implies that the dynamical behaviour of G and of H (which is in general quite complicated on the limit set) is essentially the same. Furthermore, such a conjugacy projects to a homeomorphism of the associated quotient spaces O(G)/G and O(H)/H. Notice that conversely a homeomorphism of these spaces does not necessarily lift to a map $O(G) \to O(H)$ and even if it does, it may not extend to a map of the whole sphere S^n. A topological conjugacy between the groups G and H induces an abstract algebraic isomorphism of these groups. Thus the existence of an algebraic isomorphism is a necessary condition for the existence of a topological conjugacy and so one is naturally led to consider the consequences of the existence of such an isomorphism for convergence groups.

2.1 DEFINITION. An algebraic isomorphism between convergence groups is called *type preserving* if it carries parabolic elements of one group bijectively onto the parabolic elements of the other group. In this case the isomorphism will also preserve the loxodromic elements as the elliptic elements are preserved for algebraic reasons.

Given an isomorphism $\Phi : G \to H$ between discrete convergence groups G and H, we say that Φ *induces (or is induced by) a continuous map of the limit sets* if there is a continuous map $f : L(G) \to L(H)$ such that for all $g \in G$ and $x \in L(G)$

$$(2.2) \qquad f(g(x)) = \Phi(g)(f(x)).$$

It is a rather interesting fact that a type preserving isomorphism of geometrically finite Möbius groups of S^n $(n \geq 2)$ is induced by a (quasisymmetric) homeomorphism of the limit set. In the finite volume case the parabolic fixed points have maximal rank and so are preserved for algebraic reasons so that the assumption that Φ be type preserving is unnecessary. See, for instance, §§2,3 and Thm. 3.3 of [Tu. 1].

Given an isomorphism $\Phi : G \to H$ between discrete convergence groups G and H, there is a natural densely defined function f inducing Φ. If this function is uniformly continuous on its domain of definition, then it will uniquely and continuously extend to a map of the limit sets.

For a loxodromic element g of a discrete convergence group G we denote by $a(g)$ the attractive fixed point of g (if g is parabolic, then we set $a(g)$ as the parabolic fixed point of g). It follows from (5) that the set of attractive fixed points of loxodromic elements of G is dense in the limit set. Given such an isomorphism Φ, the natural densely defined function we spoke of above is defined as follows.

$$(2.3) \qquad f(a(g)) = a(\Phi(g)),$$

To see that f satisfies (2.2) on the set of loxodromic fixed points we suppose that g' is any element of G and $x = a(g)$. Then

$$g'(x) = a(g' \circ g \circ g'^{-1})$$

so that

$$f(g'(x)) = a(\Phi(g' \circ g \circ g'^{-1})) = \Phi(g')(a(\Phi(g))) = \Phi(g')(f(x)).$$

We have established

2.4 PROPOSITION. *An isomorphism $\Phi: G \to H$ between discrete, nonelementary convergence groups is induced by a (unique) continuous map of the limit sets if and only if the map f defined above is uniformly continuous on the set of loxodromic fixed points.*

The uniqueness comes from the fact that (2.3) must always be satisfied so that any other continuous map inducing Φ will agree with f on the set of loxodromic fixed points which is dense in the limit set, and so therefore must agree with f on the whole limit set.

It is worthwhile to observe that if Φ is type preserving, then f is necessarily bijective when restricted to the loxodromic fixed point sets. To see this if $f(a(g)) = f(a(g'))$, then $\Phi(g)$ and $\Phi(g')$ are both loxodromic elements of a discrete convergence group with a common fixed point. From [G.M. I Cor. 6.9] we see that there are positive integers m and n such that

$$\Phi(g)^m = \Phi(g')^n$$

thus

$$\Phi(g^m) = \Phi(g'^n) \iff g^m = g'^n \iff a(g) = a(g').$$

Thus f is injective. The fact that f is surjective as a map of the loxodromic fixed point sets of G and H respectively (in the type preserving case) follows since $\Phi(G) = H$.

If G is a Fuchsian group of the first kind (viewed as a group acting on S^1), then there is a classical condition due to Fenchel-Nielsen, namely that of intersecting axes, which if preserved by an isomorphism, implies that this isomorphism induces a homeomorphism of the limit sets. We next show that this is also true for convergence groups of S^1 whose limit set is the whole circle, which will in turn imply our first conjugacy theorem (we will obtain a more precise result later).

We define the axis of a loxodromic element g of a discrete convergence group G acting on S^1 as

$$ax(g) = \text{the hyperbolic line joining the fixed points of g.}$$

It is important to realize that there is no natural extension of g to the closed unit disk and so we cannot think of ax(g) as an invariant line for the action of g on the disk.

An isomorphism Φ between discrete convergence groups G and H is said to preserve intersecting axes if for any pair of loxodromic elements g and g' of G

$$\text{ax}(g) \cap \text{ax}(g') \neq \emptyset \quad \Leftrightarrow \quad \text{ax}(\Phi(g)) \cap \text{ax}(\Phi(g')) \neq \emptyset.$$

If $L(G) = L(H) = S^1$, then an isomorphism which preserves intersecting axes is easily seen to be type preserving. The following theorem is proved exactly as in [Tu.2], Proposition 3.5 (the Proposition 1.4 of [Tu.2] is valid in our situation from the pairwise density of the loxodromic fixed points, see 1.4 (5)).

2.5 THEOREM. *Let G and H be discrete convergence groups of S^1 of the first kind $(L(G) = L(H) = S^1)$. Suppose that there is an isomorphism Φ of G and H which preserves the Fenchel-Nielsen intersecting axes condition. Then Φ induces a homeomorphism of the limit sets and so G and H are topologically conjugate.*

2.6 COROLLARY. *Suppose that Φ is an isomorphism between a discrete convergence group G of S^1 and a Fuchsian group Γ, both of the first kind, which preserves intersecting axes. Then G is topologically conjugate to Γ / S^1.*

§3. The map of the limit sets.

In this section we suppose that we are given a continuous map between the limit sets of discrete convergence groups G and H inducing an isomorphism and seek conditions to imply that this map must indeed be a homeomorphism. We saw previously that if the isomorphism was type preserving then such a map was a bijection between the loxodromic fixed point sets of each group. The correct formulation of the result will be in terms of points of approximation as introduced by Hedlund and Beardon-Maskit, see [B.M.].

3.1 DEFINITION. We denote by $q(x,y)$ the *spherical distance* between two points x and y of S^n. Let G be a discrete convergence group of S^n. A point w of $L(G)$ is called a *point of approximation* for G if there is a sequence $\{g_j\}$ of elements of G such that for each point x of $S^n - \{w\}$ there is a positive number $\delta = \delta(x)$ with the property that

$$(3.2) \qquad\qquad q(g_j(x), g_j(w)) \geq \delta.$$

In a convergence group, a subsequence of the sequence $\{g_n\}$ will converge locally uniformly in the complement of some point of the limit set to some other point of the limit set. Condition (3.2) implies that the convergence cannot be uniform in a neighborhood of both x and w.

We now establish the following useful characterization of points of approximation for convergence groups whose limit set is not the whole sphere S^n.

3.3 LEMMA. *Let* $w \in L(G)$. *Suppose there is a point* $x_0 \in O(G)$, *a positive number* d_0, *and a sequence* $\{g_j\}$ *in* G *such that*

$$q(g_j(x_0), g_j(w)) \geq \delta_0.$$

Then w *is a point of approximation.*

PROOF: Since G is a convergence group, there is a subsequence of the sequence $\{g_j\}$ converging to one point of the limit set locally uniformly in the complement of another (possibly the same) point in the limit set. We denote this subsequence by $\{g_j\}$. Since x_0 is an ordinary point (i.e. $x_0 \in O(G)$) we must have

$$g_j \rightarrow y_0 = \lim g_j(x_0) \quad \text{locally uniformly in } S^n - \{w\}.$$

If z is any point of $S^n - \{w\}$, then $g_j(z)$ accumulates only at y_0, where $g_j(w)$ cannot accumulate. Thus there is a positive $\delta(z)$ such that

$$q(g_j(z), g_j(w)) \geq \delta(z).$$

The lemma is proved.

In the case of Möbius groups the points of approximation are precisely the radial (or conical) limit points, see [B.M. Thm. 1]. The point to this definition is that it does not rely on the existence of an extension to the closed ball cl (B^{n+1}), which may not exist if G is merely a convergence group (and cannot exist in some cases), but is intrinsic to the action on S^n. Also one avoids having to find a topological characterization of the radial limit points of a convergence group.

3.4 LEMMA. *Let* G *and* H *be nonelementary, discrete convergence groups and let* $F:L(G) \rightarrow L(H)$ *be a continuous map of the limit sets inducing an*

121

*isomorphism. Then F is surjective and if w is a point of approximation of H,
then*

$$F^{-1}(w) = \{y \in L(G) : F(y) = w\}$$

*consists of a single point of approximation of G and any right inverse of F is
continuous at w.*

PROOF: Let x and y be points of L(G) such that $F(x) = F(y) = w$, where
$w \in L(H)$ is a point of approximation. Associated to w is a sequence $\{h_j\}$ as
in (3.2) lying in H. Let a and b be points of L(G) for which

$$F(a) \neq w, F(b) \neq w \quad \text{and} \quad F(a) \neq F(b).$$

Since G and H are nonelementary, the orbit of a single point of the limit set is
dense in the limit set, by 1.4 (6). Hence $\operatorname{Im} F = L(H)$ and since L(H) contains
more than two points the existence of such points as a and b above is clear.

Now by Lemma 3.3, there are positive numbers δ_a and δ_b such that

$$q(h_j(w), h_j(F(a))) \geq \delta_a$$

and

$$q(h_j(w), h_j(F(b))) \geq \delta_b.$$

Let Φ be the isomorphism induced by F and set $g_j = \Phi^{-1}(h_j)$, so that for each
z in L(G)

$$F(g_j(z)) = h_j(F(z)).$$

Since $F(x) = F(y) = w$ we see

$$q(F(g_j(x)), F(g_j(a))) = q(h_j(F(x)), h_j(F(a))) \geq \delta_a$$

and similarly

$$q(F(g_j(y)), F(g_j(a))) = q(h_j(F(y)), h_j(F(a))) \geq \delta_a.$$

By symmetry we also obtain

$$q(F(g_j(x)),F(g_j(b))) \geq \delta_b \quad \text{and} \quad q(F(g_j(y)),F(g_j(b))) \geq \delta_b.$$

Now F is continuous on the compact set $L(G)$ and so F is uniformly continuous. Thus there is an $\varepsilon > 0$ such that for any points $z, z' \in L(G)$,

$$q(z,z') < \varepsilon \implies q(F(z),F(z')) < \min\{\delta_a, \delta_b\}$$

This, together with the above, implies that

$$\begin{aligned}
&\text{(i)} \quad q(g_j(y),g_j(a)) \geq \varepsilon \\
&\text{(ii)} \quad q(g_j(x),g_j(a)) \geq \varepsilon \\
&\text{(iii)} \quad q(g_j(y),g_j(b)) \geq \varepsilon \\
&\text{(iv)} \quad q(g_j(x),g_j(b)) \geq \varepsilon.
\end{aligned}$$

Next, since G is a discrete convergence group, there are points p and q and a subsequence of the sequence $\{g_j\}$ (which we again relabel $\{g_j\}$ for simplicity) such that

$$g_j \to p \quad \text{locally uniformly in } S^n - \{q\}$$

and

$$g_j^{-1} \to q \quad \text{locally uniformly in } S^n - \{p\}$$

as $n \to \infty$.

Equation
    (i) implies either $y = q$ or $a = q$ and similarly
    (ii) implies either $x = q$ or $a = q$
    (iii) implies either $y = q$ or $b = q$
    (iv) implies either $x = q$ or $b = q$.

Since $a \neq b$, these equations imply that $x = y = q$ and in particular that $x = y$. Furthermore, equations (i) and (iii) imply that y is a point of approximation, in much the same way as Lemma 3.3 is proved.

The only remaining claim in the statement of the lemma is that any right inverse of F is continuous at the points of approximation of H. To see this, suppose that f is a right inverse of F. That is, $F \circ f(x) = x$, for all $x \in L(H)$. If f is not continuous at a point of approximation w, then there are sequences x_j

and y_j converging to w such that $f(x_j) \to x'$ and $f(y_j) \to y'$, with $x' \neq y'$. Since F is continuous

$$F(x') = \lim F(f(x_j)) = \lim x_j = w = \lim y_j = \lim F(f(y_j)) = F(y')$$

and we have a contradiction to the first part of the lemma.

We are now in a position to assert the validity of a number of interesting corollaries to Lemma 3.4. Firstly,

3.5 COROLLARY. *Let G and H be nonelementary discrete convergence groups and suppose that $F : L(G) \to L(H)$ is a continuous map of the limit sets induced by an isomorphism. If every point of $L(H)$ is a point of approximation, then so is every point of $L(G)$ and F is a homeomorphism. Thus*

$$F \circ G \circ F^{-1} \,/\, L(H) = H \,/\, L(H).$$

In particular, if H is of the first kind ($L(H) = S^n$), then so too is G and thus G and H are topologically conjugate.

We are especially interested in the case that H is a Möbius group. In a geometrically finite Möbius group Γ (that is, there is a finite sided fundamental polyhedron for the action of Γ on $\mathbf{B}^{n+1}$) every limit point is either a point of approximation or a parabolic fixed point, see [B.M. Thm.2.], there they deal only with the case of Kleinian groups of S^2, the general result is similar. In particular if Γ is a geometrically finite Möbius group of the first kind, then there are only countably many points of S^n which are not points of approximation.

3.6 COROLLARY. *Let G and H be nonelementary discrete convergence groups of the second kind (so $O(G)$ is open and dense). Suppose that there is a homeomorphism $F : O(G) \to O(H)$ such that*

$$F \circ G \circ F^{-1} \,/\, O(H) = H \,/\, O(H)$$

(so F projects to a homeomorphism of the quotient spaces) and that F has a continuous extension to a map of S^n. If every point of $L(H)$ is a point of approximation, then G and H are topologically conjugate. In particular, if H is a Möbius group of compact type (that is (cl $(\mathbf{B}^{n+1})$-$L(H))/H$ is compact) and F has a continuous extension, then G and H are topologically conjugate.

This last fact follows because a Möbius group of compact type is geometrically finite and contains no parabolics, so that every limit point is a point of approximation.

Notice that the loxodromic fixed points of a convergence group are always points of approximation. It is quite difficult in general to show that a given point in the limit set is actually a point of approximation. It is not difficult however to show, using standard distortion arguments, that the conical limit points of a uniformly quasiconformal group acting on cl $(\mathbf{B}^{n+1})$ are points of approximation.

§4. The Fuchsian case.

In this section we will prove that every discrete convergence group of the closed disk   cl $(\mathbf{B}^2)$ is the toplogical conjugate of a Fuchsian group. We will sometimes prefer to work in   cl $(\mathbf{R}^n)$ (the one point compactification of $\mathbf{R}^n$ rather than the conformally equivalent n-sphere $\mathbf{S}^n$.

4.1 DEFINITION. A convergence group of the closed unit ball cl $(\mathbf{B}^n)$ in $\mathbf{R}^n$ is a group of homeomorphisms of   cl $(\mathbf{B}^n)$ such that when extended by reflection via the standard Möbius inversion in $\mathbf{S}^n$ we obtain a convergence group of   cl $(\mathbf{R}^n)$.

When we speak of a convergence group of   cl $(\mathbf{B}^n)$ acting in   cl $(\mathbf{R}^n)$ we will mean this canonical extension by reflection. Furthermore, it is easy to see from the local uniform convergence of sequences that if G is a convergence group of cl $(\mathbf{B}^n)$, then the points x and y of Definition 1.1 must lie on the boundary, $\mathbf{S}^{n-1}$, of cl $(\mathbf{B}^n)$. This implies that L(G) lies in the boundary of cl $(\mathbf{B}^n)$, and that the restriction of a convergence group of cl $(\mathbf{B}^n)$ to its boundary $\mathbf{S}^{n-1}$ will be a convergence group of $\mathbf{S}^{n-1}$ (this last observation will be important to us).

To begin with we need a couple of lemmas.

4.2 LEMMA. *Let g and h be orientation preserving elements of infinite order in a discrete convergence group G of cl (R). If g and h have a common fixed point, then they lie in a cyclic subgroup of G.*

PROOF: Let us first suppose that both g and h are parabolic. Then H = <g,h> contains no loxodromic elements, since H is assumed discrete. Thus H is elementary and L(H) is a single point which we may assume to be ∞ so that the ordinary set of H is $\mathbf{R}$, see [G.M. Thm. 5.8]. There can be no elliptics as an orientation preserving elliptic transformation of cl $(\mathbf{R})$ with a fixed point is the identity. Hence every $f \in H$ is parabolic and has ∞ as its unique fixed point. The group H then acts properly discontinuously on $\mathbf{R}$ (see 1.4 (2)) so the orbit of 0 has ∞ as its only accumulation point.

We can now find a natural order on the elements of H as follows. We say

$$f_1 < f_2 \quad \text{if and only if} \quad f_1(0) < f_2(0).$$

Since $f_1 \circ f_2^{-1}$ has no fixed points other than ∞ we see

$$f_1(0) < f_2(0) \quad \Leftrightarrow \quad f_1(x) < f_2(x) \quad \text{for all } x \in \mathbf{R}.$$

Let f be the unique element of H with

$$\text{Identity} < f < f' \qquad \text{for all } f' \in H\text{-}\{\text{Identity}\} \text{ with } f'(0) > 0.$$

We claim H is generated by f. If this is not the case then there is $f' \in H$ and a positive integer n such that

$$f^n < f' < f^{n+1}.$$

We would then have (since f is increasing)

$$\text{Identity} < f^{-n} \circ f' < f,$$

which is impossible by the choice of f.

This proves the lemma in the case that H is parabolic. If either g or h is loxodromic, then from [G.M. Thm. 6.7], there is an integer k such that either $h \circ g^k = g^k \circ h$ or $h^k \circ g = g \circ h^k$ (the result is stated there only for dimension at least two, although the proof works just as well in dimension one). This relation implies that g and h have common fixed points. The fact that $H = \langle g,h \rangle$ is cyclic then follows as in the parabolic case. The lemma is proved.

If G is a discrete convergence group and $x \in \mathbf{R}^n$, *the stabilizer of* x, which we denote by G_x, is the subgroup of G which fixes the point x. That is

$$G_x = \{ g \in G : g(x) = x \}.$$

4.3 LEMMA. *Let G be a discrete convergence group of $cl\,(\mathbf{R}^2)$. If $x \in O(G)$, then the stabilizer G_x of x is either the identity, a cyclic group generated by a single elliptic element or a dihedral group. (A convergence group is dihedral if it is topologically conjugate to a group generated by two reflections across lines through the origin whose angle of intersection is a rational multiple of π.) In all cases G_x is topologically conjugate to a finite Möbius group.*

PROOF: Assume first that G_x is an orientation preserving group. Since $x \in O(G)$, each $g \in G$ is either the identity or is elliptic and the group G_x is finite. Hence we can find a neighborhood U of x such that $G_x(U) = U$ and that every $g \in G\text{-}\{\text{identity}\}$ fixes only x in U. Then

$$\Pi : U\text{-}\{x\} \;\to\; (U\text{-}\{x\})\,/\,G_x$$

is a covering projection onto a surface which can be triangulated. Thus we may obtain a G_x-invariant triangulation of $U\text{-}\{x\}$. Let C be a P.L. (piecewise linear) circle in this triangulation separating x from the boundary of U and bounding a disk V containing x. Now

$$V^* = \cap \{g(V) : g \in G_x\}$$

is a neighbourhood of x. The boundary of the component of V^* containing x is a G_x-invariant circle C^* (recall G_x is finite). On C^* the group G_x is a finite group acting without fixed points and so cyclic. The lemma then follows in the orientation preserving case since G_x is generated by a single orientation preserving periodic homeomorphism of $\mathbf{S}^2$ which must be topologically conjugate to rotation, see [G.M. I §7.25].

If G_x also contains orientation reversing elements, we let H denote the normal, orientation preserving subgroup of index two in G_x. From the first part of the lemma we may assume that H is a Möbius group. We know that $cl\,(\mathbf{R}^2)/H = S$ is also a two dimensional sphere, and the map $cl\,(\mathbf{R}^2) \to S$ is a branched covering with two branch points. Since H is normal in G_x, the group G_x/H acts on S. Let f be an orientation reversing element of G_x. Regarded as a map of S, f is an orientation reversing map of order two and hence conjugate to a Möbius reflection, (again see [G.M. I §7.25]). Lifting this conjugation to $cl\,(\mathbf{R}^2)$ we obtain the lemma.

We are now in a position to establish

4.4 THEOREM. *Any discrete convergence group G of cl $(\mathbf{B}^2)$ is topologically conjugate to a Fuchsian group.*

PROOF: In order to prove the theorem we must first find a Fuchsian group Γ and a homeomorphism $F : O(G) \to O(\Gamma)$ conjugating $G \mid O(G)$ to $\Gamma \mid O(\Gamma)$. If $\mathbf{B}^2 / G$ is compact, the simplest way to do this is to give the subset $O(G)$ of cl $(\mathbf{B}^2)$ a conformal structure so that every $g \in G$ is conformal in this structure. To define it, we find a G-invariant triangulation of $O(G)$, by lifting a finite triangulation of the compact quotient space (to find this we must apply Lemma 4.3 which implies that the quotient is a (bordered) surface with the same kind of branching as that which occurs for Fuchsian groups). Then the trick of Ahlfors and Sario, [A.S. pg.127] gives the conformal structure. One must then check that Γ is actually a Fuchsian group and not a conformal group of $\mathbf{R}^2$. In this latter case Γ would have an abelian subgroup of rank two and of finite index since the quotient is compact. This is impossible from Lemma 4.2. Unfortunately this simple trick may not work in the noncompact case. For instance Γ may be of the second kind, even if G is of the first kind.

Thus we assume that $S_G = O(G) / G$ is not compact and seek the appropriate Fuchsian group Γ. The projection map $P : O(G) \to S_G$ gives some additional structure to S_G. First of all we define a map

$N_G : S_G \to \mathbf{Z}_+$ (the positive integers), by the rule

$$N_G(x) \text{ is the degree of branching of } P \text{ at } x.$$

If no $y \in P^{-1}(x)$ is fixed by a reflection, then $N_G(x)$ is just the order of the stabilizer G_y of y, where $P(y) = x$ Otherwise we define $N_G(x) = \#G_y /2$, where $P(y) = x$. Notice that if G_y contains a reflection, then it has even order as it is a finite group with a subgroup of index two. Since G acts properly discontinuously in $O(G)$ it is clear that $\{ x \in S_G : N_G(x) > 1 \}$ is a discrete subset of S_G.

We note in addition that S_G has two kinds of boundary: one part comes from $(S^1 - L(G)) / G$ and the other part comes from points fixed by reflections in G. We set $B_G = (S^1 - L(G))/G$ as this first part of the boundary. We now have the triple

$$(S_G, B_G, N_G)$$

which was considered in [Tu.3. pg. 5]. This triple contains all the information we need to find Γ.

If G is of the second kind, or if $O(G) / G$ is not compact, then Theorem 2 of [Tu.3] gives a Fuchsian group Γ such that the triples (S_G, B_G, N_G) and $(S_\Gamma, B_\Gamma, N_\Gamma)$ are homeomorphic. Actually, for technical reasons, this theorem considers Γ as acting on an open set E of cl $(\mathbf{B}^2)$ - $L(\Gamma)$ such that $\mathbf{B}^2$ lies in E and that $E \cap \mathbf{S}^1$ is dense in $\mathbf{S}^1$ if $E \neq$ cl $(\mathbf{B}^2)$. Thus here we have

$$S_\Gamma = E / \Gamma \quad \text{and} \quad B_\Gamma = (E - \mathbf{B}^2) / \Gamma.$$

Note too, that $N_G(x) = 1$ at all points corresponding to points of $\mathbf{S}^1$.

Once we have the homeomorphism F' of triples we can lift it to a homeomorphism $F : O(G) \to E$ conjugating G to Γ. Here, either $E = \mathbf{B}^2 = O(\Gamma)$ or cl $(\mathbf{B}^2)$ - E is a totally disconnected subset of $\mathbf{S}^1$. This follows since F' respects branching and the separation of the boundary into the pieces coming from $\mathbf{S}^1$ and from reflections, see [Tu.2 Prop.2.4] (this proposition considers only Fuchsian groups and only the situation in $\mathbf{B}^2$, so that all the boundary comes from reflections, but the proof given there applies in our situation with only very slight modification as the local behaviour of the projection map P is the same for both convergence groups and for Fuchsian groups by Lemma 4.3 (note that it can be applied since we can extend G canonically to a group of cl $(\mathbf{R}^2)$ via Möbius inversion). Notice in particular that the local behaviour is the same also on the boundary of cl $(\mathbf{B}^2)$.

Having obtained our map F we must show that it can be homeomorphically extended to the limit set. If G is of the second kind this is immediate as $L(G)$ and $\mathbf{S}^1$ - E will be totally disconnected subsets of $\mathbf{S}^1$ ($B_G \neq \emptyset \Leftrightarrow B_\Gamma \neq \emptyset$ as they are homeomorphic, and so G is of the second kind if and only if Γ is as well). If G is of the first kind, then so too is Γ and a little more is required to obtain the desired extension.

For $\gamma \in \Gamma$, we denote by $F[\gamma]$ the element $F^{-1} \circ \gamma \circ F$ of G. We first show that if $\gamma \in \Gamma$ is loxodromic, then so too is $F[\gamma]$ and conversely. By the pairwise density of the loxodromic fixed points in $\mathbf{S}^1 \times \mathbf{S}^1$, see 1.4 (5), there is another loxodromic $\gamma' \in \Gamma$ such that the axes $ax(\gamma)$ and $ax(\gamma')$ intersect. We can easily extend $F \mid ax(\gamma)$ and $F \mid ax(\gamma')$ to the fixed points of γ and γ'. If $F[\gamma]$ is parabolic, then $F(ax(\gamma))$ is a topological circle which meets $\mathbf{S}^1$ only at the fixed point of $F[\gamma]$.

Since $F(ax(\gamma)) \cap F(ax(\gamma'))$ is a single point and since γ and γ' do not lie in a cyclic subgroup (neither therefore do $F[\gamma]$ and $F[\gamma']$) we have a contradiction to Lemma 4.2.

Thus F preserves loxodromic elements and from the above argument we see it also preserves intersecting axes. These properties imply that there is a unique homeomorphism of S^1 induced by F. The argument of [Tu.2. pp 30-32] is now valid in our situation, as it is based on the existence of a neighborhood basis for each point of S^1 consisting of sets U_i lying in cl $(\mathbf{B}^2)$, and whose boundary is an axis of a hyperbolic $\gamma \in \Gamma$. The theorem is proved.

4.5 COROLLARY. *Let G be a discrete convergence group of S^1. Then any extension of G to a convergence group of cl $(\mathbf{B}^2)$ is essentially unique. That is if G_1 and G_2 are any such extensions, then there is a self homeomorphism h of cl $(\mathbf{B}^2)$ such that*

$$h \circ G_1 \circ h^{-1} = G_2$$

and $h \mid S^1 = Identity$.

PROOF: Firstly, find Fuchsian groups Γ_i and self homeomorphisms h_i of cl $(\mathbf{B}^2)$ such that

$$h_i \circ G_i \circ h_i^{-1} = \Gamma_i \qquad i = 1,2.$$

Then, since $G_1 \mid S^1 = G_2 \mid S^1$, $h_3 = h_2 \circ h_1^{-1} \mid S^1$ is a homeomorphism of S^1 conjugating $\Gamma_1 \mid S^1$ to $\Gamma_2 \mid S^1$. From [Tu.3. Thm. 3], h_3 has a homeomorphic extension to cl $(\mathbf{B}^2)$ (which we still call h_3) conjugating Γ_1 to Γ_2. The desired map h of the corollary is then easily seen to be $h = h_2^{-1} \circ h_3 \circ h_1$.

4.6 COROLLARY. *Let G be a discrete convergence group of S^2 with an invariant closed topological disk. Then G is topologically conjugate to a Kleinian group.*

PROOF: We first conjugate G by a homeomorphism from S^2 to cl $(\mathbf{R}^2)$ such that the invariant topological disk in S^2 is now the unit disk of $\mathbf{R}^2$. We denote this new group by G as well.

Let r : cl $(\mathbf{R}^2)$ → cl $(\mathbf{R}^2)$ $(r^2$ = identity) be the Möbius inversion in the unit cir-cle, $\mathbf{S}^1$. From Theorem 4.4, there is a homeomorphism h conjugating G | cl $(\mathbf{B}^2)$ to a Fuchsian group Γ. We extend h to $\mathbf{S}^2$ via reflection and set

$$H = h \circ G \circ h^{-1}.$$

So H | cl $(\mathbf{B}^2)$ = Γ. Then both H and the natural extension of Γ are convergence groups which are extensions of the convergence group Γ | $\mathbf{S}^1$ = H | $\mathbf{S}^1$. Thus from Corollary 4.5, there is a homeomorphism f of cl $(\mathbf{R}^2)$ - $\mathbf{B}^2$, which is the identity on $\mathbf{S}^1$, conjugating H to Γ. Extending this map f via the identity, we find that

$$\Gamma = f \circ h \circ G \circ h^{-1} \circ f^{-1}$$

is the desired Kleinian group and the required conjugacy.

We remark that if there is only an invariant topological circle for the convergence group G, then we can find a normal index two subgroup of G with an invariant closed disk D. From the above corollary we may conjugate G so that this subgroup becomes a Möbius group Γ and D becomes cl $(\mathbf{B}^2)$. Suppose that $\mathbf{B}^2 / \Gamma$ is a surface of finite type, that is if Γ is finitely generated. Let h ∈ G be an element such that ⟨Γ,h⟩ = G. We wish to show that h (and hence G since Γ is normal and finite index) can be assumed to be quasiconformal. The best way to see this is to consider the surface S = O(Γ) / Γ ∪ {punctures} and the group H of order two induced by h, acting on S. Now S is a Riemann surface with some additional structure corresponding to the punctures and branch points. We may find an H-invariant triangulation respecting this extra structure and then change this triangulation slightly so as to become a quasiconformal triangulation and in this way h will become a quasiconformal self homeomorphism of O(Γ) = O(G) after lifting. We denote by h' the mapping so obtained. Since Γ is geometrically finite and normal in G, we see h' ∘ Γ ∘ h'$^{-1}$ is equal to Γ so that h' is a Γ compatible quasiconformal homeomorphism of O(Γ). Such compatible quasiconformal homeomorphisms extend to quasiconformal homeomorphisms of cl $(\mathbf{R}^2)$ by [Tu.1 Thm. 4.2]. We have now shown that there is a homeomorphism f of O(G) which conjugates

G = ⟨Γ,h⟩ to the group ⟨Γ,h'⟩ and such that f ∘ Γ ∘ f^{-1} = Γ. Again appealing to Thm. 4.2 of [Tu.1] we may extend f to a homeomorphism f' of cl $(\mathbf{R}^2)$. Thus G is topologically conjugate via f' to the group ⟨Γ,h'⟩ which is uniformly quasiconformal, since h' is quasiconformal and Γ is a Fuchsian subgroup of index two. Since uniformly quasiconformal groups of cl $(\mathbf{R}^2)$ are quasiconformally conjugate to Kleinian groups (see, Sullivan, Tukia, in particular [Tu.4 Thm. F and Remark F2]), we obtain the following corollary.

4.7 COROLLARY. *Let G be a finitely generated convergence group of S^2 with an invariant topological circle. Then G is topologically conjugate to a Kleinian group.*

Combining the above results with some of the results of [G.M. I] we obtain the following characterization of certain groups of homeomorphisms of S^2.

4.8 COROLLARY. *Let G be a finitely generated group of homeomorphisms of S^2 which acts properly discontinuously in the complement of some totally disconnected set E. Suppose that there is a G-invariant topological circle C of S^2. Then G is topologically conjugate to a Kleinian group. Furthermore if G preserves the components of S^2 - C, the assumption that G be finitely generated is unnecessary.*

PROOF: The hypothesis that G acts properly discontinuously in the complement of a totally disconnected set implies that G is a discrete convergence group, see [G.M. Thm. 7.8]. The result then follows from Corollary 4.6.

Actually in Corollary 4.7 some restrictions are necessary. This is because there are discrete convergence groups G of S^2 whose limit sets are totally disconnected Cantor sets and which are not the topological conjugates of any Kleinian group, see [M.G.] and [G.M. II]. G will act properly discontinuously in the complement of L(G), thus from Corollary 4.7 there can be no invariant topological circle. This is clear in the examples cited above, as the reason that they are not the topological conjugates of any Kleinian group is that the stabilizer of some point in the limit set is a free group on at least two generators (which cannot occur in the Kleinian case). If there were an invariant circle for the action of G, then an argument analogous to that of Lemma 4.2 would imply that the stabilizer of any point in the limit set (which must lie on this topological circle as L(G) is the smallest, nonempty, closed G-invariant set) contains a cyclic subgroup of finite index, and this excludes the possibility that the stabilizer contains a free group of rank two or more.

Finally, there is evidence to suggest that a group of homeomorphisms of S^1 is a topological conjugate of a Fuchsian group if and only if it is a discrete convergence group. The general case seems to be quite difficult but the second named author has shown, see [Tu. 5], that either a discrete convergence group is the topological conjugate of a Fuchsian group or is of such a special nature that it might be presumed not to exist. For more general results on the conjugacy of finitely generated convergence groups of S^2 to Kleinian groups, see [M.S.].

§5. 3-dimensional convergence groups of compact type.

In this section we will show that a torsion free convergence group of compact type acting on the closed unit ball $cl\,(\mathbf{B}^3)$ of $\mathbf{R}^3$ is topologically conjugate to a Möbius group on its ordinary set. We will do this by showing that the associated 3-manifold contains no incompressible tori and then use some of the three manifold theory developed by Thurston, see [M.B.], to see that the quotient has a hyperbolic structure, lifting to the universal covers we obtain the desired Möbius group and the required conjugacy. We then turn to consider when this conjugacy can be extended homeomorphically to the limit set, thus obtaining a conjugacy on all of $cl\,(\mathbf{B}^3)$.

5.1 DEFINITION. A convergence group of $cl\,(\mathbf{B}^3)$ is said to be of *compact type*, if $O(G)/G$ is compact, where $O(G) = cl\,(\mathbf{B}^3) - L(G)$.

A continuous map $f\colon F^2 \to M^3$ from a surface F^2 into a three manifold M^3 is said to be incompressible if the associated map of the fundamental groups is injective. If F^2 lies in M^3 and is a properly embedded surface (that is, ∂F is contained in ∂M), then we say that F^2 is incompressible if the inclusion map is incompressible. It follows from the stong torus theorem, see, for instance, [Sc. §3], that if M is a compact, orientable, irreducible 3-manifold with incompressible boundary and which admits an incompresible map of the torus T^2, then there is an embedded incompressible torus in M or otherwise the fundamental group of M contains an infinite normal cyclic subgroup. This latter case cannot occur if M is the quotient of a convergence group of $\mathbf{B}^3$ as the convergence group will be nonelementary, see [M.G.]. Note too, that all such quotients will be irreducible, that is every topologically flat two sphere will bound a three ball.

5.2 THEOREM. *Let G be a torsion free discrete convergence group of $cl\,(\mathbf{B}^3)$. Suppose that*

$$M = O(G)\,/\,G$$

is compact. Then there is no incompressible map of a torus into M.

PROOF: Note that G acts properly discontinuously in $O(G)$ so that the quotient space M is a compact three manifold, possibly with boundary.

Suppose that $f : T^2 \to M$ is incompressible. Since finite covers of tori are still tori we may assume by passing to suitable finite covers of M and T^2, that M is orientable. From the remarks after Definition 5.1 we may suppose that there is an embedded incompressible torus T contained in M. Let $H = i_*(T) < G = \pi_1(M)$, where i is the inclusion map. Since T is incompressible, H is isomorphic to $\mathbf{Z} + \mathbf{Z}$. Let $p : O(G) \to M$ be the canonical projection map and let $P_1, P_2,\ldots$ be the components of $p^{-1}(T)$ in $O(G)$. If H_i is the stabilizer of P_i, then we see as above that $H_i \cong \mathbf{Z} + \mathbf{Z}$. Now H_i is an abelian group and so elementary, see 1.4 (3). It has one end and therefore at most one limit point, see [M.G.], and thus H_i is generated by a pair of parabolic elements (in fact every element of H_i other than the identity is parabolic). We denote the limit point of H_i by $z_i \in S^2 = \partial B^3$. Now as H_i is a discrete convergence group with a single limit point z_i, H_i acts properly discontinuously and freely on $cl\,(B^3) - \{z_i\}$. We claim firstly that

(5.3) $P_i \cup \{z_i\}$ is topologically a two sphere.

Clearly P_i is homeomorphic to $\mathbf{R}^2$ as P_i is a universal cover of T. The claim now follows as P_i is H_i-invariant and any sequence $\{h_i\}$ of distinct elements of H_i, converges locally uniformly in P_i to the value z_i.

Let B_i be the component of $cl\,(B^3) - (P_i \cup \{z_i\})$ contained in $\mathbf{B}^3$. Since the components of $cl\,(B^3) - (P_i \cup \{z_i\})$ are not homeomorphic, it is clear that B_i is left invariant by H_i. It is not clear that B_i is a ball. Since p is a local homeomorphism, it is clear that P_i is locally flat, however it might be possible that P_i is wild near z_i (that this is not the case follows from our later results). In any case all we need is the following claim

(5.4) $\qquad (cl(B_i) - \{z_i\})/H_i \qquad$ is compact.

To see this we let G_i be the subgroup of G which stabilizes the point z_i. From [G.M. I Thm.6.10] we see that G_i is elementary. Furthermore H_i lies in G_i, so that G_i contains at least one parabolic element. From the classification of the elementary convergence groups, recalling that G is torsion free, we find that every element of G_i is parabolic. Again we see that G_i acts freely and properly discontinuously in $cl\,(B^3) - \{z_i\}$ and so in particular on $S^2 - \{z_i\}$.

Let

$$S_i = (S^2 - \{z_i\}) / G_i$$

and

$$S_i' = (S^2 - \{z_i\}) / H_i.$$

As G is a convergence group both S_i and S_i' are surfaces because G_i and H_i act as covering transformations. Now, as $H_i \cong \mathbf{Z}+\mathbf{Z}$ we see S_i' is a torus, for the only surface with fundamental group $\mathbf{Z}+\mathbf{Z}$ is T^2 (compact or otherwise). Also G_i is a suface group which contains a subgroup isomorphic to $\mathbf{Z}+\mathbf{Z}$. It is not difficult to see that then $G_i \cong \mathbf{Z}+\mathbf{Z}$ as well, since it is the fundamental group of a compact surface containing a subgroup isomorphic to $\mathbf{Z}+\mathbf{Z}$. Hence G_i is abelian and so H_i is normal and of finite index in G_i as S_i' covers S_i and both are compact surfaces.

Next let $g \in G$ and consider $g(\text{cl}\,(B_i))$. If $g(z_i) \neq z_i$, then we see (because either $g(P_i) = P_i$ or

$g(P_i) \cap P_i = \emptyset$ as G is the group of cover transformations and P_i is a component of $p^{-1}(T)$) that

$$(5.5) \qquad g(\text{cl}\,(B_i)) \cap \text{cl}\,(B_i) = \emptyset.$$

Thus $g(\text{cl}\,(B_i)) \cap \text{cl}\,(B_i) \neq \emptyset$ implies that $g \in G_i$, the stabilizer of z_i.
Let

$$C_i = \cup \{ g(\text{cl}\,(B_i)) : g \in G_i \}$$

Since B_i is H_i invariant and since G_i / H_i is a finite group, C_i is closed and G_i-invariant. From (5.5)

$$g(C_i) \cap C_i = \emptyset \qquad \text{if and only if} \qquad g \in G - G_i.$$

We now show that

$$C = \{ g(C_i\text{-}\{z_i\}) : g \in G \}$$

is locally finite in $O(G)$ and so in fact closed in this set. To see this, pick any point $z \in O(G)$ and let U be a small neighborhood of z on which the projection p is injective and such that either $p(U) \cap T$ is empty or connected, the existence of such neighborhoods is clear. Then U intersects the boundary of at most one $g(C_i\text{-}\{z_i\})$, since $p(\partial C_i) = T$, and hence meets at most one $g(C_i\text{-}\{z_i\})$.

It now follows that

$$(C\text{-}L(G))\,/\,G = (C_i\text{-}\{z_i\})\,/\,G_i$$

is compact, being a closed subset of the compact 3-manifold M. This finally establishes the claim (5.4).

Consider now

$$M_i \;=\; (cl\,(B_i)\text{-}\{z_i\})\,/\,H_i.$$

We have shown that M_i is a compact orientable three manifold with incompressible boundary F such that the map $\pi_1(F) \rightarrow \pi_1(M_i)$ induced by the inclusion is an isomorphism.

From [He. Thm.10.2 p. 89] we find there are no such three manifolds. This final contradiction implies that M can have no incompressible tori and the theorem is proved.

We can now appeal to the appropriate three manifold theory to obtain the following corollary. It follows from Theorem A' p. 70 of [M.B.], also note that the discussion on pp. 62 and 63 show we need only consider the convex core of the Kleinian group Γ conjugate to G (instead of all of $cl\,(\mathbf{B}^3) - L(\Gamma)$). The theorem mentioned above asserts that in our situation (compact quotient, irreducible, no incompressible tori) the quotient space $(cl\,(\mathbf{B}^3) - L(G))/G$ is homeomorphic to a hyperbolic space form. That is the quotient of a Möbius group acting on $\mathbf{B}^3$.

5.6 COROLLARY. *Let G be a torsion free convergence group of compact type acting on $\mathbf{B}^3$. If either*

(1) G is of the second kind or

(2) $\mathbf{B}^3 \,/\, G$ contains an incompressible surface (that is, M is Haken),

then there is a Möbius group Γ and a homeomorphism

$$F\colon\; cl\,(\mathbf{B}^3) - L(G) \;\rightarrow\; cl\,(\mathbf{B}^3) - L(\Gamma)$$

such that

136

$$F \circ G \circ F^{-1} / (cl(\mathbf{B}^3) - L(\Gamma)) = \Gamma / (cl(\mathbf{B}^3) - L(\Gamma)).$$

Notice from the results of section three, that if F has a continuous extension to $cl(\mathbf{B}^3)$, then this extension is actually a homeomorphism and so

$$F \circ G \circ F^{-1} = \Gamma$$

as groups of $cl(\mathbf{B}^3)$. Finding this continuous extension is quite difficult in general. However, if $L(G)$ is a Cantor set (perfect and totally disconnected) and G contains no parabolics, then G is isomorphic to a finitely generated free group (see [M.G.] and recall the quotient is compact and so the fundamental group is finitely generated). So too then is Γ. And hence, since Γ contains no parabolics (it is a Möbius group of compact type) $L(\Gamma)$ is also a Cantor set ($\Gamma \mid S^2$ is a Schottky group of compact type). Hence F does indeed have a unique extension to a homeomorphism of $cl(\mathbf{B}^3)$, (Cantor sets in S^n are determined by their complements, this is the uniqueness of Freudenthal compactifications).

There is an important special case when the map F will automatically extend homeomorphically to the boundary. This is when the group G is uniformly quasiconformal. Since M is a compact three manifold we can replace the homeomorphism of the quotient spaces by a quasiconformal homeomorphism, this is because of the Hauptvermutung for quasiconformal mappings due to Sullivan, see [T.V] (actually M inherits a natural quasiconformal structure from G). When we lift this quasiconformal mapping, we will obtain (since G is uniformly quasiconformal) a quasiconformal conjugacy on the ordinary sets. In particular the conjugacy is defined and quasiconformal on $\mathbf{B}^3$. Quasiconformal maps of $\mathbf{B}^n$ extend to quasiconformal maps of $cl(\mathbf{B}^n)$, see [Ge.]. This extension will preserve the conjugacy by continuity. Thus we find that torsion free quasiconformal groups of compact type and the second kind (or even the first kind if we assume that the quotient is Haken) are conjugate to Möbius groups in dimension three. We record this as

5.7 THEOREM. *Let G be a torsion free, uniformly quasiconformal group acting on $cl(\mathbf{B}^3)$, of compact type and the second kind. Then there is a Möbius group Γ and a quasiconformal self homeomorphism F of $cl(\mathbf{B}^3)$ such that*

$$F \circ G \circ F^{-1} = \Gamma.$$

We also mention that there is some hope of extending the above results to the geometrically finite case. That is the case where the quotient space associated to the discrete convergence group G acting on $cl(\mathbf{B}^3)$ is a 3-manifold with rank one or rank two cusped ends. Essentially this means there is a compact submanifold M' of M such that $M - M' = C_1 U C_2 U...U C_n$ and each C_i is homeomorphic to one of

$$T \times [0,\infty) \quad \text{or} \quad S^1 \times [0,1] \times [0,\infty)$$

and in the latter case $\partial C_i \cap \partial M = S^1 \times S^0 \times [0,\infty)$ so that $T \times \{0\}$ or $S^1 \times [0,1] \times \{0\}$ represent incompressible surfaces in M.

We would like to show in such a case that M is homeomorphic to a geometrically finite hyperbolic space form. In order to apply Theorem B' pg. 70 of [M.B.] we should show that the pair (M, P), where $P = cl(\partial M'-\partial M)$, is a pared manifold (see pg. 58). The conditions (i) and (ii) of pp. 58 and 59 are easily seen to be true provided one can show that if $C_{i,k}$ are the closures of the components of $p^{-1}(C_i)$, then for every i, the $C_{i,k}$ are disjoint. This seems to be true, however the details would be rather long and so we will not proceed further here.

Finally we outline the proof of the following theorem. It should be compared with the examples due to Freedman and Skora [F.S.] of a convergence group G of S^3 with $L(G)$ a Cantor set and $(S^3-L(G))/G$ compact, such that G is not conjugate to any Möbius group. A consequence is, for instance, that there cannot be an invariant toplogically flat two sphere in their example (this is not difficult to see for other reasons as well, such as the wildness of the limit Cantor set).

5.8 THEOREM. *Let G be a torsion free group of homeomophisms of S^3 acting properly discontinuously in the complement of a Cantor set E and such that $(S^3-E)/G$ is compact. Suppose that there is a G-invariant topologically flat two sphere S and that G stabilizes each component of $S^3 - S$. Then G is topologically conjugate to a Schottky group of S^3 and so isomorphic to a finitely generated free group.*

PROOF: The assumption that G acts properly discontinuously in the complement of a Cantor set implies that G is a nonelementary discrete convergence group and that $L(G)$ is contained in $E \cap S$. By conjugating G with a homeomorphism of S^3 we may assume that $S = S^2$, the equatorial two sphere in S^3. Let U and V denote the components of S^3-S^2. The assumptions imply that

$$(S^2-L(G))/G = F$$

is a closed surface of genus at least two and that both

$$M = (cl\,(U)\text{-}L(G))\,/\,G \quad \text{and} \quad M' = (cl\,(V)\text{-}L(G))\,/\,G$$

are compact three manifolds whose only boundary component (since E is zero dimensional) is the closed surface F. As above, see [He. Thm. 10.2], the natural map $\pi_1(F) \to \pi_1(M)$ cannot be an isomorphism. Thus there is a nontrivial loop on F spanning a disk in M (by Dehn's Lemma). If we cut open along this disk we obtain either another compact manifold with boundary a single surface of genus strictly less than F, or two such manifolds. Proceeding inductively, we cut M into disjoint balls. This shows that M is a handlebody and in particular that $G = \pi_1(M)$ is a finitely generated free group

From Corollary 5.6, there is a Möbius group Γ (acting on U) and a homeomorphism

$$\Psi : U\text{-}L(G) \to U\text{-}L(\Gamma) \qquad \text{such that}$$

$$f^{-1} \circ (\Gamma \mid (cl\,(U)\text{-}L(\Gamma))) \circ f = G \mid (cl\,(U)\text{-}L(G)).$$

The Kleinian group Γ naturally acts on U, and the quotient $(S^2\text{-}L(\Gamma))\,/\,\Gamma$ is the closed surface F. Since Γ is of compact type it contains no parabolic elements, [Tu. 1 Cor. 2.5], and since Γ is a free group (as it is isomorphic to G) we find from [Ma. Thm 1] that Γ is a Schottky group. Alternatively one can conclude that Γ is a Schottky group from the classification of function groups together with the fact that, since Γ is of compact type, there are no parabolics and no degenerate factor subgroups. Since Γ is a Schottky group its limit set is a Cantor set. Thus Ψ will extend to a homeomorphism of $cl\,(U)$ conjugating G to Γ and then by reflection to a homeomorphism of S^3.

Thus, after conjugating G by Ψ we are in the situation that G is a discrete convergence group of S^3 whose limit set is a Cantor set and such that $G \mid cl\,(U)$ is a Möbius group. The above arguments have shown that M is a compact hyperbolic manifold with boundary a single surface F and that M' is a compact three manifold also with boundary F. In particular both M and M' are irreducible Haken manifolds with boundary. Notice too that from Cor 5.6, $G \mid (\,cl\,(V)\text{-}L(G))$ is conjugate to a Möbius group and so applying the same reasoning as above we find $G \mid V$ is conjugate to a Möbius group.

Identifying U with V via the reflection r and defining $G \mid cl\,(U) = \Gamma$ and $r \circ G \circ r \mid cl\,(U) = H$, we see $\Gamma \mid S^2 = H \mid S^2$. Now as we noted above H is conjugate to a Kleinian group Γ' via a homeomorphism, say $f : cl\,(U) \to cl\,(U)$. Thus $f \mid S^2$ is a Γ compatible homeomorphism of S^2 and so has a homeomorphic extension f' to a Γ compatible homeomorphism of U, see Thm. 4.2 of [Tu.1].

Hence $h = f^{-1} \circ f'$ gives an extension of the Identity map of $\mathbf{S}^2$ which conjugates Γ to H. Finally $r \circ h$ extended via the identity produces the desired conjugacy of G to the Möbius group Γ. This then establishes the theorem.

Yale University, New Haven, CT 06520

REFERENCES

[AS] L. V. Ahlfors and L. Sario. "Riemann surfaces", Princeton Univ. Press, (1960).

[BM] A. F. Beardon and B. Maskit. *Limit points of Kleinian groups and finite sided fundamental polyhedra,* Acta. Math. **132** (1974).

[FS] M. Freedman and R. Skora. *Strange actions of groups on spheres,* J. Diff. Geom. **26** (1987) 75-98.

[Ge] F. W. Gehring. *Rings and quasiconformal mappings in space,* Trans. A.M.S. **103** (1962).

[GM] F. W. Gehring and G.J. Martin. *Discrete quasiconformal groups I, II,* Proc. London Math. Soc. (3) **55** (1987) 331-358.

[He] J. Hempel. "3-manifolds", Ann. Math. Stud. 86, Princeton Univ. Press, (1976).

[MB] J. W. Morgan and H. Bass. "The Smith conjecture", Academic Press, (1984).

[MG] G. J. Martin and F. W. Gehring. *Generalizations of Kleinian groups* M.S.R.I. Preprint. (1986).

[MS] G. J. Martin and R. Skora. *Group actions on* S^2, To appear.

[Ma] B. Maskit. *A characterization of Schottky groups,* J. D'Analyse Math **19**, (1967).

[Sc] G. P. Scott. *Strong annulus and torus theorems and the enclosing property of characteristic submanifolds of three manifolds,* Quart. J. Math. Oxford (2) **35** (1984).

[Tu1] P. Tukia. *On isomorphisms of geometrically finite Möbius groups,* I.H.E.S. Publ. No. 61 (1985).

[Tu2] P. Tukia. *On discrete groups of the unit disk and their isomorphisms,* Ann. Acad. Sci. Fenn. Ser. A.I. Math **504** (1972).

[Tu3] P. Tukia. *Extension of boundary homeomorphisms of discrete groups of the unit disk,* Ann. Acad. Sci. Fenn. Ser. A.I. **548** (1973)

[Tu4] P. Tukia. *On quasiconformal groups,* To appear, J. D'Analyse Math.

[Tu5] P. Tukia. *Homeomorphic conjugates of Fuchsian groups,* To appear.

[TV] P. Tukia and J. Väisälä. *Lipschitz and quasiconformal approximation and extension,* Ann. Acad. Sci. Fenn. Ser. A.I. **6** (1981).

The limit set of a discrete group
of hyperbolic motions

BY PETER J. NICHOLLS

1. Introduction.

Consider a discrete group Γ of Mobius transformations acting in the unit ball of R^n $(n \geq 2)$. The limit set of Γ is that subset of the unit sphere where Γ orbits accumulate and, as such, is the set of points where Γ fails to act discontinuously. Over the last several years much work has been done on the classification of limit points—a major impetus in this direction has been provided by the application of ergodic theory to discrete groups. Put simply, in order to understand the dynamics of the flow along a geodesic, for example, one must have information about the (limit) point situated at the "end" of the geodesic. In order to obtain general theorems about the classes of groups for which certain flows exhibit ergodic properties, one needs information on the size of various subsets of the limit set. Similarly the existence, or otherwise, of wandering sets under the group action on the sphere depends solely on the size of some other subsets of the limit set.

Many subclasses of the limit set have been identified and their properties explored. The literature is now very extensive and, due to the differing definitions and notations, is somewhat confusing.

The main aim of this paper is to give an account of the development and applications of some important subclasses of the limit set. In each case the definition will be followed by analytic and geometric interpretations and then by statements of the major theorems relating to their existence and applications. We conclude with a section covering results on the size of the limit set.

In the next section we collect some preliminary results necessary for the discussion.

2. Preliminaries.

Throughout we denote by B the unit ball in R^n and by S the unit sphere

$$B = \{x : |x| < 1\} \quad S = \partial B.$$

Research supported in part by the National Science Foundation.

The hyperbolic metric is denoted by ρ and is derived from the differential

$$d\rho = \frac{2|dx|}{1 - |x|^2}.$$

We shall be concerned with the measure of certain subsets of the unit sphere. We use w for the Lebesgue surface area on S and shall have occasion to measure the surface area of a subset of S interior to a ball. The following lemma is proved in [24].

LEMMA 2.1. *For* $\eta \in S$ *and* $\lambda > 0$, *set* $A = \{x \in S : |x| - \eta| < \lambda\}$. *Then*

$$w(A) = M \int_0^\mu (\sin \theta)^{n-2} d\theta$$

where $\mu = \arccos(1 - \lambda^2/2)$ *and* M *is an absolute constant.*

If γ is a Mobius transform preserving B then the Jacobian $\gamma'(x)$ is a positive multiple of an orthogonal matrix and this multiple, which measures the change in linear scale (the same in all directions), will be denoted by $|\gamma'(x)|$. The following result is to be found in [3].

THEOREM 2.2. *Let* γ *be a Mobius transform preserving* B. *Then*

1. *For any* $x, y \in B \cup S$,

$$|\gamma(x) - \gamma(y)| = |\gamma'(x)|^{1/2}|\gamma'(y)|^{1/2}|x - y|.$$

2. *For any* $x \in B$,

$$1 - |\gamma(x)|^2 = |\gamma'(x)|(1 - |x|^2).$$

3. *For any* $\xi \in S$,

$$|(\gamma^{-1})'(\xi)| = (1 - |\gamma(0)|^2)|\xi - \gamma(0)|^{-2}.$$

In order to measure the size of various subsets of the unit sphere the following definition is useful. If $a \in B$ and $k, \alpha > 0$ we define

$$(2.1) \qquad I(a : k, \alpha) = \{x \in S : |x - \frac{a}{|a|}| < k(1 - |a|)^\alpha\}.$$

Thus $I(a : k, \alpha)$ is that part of S which lies within a distance $k(1 - |a|)^\alpha$ of the projection of a from the origin onto S.

We next need a formula for the hyperbolic distance from a point to a line in B. Such a formula is to be found in [13, p. 162] but not in the form best suited to our purpose. The proof of the following is a routine calculation which is not repeated here.

THEOREM 2.3. *Suppose $a \in B$ and $\xi, \eta \in S$, $\xi \neq \eta$. If s denotes the hyperbolic distance from a to the geodesic joining ξ and η then*

$$\cosh s = \frac{2|a - \xi||a - \eta|}{|\xi - \eta|(1 - |a|^2)}.$$

We next consider cones at a point $\xi \in S$. If $x \in B$, $\xi \in S$ and λ satisfies $0 < \lambda < \pi/2$ then we say x belongs to the cone at ξ of opening λ if the angle between the vectors ξ and $\xi - x$ is at most λ and, further, $|x - \xi| < 2\cos\lambda$. The cosine of the angle between ξ and $\xi - x$ is calculated to be

$$\frac{\xi \cdot (\xi - x)}{|\xi||\xi - x|} = \frac{2 - 2\xi \cdot x}{2|\xi - x|} = \frac{(\xi - x) \cdot (\xi - x) + 1 - |x|^2}{2|\xi - x|} = \frac{|\xi - x|^2 + 1 - |x|^2}{2|\xi - x|}$$

and we have proved the following.

LEMMA 2.4. *If $x \in B$, $\xi \in S$ and λ satisfies $0 < \lambda < \pi/2$ then x belongs to the cone at ξ of opening λ and if and only if $|x - \xi| < 2\cos\lambda$ and*

$$\frac{|\xi - x|^2 + 1 - |x|^2}{2|\xi - x|} > \cos\lambda.$$

The following result characterizes conical approach to the boundary in a number of ways. Its proof is a straightforward application of Theorem 2.3 and Lemma 2.4.

THEOREM 2.5. *Suppose $\xi \in S$ and $\{x_n\}$ is a sequence of points of B with $|x_n| \to 1$ as $n \to \infty$. The following are equivalent*

1. *There exists $a > 0$ such that, for n large enough, x_n lies in the cone of opening a at ξ.*

2. *There exists $b > 1$ such that, for n large enough,*

$$|x_n - \xi| < b(1 - |x_n|).$$

3. *There exists $c > 0$ such that, for n large enough,*

$$\xi \in I(x_n : c, 1).$$

4. *There exists $d > 0$ such that if l is any geodesic ending at ξ then, for n large enough, $\rho(x_n, l) < d$.*

It can be shown that the constants a, b, c, d are related by

$$b \approx \frac{1}{\cos a} \approx \cosh d \approx (1 + c^2)^{1/2}.$$

In other words, (1) implies (2) with $b = \dfrac{1}{\cos a} + \epsilon$ for any $\epsilon > 0$ and (2) implies (1) with $a = \arccos(1/b) + \epsilon$ for any $\epsilon > 0$. Similar remarks hold for the relations between b, c and d. We next consider horospheres. A **horosphere** at $\xi \in S$ is a sphere in R^n which is internally tangent to the unit sphere S at ξ. A **horoball** is the interior of a horosphere.

THEOREM 2.6. *If $\xi \in S$, $x \in B$ and $0 < k < 1$ then x is on the horosphere at ξ of euclidean radius k if and only if*

$$(1 - |x|^2)|x - \xi|^{-2} = \frac{1 - k}{k}.$$

The point x is in the horoball at ξ of radius k if and only if

$$(1 - |x|^2)|x - \xi|^{-2} > \frac{1 - k}{k}.$$

PROOF: Suppose $x \in B$ with $(1 - |x|^2)|x - \xi|^{-1} = \dfrac{1 - k}{k}$ then

$$1 - |x|^2 = \frac{1 - k}{k}(x - \xi) \cdot (x - \xi) = \frac{1 - k}{k}(|x|^2 + 1 - 2x \cdot \xi)$$

and so

$$(2.2) \qquad 2x \cdot \xi = \frac{(|x|^2 + 1 - 2k)}{(1 - k)}.$$

Now the square of the distance of x from the center of the horosphere is

$$\begin{aligned}
|x - (1 - k)\xi|^2 &= |x|^2 + (1 - k)^2 - 2(1 - k)x \cdot \xi \\
&= |x|^2 + (1 - k)^2 - (|x|^2 + 1 - 2k) \\
&= k^2
\end{aligned}$$

where we substituted for $x \cdot \xi$ from (2.2) above. Thus x lies on the horosphere. Our argument is clearly reversible and we have the if and only if condition. The statement concerning the horoball is an easy modification of this. $\qquad\qquad\square$

We consider next a discrete group Γ preserving the unit ball B. For $a \in B$ we define the orbit of a, $\Gamma(a)$, to be the set $\{\gamma(a) : \gamma \in \Gamma\}$. A point $\xi \in S$ is a **limit point** for the discrete group Γ if for one, and hence every, point $x \in B$ the orbit $\Gamma(x)$ accumulates at ξ. The set of limit points is denoted by $\Lambda(\Gamma)$ or simply Λ. For the time being we merely remark that Λ is a closed subset of S and its complement is the set of **ordinary points**. The group Γ is said to be of the **first kind** if $\Lambda = S$ and of the **second kind** otherwise.

If $a \in B$ and Γ is a discrete group containing no element which fixes a (except the identity) then the Dirichlet region centered at a is defined as follows

$$D_a = \{x \in B : \rho(x,a) < \rho(x,\gamma(a)) \quad \gamma \in \Gamma - I\}.$$

In other words, from each orbit we select the point closest to a.

From the differential $d\rho$ defined in B we may construct a hyperbolic volume element as follows

$$dV = \frac{2^n dx_1 dx_2 \ldots dx_n}{(1 - |x|^2)^n}$$

and use this to measure the volume of a Dirichlet region D_a. The discrete group Γ is said to be of **finite volume** if for one (and hence every) non-fixed point $a \in B$, $V(D_a) < \infty$. As examples, one can consider any group with D_a compact in B or a geometrically finite Fuchsian group of the first kind.

We are interested in how fast the points in an orbit tend to S. The first observation is that any two orbits $\Gamma(a)$, $\Gamma(b)$ are comparable in the sense that the ratios

$$\frac{1 - |\gamma(a)|}{1 - |\gamma(b)|} \qquad \text{for } all \; \gamma \in \Gamma$$

lie between finite limits. To see this note that $\rho(0, \gamma b) \leq \rho(0, \gamma a) + \rho(a, b)$ and so

$$\log \left(\frac{1 + |\gamma(b)|}{1 - |\gamma(b)|} \right) \leq \log \left(\frac{1 + |\gamma(a)|}{1 - |\gamma(a)|} \right) + \rho(a, b)$$

from which

$$1 - |\gamma(a)| \leq 2e^{\rho(a,b)}(1 - |\gamma(b)|)$$

for all γ.

A good way to study the rate at which orbits tend to S is to consider the convergence of the series

$$\sum_{\gamma \in \Gamma}(1 - |\gamma(a)|)^\alpha$$

for various $\alpha > 0$. From our remarks above, the convergence or otherwise of this series is independent of $a \in B$. Thus in general we will consider the series

$$(2.3) \qquad \sum_{\gamma \in \Gamma} (1 - |\gamma(0)|)^{\alpha}.$$

In many ways it is more natural to look instead at the series

$$(2.4) \qquad \sum_{\gamma \in \Gamma} e^{-\alpha \rho(0, \gamma(0))}$$

and in view of the fact that $\rho(0, \gamma(0)) = \log \left(\dfrac{1 + |\gamma(0)|}{1 - |\gamma(0)|} \right)$ it is immediate that (2.3) and (2.4) converge or diverge together. For the group Γ we define the exponent of convergence, or critical exponent, $\delta(\Gamma)$ to be the infimum of those positive reals α for which the series (2.4) converges. The following result is based on fairly elementary volume estimates. See [**3**, Chapter 7] for the proofs of the first two parts and [**41**, p. 518] for the proof of the third part in dimension 2 (the proof in higher dimensions is analogous).

THEOREM 2.7. *Let Γ be a discrete group preserving B and, for $\alpha > 0$, consider the series*

$$\sum_{\gamma \in \Gamma} e^{-\alpha \rho(0, \gamma(0))}.$$

1. *If $\alpha > n - 1$ the series converges.*
2. *If Γ is of the second kind and $\alpha = n - 1$, the series converges.*
3. *If Γ is of finite volume and $\alpha = n - 1$, the series diverges.*

For our later results concerning the size of various limit point classes we need a further definition. If $a \in B$, $k, \alpha > 0$ and Γ is a discrete group enumerated by $\Gamma = \{\gamma_n : n = 0, 1, 2, \ldots\}$ we define the following subset of the unit sphere S

$$(2.5) \qquad L(a : k, \alpha) = \bigcap_{N=1}^{\infty} \bigcup_{n > N} I(\gamma_n(a) : k, \alpha).$$

Thus $L(a : k, \alpha)$ comprises those points of S which lie in infinitely many of the neighborhoods $I(\gamma(a) : k, \alpha)$, $\gamma \in \Gamma$. Since, on any sequence $\{\gamma_m\}$, $|\gamma_m(a)| \to 1$ we see that, for any $k, \alpha > 0$, $L(a : k, \alpha)$ comprises limit points. Note further that the size of k and α regulate the rate at which the orbit of a approaches $\xi \in L(a : k, \alpha)$.

Our next result is a fairly immediate consequence of the Borel Canteli lemma from probability theory.

THEOREM 2.8. *Let Γ be a discrete group acting in B for which the series*

$$\sum_{\gamma \in \Gamma}(1 - |\gamma(a)|)^{(n-1)\alpha}$$

converges. Then the set $\cup_{k>0}L(a : k, \alpha)$ has zero w-measure as a subset of S.

PROOF: Fix $k > 0$. From Lemma 2.1 we observe that

$$w(I(\gamma(a) : k, \alpha)) \leq \frac{M}{n-1}\mu^{n-1}$$

where $\mu = \arccos(1 - k^2(1 - |\gamma(a)|)^{2\alpha}/2)$. Thus for $|\gamma(a)|$ close enough to 1 we will have $\mu < 2k(1 - |\gamma(a)|)^\alpha$, and so, except for finitely many $\gamma \in \Gamma$, we will have

$$(2.6) \qquad w(I(\gamma(a) : k, \alpha)) \leq A(1 - |\gamma(a)|)^{(n-1)\alpha}$$

where A is a constant depending only upon k and the dimension n. If $w(L(a : k, \alpha)) = \epsilon > 0$ then infinitely many of the sets $I(\gamma(a) : k, \alpha)$ would have w-measure at least ϵ and we deduce from (2.6) that

$$\sum_{\gamma \in \Gamma}(1 - |\gamma(a)|)^{(n-1)\alpha} = \infty.$$

This contradiction shows that $w(L(a : k, \alpha)) = 0$ for any $k > 0$. Now from the definition (2.1) we see that if $k' > k > 0$ then $I(a : k, \alpha) \subset I(a : k', \alpha)$ and so the set $\cup_{\gamma \in \Gamma}L(a : k, \alpha)$ may be written as a countable union of sets of zero w-measure and this completes the proof of the Theorem. $\qquad\square$

Following ideas of Sprindzuk [34, p. 21] we will prove

THEOREM 2.9. *Let Γ be a discrete group acting in B. Fix $\alpha > 0$ and k_1, k_2 satisfying $k_1 > k_2 > 0$ then, for any $a \in B$,*

$$w(L(a : k_1, \alpha)) = w(L(a : k_2, \alpha)).$$

PROOF: Note that for any $\gamma \in \Gamma$, $I(\gamma(a) : k_2, \alpha) \subset I(\gamma(a) : k_1, \alpha)$. If we write $\Gamma = \{\gamma_m : m = 0, 1, 2, \ldots\}$ then, as $m \to \infty$,

$$w(I(\gamma_m(a) : k_1, \alpha)) \approx \frac{Mk_1^{n-1}}{n-1}(1 - |\gamma_m(a)|)^{(n-1)\alpha}$$

from Lemma 2.1. It follows that there exists $\delta > 0$ such that, for m large enough,

$$(2.7) \qquad \frac{w(I(\gamma_m(a) : k_2, \alpha))}{w(I(\gamma_m(a) : k_1, \alpha))} > \delta.$$

Now define

$$J = \bigcap_{l=1}^{\infty} \bigcup_{m=l}^{\infty} I(\gamma_m(a) : k_1, \alpha) \quad \text{and} \quad B_l = \bigcup_{m=l}^{\infty} I(\gamma_m(a) : k_2, \alpha)$$

and set $D_l = J - B_l$. To prove the theorem it suffices to show that every D_l is of w-measure zero. If this is not the case then D_l contains a point of metric density—say ξ. Since $\xi \in J$ then $\xi \in I(\gamma_m(a) : k_1, \alpha)$ for infinitely many m and, for such m,

$$(2.8) \qquad w[D_m \cap I(\gamma_m(a) : k_1, \alpha)] \approx w(I(\gamma_m(a) : k_1, \alpha)) \quad \text{as} \quad m \to \infty$$

since $w(I(\gamma_m(a) : k_1, \alpha)) \to 0$ as $m \to \infty$. On the other hand, the sets D_m and $I(\gamma_m(a) : k_2, \alpha)$ do not intersect if $m \geq l$, and hence $D_m \cap I(\gamma_m(a) : k_1, \alpha)$ and $I(\gamma_m(a) : k_2, \alpha)$ are non-intersecting subsets of $I(\gamma_m(a) : k_1, \alpha)$. Therefore

$$w[I(\gamma_m(a) : k_1, \alpha)] \geq w[I(\gamma_m(a) : k_2, \alpha)] + w[D_m \cap I(\gamma_m(a) : k_1, \alpha)]$$

$$\geq \delta w[I(\gamma_m(a) : k_1, \alpha)] + w[D_m \cap I(\gamma_m(a) : k_1, \alpha)]$$

and so $w[D_m \cap I(\gamma_m(a) : k_1, \alpha)] \leq (1 - \delta)w[I(\gamma_m(a) : k_1, \alpha)]$ which contradicts (2.8). This completes the proof of the Theorem. $\qquad \square$

The following corollary is immediate.

COROLLARY 2.10. *Let Γ be a discrete group acting in B. Fix $\alpha > 0$ and $a \in B$ then*

$$w(\bigcup_{k>0} L(a : k, \alpha)) = w(\bigcap_{k>0} L(a : k, \alpha)).$$

Our analysis of the limit set will be based on the rate at which orbits approach the point in question. We will start by considering the most rapid rate possible and then successively weaken the required rate of approach.

3. The Line Transitive Set.

Given a discrete group Γ acting in B and a point $\xi \in \Lambda(\Gamma)$ then, for any $\gamma \in \Gamma$ and any $a \in B$ we clearly have $1 - |\gamma(a)| \leq |\xi - \gamma(a)|$.

In terms of orbital approach, the best we can hope for is that, on a sequence $\{\gamma_n\} \subset \Gamma$,

$$\frac{|\xi - \gamma_n(a)|}{1 - |\gamma_n(a)|} \to 1 \quad \text{as} \quad n \to \infty.$$

We could even ask that for any $a \in B$ such a sequence $\{\gamma_n\}$ exist. In fact we start by asking even more than this.

DEFINITION: The point $\xi \in \Lambda(\Gamma)$ is said to be a line transitive point for Γ if for every pair $a, b \in B$ there exists a sequence $\{\gamma_n\} \subset \Gamma$ such that

$$\lim_{n\to\infty} \frac{|\xi - \gamma_n(a)|}{1 - |\gamma_n(a)|} = 1 \quad \text{and} \quad \lim_{n\to\infty} \frac{|\xi - \gamma_n(b)|}{1 - |\gamma_n(b)|} = 1.$$

Suppose ξ is line transitive and σ is a geodesic ending at ξ (with η the other end point of σ) we have

$$\cosh \rho(\gamma_n(a), \sigma) = \frac{2|\gamma_n(a) - \xi||\gamma_n(a) - \eta|}{|\xi - \eta|(1 - |\gamma_n(a)|^2)}$$

(from Theorem 2.3) and so, on the sequence $\{\gamma_n\}$, $\rho(\gamma_n(a), \sigma) \to 0$ and, similarly, $\rho(\gamma_n(b), \sigma) \to 0$. By the invariance of the hyperbolic metric we have

$$\rho(a, \gamma_n^{-1}(\sigma)) \to 0 \quad \text{and} \quad \rho(b, \gamma_n^{-1}(\sigma)) \to 0$$

as $n \to \infty$. Thus, for any pair of points $a, b \in B$ there is a sequence of images of the geodesic σ coming arbitrarily close to both points. We have proved:

THEOREM 3.1. *If the point $\xi \in \Lambda(\Gamma)$ is a line transitive point and σ is an arbitrary geodesic ending at ξ then the Γ-images of σ are dense in the set of all geodesics.*

This result explains the name "Line Transitive"—the set of line transitive points is denoted by T_l.

The class T_l was the first special class of limit points to be isolated. In 1923 Artin [9] characterized T_l for the modular group acting in the upper half of the complex plane—he showed that T_l comprises those real numbers whose continued fraction representation contains each finite sequence of integers. Myrberg [21] later showed that for finitely generated Fuchsian groups of the first kind the set T_l has full measure on the circle. Other early work on the set T_l (in dimension 2) is to be found in the work of: Koebe [18], Lobell [20], Myrberg [21], and Shimada [33]. The papers of Koebe and Lobell contain a proof of the following result. We include a proof since the original papers are hard to find.

THEOREM 3.2. *If Γ is of the first kind then $T_l \neq \Phi$.*

PROOF: Since Γ is of the first kind then $\Lambda(\Gamma) = S$ and we know that the set of hyperbolic fixed point pairs is dense in $S \times S$ [**16**, p. 122]. Following Hedlund's methods [**16**, p. 123] it may then be shown that if A, B, C, D are four open neighborhoods in S then there exists $\gamma \in \Gamma$ with $\gamma(A) \cap C \neq \Phi$ and $\gamma(B) \cap D \neq \Phi$.

Now for integer n we may partition S into n regions of equal w-measure, say $E_1, \ldots, E_n$. Choose A, B open neighborhoods in S and select E_i, E_j. By the continuity of Mobius transforms we may find open sub-neighborhoods say A', B' of A, B and a Mobius $\gamma_{i,j} \in \Gamma$ such that if $a \in A'$, $b \in B'$ then $\gamma_{i,j}(a) \in E_i$, $\gamma_{i,j}(b) \in E_j$. This procedure may be repeated for all pairs E_i, E_j and we have two open neighborhoods say A, B of S and a collection $\gamma_{i,j}$ of Mobius transforms in Γ such that for any $a \in A$, $b \in B$, $\gamma_{i,j}(a) \in E_i$, $\gamma_{i,j}(b) \in E_j$.

We repeat this procedure with the integer $n + 1$, starting with the neighborhoods A, B just obtained and find ultimately that there exists a geodesic whose Γ-images are dense in the set of all geodesics. One end point of this geodesic must be in T_l and the Theorem is proved. $\square$

What is the geometric significance of the line transitive set? Imagine in the unit ball a point together with a direction. We call this object a line element—say, l_1. It determines a directed geodesic which we suppose ends at a line transitive point ξ. As the line element slides along this geodesic it comes arbitrarily close to group images of any other line element—say l_2.

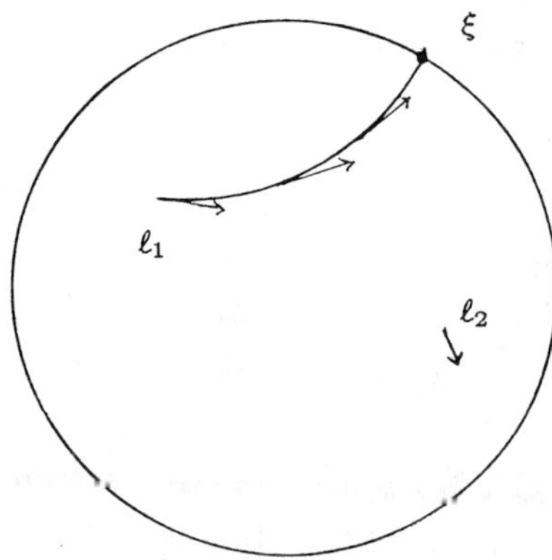

On the quotient space B/Γ these ideas give rise to the geodesic flow and the existence of a line transitive point for Γ simply means that there is a point with a dense trajectory under the flow.

4. The Point Transitive Set.

If we weaken the requirement for a line transitive point and require only that for every $a \in B$ a sequence of Γ-images of a approach the limit point almost radially, then the limit point is said to be point transitive.

DEFINITION: The point $\xi \in \Lambda(\Gamma)$ is said to be a point transitive point for Γ if for every $a \in B$ there exists a sequence $\{\gamma_n\} \subset \Gamma$ such that

$$\lim_{n \to \infty} \frac{|\xi - \gamma_n(a)|}{1 - |\gamma_n(a)|} = 1.$$

The argument used in section 3 easily yields the following result which explains the name "point transitive".

THEOREM 4.1. *If the point $\xi \in \Lambda(\Gamma)$ is a point transitive point and σ is an arbitrary geodesic ending at ξ then the Γ-images of σ are dense in B.*

The set of point transitive points will be denoted T_p and clearly $T_l \subset T_p$ for any discrete group Γ. The following result of Sheingorn [31] shows that in general these sets will not be equal.

THEOREM 4.2 (Sheingorn). *If Γ is the modular group acting in the upper half of the complex plane then $T_p \neq T_l$.*

Whereas, for groups of the first kind, the set T_l is always nonempty, for groups of the second kind the set T_p is always empty.

THEOREM 4.3. *If Γ is of the second kind then $T_p = \Phi$.*

PROOF: Let $\eta \in \partial B$ be an ordinary point for Γ. Since the ordinary set is open there exists a neighborhood N of η in ∂B which comprises ordinary points. It is geometrically evident that a point a of B may be chosen so close to η that any geodesic passing through a has one end point in N. Such a geodesic clearly cannot be approximated by images of any other geodesic since this would make the end point in N a limit point for the group. $\qquad\square$

The class T_p has been used extensively in connection with number theory and the boundary behavior of automorphic functions and forms—see, for example, Lehner [19], Nicholls [23], and Sheingorn [32].

5. The Conical Limit Set.

We next weaken the requirement that orbits approach a limit point almost radially and require instead that they approach within a cone. In view of Theorem 2.5 we see that the property given in the following definition is equivalent to conical approach.

DEFINITION: The point $\xi \in \Lambda(\Gamma)$ is said to be a conical limit point for Γ if for every $a \in B$ there exists a sequence $\{\gamma_n\} \subset \Gamma$ on which the sequence $\dfrac{|\xi - \gamma_n(a)|}{1 - |\gamma_n(a)|}$ remains bounded.

It is immediate from the definition that the conical limit set—denoted by C—is a subset of T_p. The following result is an immediate consequence of Theorem 2.5.

THEOREM 2.5. *The point $\xi \in S$ is a conical limit point for Γ if and only if there is a geodesic σ ending at ξ such that for any point $a \in B$ there are infinitely many Γ-images of σ within a bounded hyperbolic distance of a.*

COROLLARY 5.2. *If ξ is fixed by a hyperbolic element of Γ then ξ is a conical limit point.*

It is well known that the closure of the set of group images of the axis of a hyperbolic element in the group comprises the set of images of the axis and so, from Theorem 4.1, a hyperbolic fixed point is not in T_p. Thus $T_p - C$ is not empty, however, this difference is fairly small.

THEOREM 5.3. *If Γ is a discrete group then $w(C) = w(T_p)$.*

PROOF: Recalling the set $L(x : k, \alpha)$ defined in (2.5) and using Theorem 2.5 we see that if $\{x_n\}$ is a countable dense subset of B we have

$$T_p = \bigcap_{n \geq 1} \bigcap_{k > 0} L(x_n : k, 1) \quad \text{and} \quad C = \bigcap_{n \geq 1} \bigcup_{k > 0} L(x_n : k, 1)$$

and so $w(T_p) = w(C)$ from Corollary 2.10. $\square$

Using the fact that

$$C = \bigcap_{x \in B} \bigcup_{k > 0} L(x : k, 1) = \bigcup_{k > 0} L(x : k, 1)$$

for any $x \in B$, the following result is a corollary of Theorem 2.8.

THEOREM 5.4. *Let Γ be a discrete group acting in B for which the series*

$$\sum_{\gamma \in \Gamma} (1 - |\gamma(a)|)^{n-1}$$

converges. Then $w(C) = 0$.

Conical limit points were introduced (in dimension 2) in 1936 by Hedlund [17] and were used by him in his study of horocyclic transitive points. The conical limit set has been studied over the years by a number of authors. Particular mention should be made of: Lehner [19, Chapter 10] who observed the connection with Diophantine approximation; Beardon and Maskit [14] who first gave the characterization (2) of Theorem 2.5 and generalized Hedlund's results to the three-dimensional case—they called such points "points of approximation". The conical limit set plays a critical role in the ergodic theory of discrete groups—see for example the work of Sullivan [36]—and are of importance in the development of general rigidity theorems—see the work of Tukia [43] and Agard [1].

The following result—the converse of Theorem 5.4—was first proved, for Fuchsian groups, by Tsuji [41, p. 530]. His proof depends heavily on classical complex analysis and does not extend to higher dimensions. It follows (in all dimensions) from a deep ergodic result of Sullivan [36, p. 483]. However, there is also available an elementary proof due to Thurston—see [3, p. 97].

THEOREM 5.5. *Let Γ be a discrete group acting in B for which the set of conical limit points has zero w-measure. Then the series*

$$\sum_{\gamma \in \Gamma} (1 - |\gamma(a)|)^{n-1}$$

converges for all $a \in B$.

For certain groups we may specify completely the nature of the conical limit set. The following result is an easy consequence of Theorem 2.5.

THEOREM 5.6. *Let Γ be a discrete group preserving B. If Γ is convex co-compact then every limit point of Γ is a conical limit point.*

Beardon and Maskit [14] have proved the following result.

THEOREM 5.7. *If Γ has a Dirichlet region D_a with a finite number of faces then every limit point of Γ is either a conical limit point or a cusped parabolic fixed point.*

For a group which diverges at the exponent $n - 1$, the group action on S is ergodic [3, p. 91]. Since the conical limit set is clearly group invariant and, by Theorem 5.5, has positive measure it must then have full measure. Now using Theorem 5.3 we see that T_p also has full measure. In fact one can obtain more—if the group diverges at the exponent $n-1$ then Sullivan's theorem [36, p. 476] tells us that the geodesic flow is ergodic and from this one can deduce that T_l has full measure—see Tsuji [41, p. 530] for the proof in the Fuchsian case.

Combining our results to date on the size of the sets T_l, T_p, C we have the following result.

THEOREM 5.8. *For any discrete group Γ, $T_l \subset T_p \subset C$. If Γ diverges at the exponent $n - 1$ then $w(T_l) = w(S)$. If Γ converges at the exponent $n - 1$ then $w(C) = 0$.*

To see the geometric significance of the conical limit set imagine a line element l_1 determining a geodesic ending at a conical limit point. As the line element slides along the geodesic it keeps meeting images of some compact portion of B.

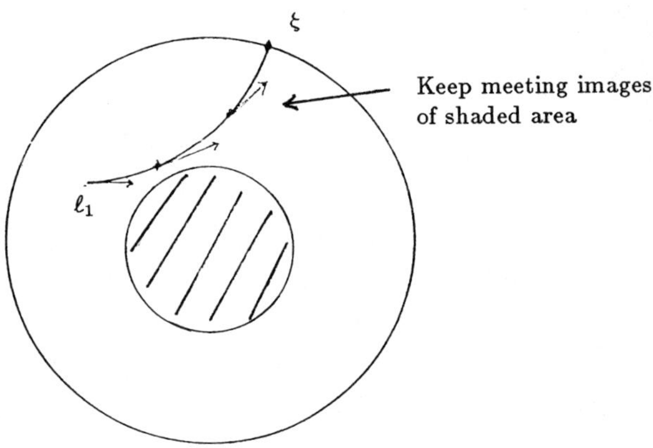

On the quotient space this geodesic flow keeps returning to a compact part of the manifold. It is this notion of recurrence of the geodesic flow which, coupled with Theorem 5.5, leads to the fact that discrete groups diverging at the exponent $n-1$ have many interesting ergodic properties— see the work of Sullivan [36] for an account.

6. The Horospherical Limit Set.

Analogous to the notion of orbits approaching the boundary in a conical region is that of an orbit approaching the boundary in a horosphere. We first give the analytic definition.

DEFINITION: Let Γ be a discrete group acting in B. A point $\xi \in S$ is a horospherical limit point for Γ if for every $a \in B$ there exists a sequence $\{\gamma_n\} \subset \Gamma$ such that

$$\frac{|\xi - \gamma_n(a)|^2}{1 - |\gamma_n(a)|} \to 0 \quad \text{as} \quad n \to \infty.$$

We see from Theorem 2.6 that at a horospherical limit point ξ the orbit of every point of B enters every horoball at ξ.

The horospherical limit set is denoted by H and the following result is an immediate consequence of the definition.

THEOREM 6.1. *Let Γ be a discrete group acting in B then*

$$H = \bigcap_{k>0} L(a : k, 1/2).$$

In terms of derivatives we note from Theorem 2.2 that $\xi \in H$ if and only if $\{|\gamma'(\xi)| : \gamma \in \Gamma\}$ is an unbounded set of reals.

We next consider image horospheres. The next result is a routine calculation using Theorems 2.2 and 2.6.

LEMMA 6.2. *Suppose $\xi \in S$ and H is a horosphere at ξ of euclidean radius k. If γ is a Mobius transform preserving B then $\gamma(H)$ is a horosphere at $\gamma(\xi)$ of euclidean radius*

$$\frac{k|\gamma'(\xi)|}{1 - k + k|\gamma'(\xi)|}.$$

It follows from Lemma 6.2 that any horosphere at a horospherical limit point has images of radius arbitrarily close to one. If the group is of the first kind then an argument of Hedlund [17, p. 537] may be applied to show that images of any such horosphere approximate any horosphere and again using Lemma 6.2 we obtain

THEOREM 6.3. *Let Γ be a discrete group acting in B. If Γ is of the first kind then $\xi \in H$ if and only if the set $\{|\gamma'(\xi)| : \gamma \in \Gamma\}$ is dense in the positive reals.*

The notion of a horosphere having images approximating every horosphere is analogous to the notion of a line transitive point and Hedlund thus described this situation (in dimension 2) by saying that the point at infinity of the horosphere was a "horocyclic transitive point" [17]. It was his work on the characterization of such points which led him to consider what we now call conical limit points. His results were later extended to the three-dimensional case by Tuller [44].

In dimension 2 consider a line element l_1 determining a geodesic ending at a horospherical limit point ξ. Construct a horocycle at ξ passing through the carrier point of l_1. As the line element moves around the horocycle (still pointing at ξ) it comes arbitrarily close to group images of any other line element—say l_2.

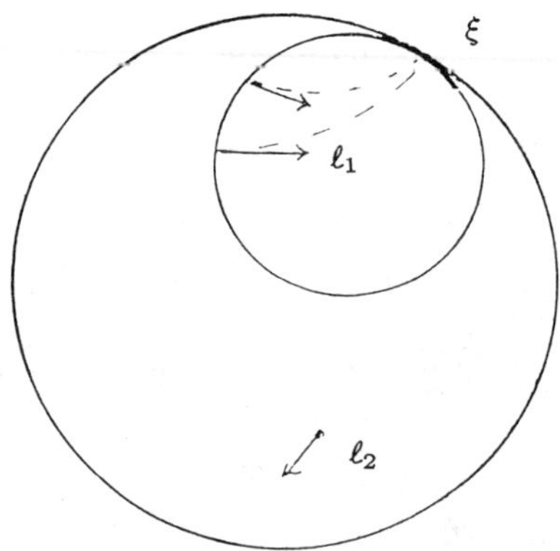

On the quotient space B/Γ these ideas give rise to the horocyclic flow and the existence of a horospherical limit point (for a group of the first kind) means that there is a point with a dense trajectory under the flow.

In terms of the group action on S, Sullivan has shown that the horospherical limit set H is the conservative piece. For details on this and related topics see [36], [3], and [26].

7.   The Dirichlet Set.

Our last class of boundary points is defined as follows.

DEFINITION: For a discrete group Γ the point $\xi \in S$ is a **Dirichlet point** if for every $a \in B$ the set

$$\left[\frac{|\gamma(a) - \xi|^2}{1 - |\gamma(a)|} : \gamma \in \Gamma \right]$$

has an attained minimum.

The set of Dirichlet points is denoted by D and in order to understand the name we need two lemmas.

LEMMA 7.1. *Let σ be the hyperbolic ray connecting $a \in B$ and $\xi \in S$. Then the horoball H at ξ through a may be written*

$$H = \bigcup_{y \in \sigma} \{x : \rho(x,y) < \rho(a,y)\}.$$

PROOF: We need only observe that, for any $y \in \sigma$, the sphere $\{x : \rho(x,y) = \rho(a,y)\}$ is internally tangent to the horosphere ∂H at a. $\qquad\square$

LEMMA 7.2. *Let Γ be a discrete group acting in B and $\xi \in S$. If $a \in B$ then $\xi \in \partial D_a$ if and only if the horoball at ξ through a contains no Γ image of a.*

PROOF: Suppose $\xi \in \partial D_a$ then by convexity the geodesic ray σ joining a to ξ is in D_a and so for any $y \in \sigma$ the ball $\{x : \rho(x,y) < \rho(a,y)\}$ contains no Γ images of a and the conclusion follows from Lemma 7.1. The proof in the reverse direction is entirely similar. $\qquad\square$

From Lemma 7.2 we see that $\xi \in D$ if and only if for every $a \in B$ there exists $\gamma \in \Gamma$ with the property that the horoball at ξ through $\gamma(a)$ contains no Γ image of a. In view of Theorem 2.6 we have the following result.

THEOREM 7.3. *Let Γ be a discrete group acting in B and $\xi \in S$ then ξ is a Dirichlet point for Γ if and only if, for every $a \in B$ there exists $\gamma \in \Gamma$ with $\gamma(\xi) \in \partial D_a$.*

Thus the Dirichlet set is precisely the set of points of S which are represented on the boundary of every Dirichlet region. This set clearly includes ordinary points and cusped parabolic fixed points. The set D of Dirichlet points has an interesting property. If $\xi \in D$ is not a fixed point then from each orbit $\Gamma(a)$ we may select a representative which minimizes $|\gamma(a) - \xi|^2 (1 - |\gamma(a)|)^{-1}$ and it is possible to do this in such a way as to obtain a convex fundamental region for Γ. This region has a very natural interpretation as a Dirichlet region centered at ξ in a topological sense. The details of the construction are to be found in Beardon and Nicholls [15] and Nicholls [24].

In terms of derivatives we note from Theorem 2.2 and 2.6 that $\gamma(0)$ belongs to the horoball at ξ of euclidean radius k if and only if

$$|(\gamma^{-1})'(\xi)| > \frac{1-k}{k}$$

and as an immediate consequence we see that if $\xi \in D$ then the sequence $\{\gamma'(\xi)\}$ $\gamma \in \Gamma$ accumulates only at zero. In fact Pommerenke [29] has shown much more.

THEOREM 7.4. *Let Γ be a discrete group acting in B. For almost all (w) $\xi \in D$ the series $\sum_{\gamma \in \Gamma} |\gamma'(\xi)|$ converges.*

PROOF: Select $a \in B$ and write $e_a = \partial D_a \cap S$. We first prove that if $\gamma \neq I$ then $e_a \cap \gamma(e_a)$ is countable. To see this suppose $\xi \in e_a \cap \gamma(e_a)$, then the rays joining a to ξ and a to $\gamma^{-1}(\xi)$ lie in D_a. Thus the ray from $\gamma(a)$ to ξ lies in $\gamma(D_a)$ and we deduce that the hyperbolic bisector of the segument joining a to $\gamma(a)$ ends at ξ. There are only countably many such bisectors and hence countably many such ξ. It follows that the sets $\{\gamma(e_a) : \gamma \in \Gamma\}$ overlap in at most a countable set and so the sum $\sum_{\gamma \in \Gamma} w(\gamma(e_a))$ converges. Thus

$$\sum_{\gamma \in \Gamma} \int_{e_a} |\gamma'(\xi)| dw = \int_{e_a} \sum_{\gamma \in \Gamma} |\gamma'(\xi)| dw$$

converges and it follows that the series $\sum_{\gamma \in \Gamma} |\gamma'(\xi)|$ converges for almost every point of e_a. Clearly then the series converges for almost every point of $\cup_{\gamma \in \Gamma} \gamma(e_a)$. But this latter set includes D and the Theorem is proved.

$\square$

Contrasting D with H we observe that $D \cap H = \Phi$ for any group Γ but there is a possibility that $D \cup H \neq S$ and we will consider this situation later. However, the following result (implicit in the work of Pommerenke [29] in two dimensions and due to Sullivan in higher dimensions) shows that $D \cup H$ comprises most of S.

THEOREM 7.5. *Let Γ be a discrete group acting in B. The sphere S may be written as the disjoint union*

$$S = H \cup D \cup Q$$

where $w(Q) = 0$.

PROOF: Using Theorem 6.1 and Corollary 2.10 we observe that for any discrete group the set H has the same measure (w) as the set

$$\bigcup_{k>0} L(a : k, 1/2)$$

but this latter set comprises those points ξ of S with the property that an orbit enters **some** horoball at ξ infinitely often. The complement of this set comprises those points η of S with the property that **every** horoball at η meets every orbit finitely often. Such points, in view of Lemma 7.2, are points of D and the Theorem is proved. $\qquad\square$

We have already remarked that H is the conservative piece of the group action on S and we see from Theorem 7.5 that D is the dissipative piece. In fact, the set $\{\xi : \xi \in S, |\gamma'(\xi)| < 1 \textit{ for all } \gamma \in \Gamma - I\}$ is a fundamental region for the group action on the complement of H in S—see [**26**].

What can be said about the set Q of Theorem 7.5? If $\xi \in Q$ then for some $a \in B$ the set

$$\frac{|\gamma(a) - \xi|^2}{1 - |\gamma(a)|} : \gamma \in \Gamma$$

is bounded away from zero and does not have an attained minimum. Geometrically this means that there is a critical horoball based at ξ containing no Γ-equivalents of a but with the property that **any** larger horoball contains infinitely many such equivalents. Such a limit point is called a **Garnett point** and such objects are known to exist in all dimensions—see [**25**] for example. They can arise as limit points which are represented on the boundary of some but not all Dirichlet regions for the group.

8. Size of the Limit Set.

A good way to measure the size of the limit set is to use Hausdorff measure and we briefly review this notion.

Suppose E is a subset of R^n then for α and ϵ positive we define

$$m_{\alpha,\epsilon}(E) = \inf \sum_i |I_i|^\alpha$$

where the infimum is taken over all coverings of E by sequences $\{I_i\}$ of open sets I_i with diameter $|I_i|$ less than ϵ. The Hausdorff α-dimensional measure of E is defined to be

$$m_\alpha(E) = \lim_{\epsilon \to 0} m_{\alpha,\epsilon}(E).$$

The Hausdorff dimension of E is defined to be the unique non-negative number $d(E)$ such that

$$m_\alpha(E) = 0 \text{ if } \alpha > d(E) \quad \text{and}$$

$$m_\alpha(E) = +\infty \text{ if } 0 \le \alpha < d(E).$$

We start by considering non-elementary discrete groups preserving the complex plane—Kleinian groups—and note that these are conjugate to discrete groups preserving the unit ball of R^3.

In 1941 Myrberg [22] showed that $\Lambda(\Gamma)$ has positive logarithmic capacity. The first results concerning Hausdorff dimension of $\Lambda(\Gamma)$ for arbitrary Γ are due to Beardon [10,11] who improved Myrberg's result to show that the limit set has positive Hausdorff dimension. Beardon considers a non-cyclic Schottky group in the complex plane and shows that its limit set is a finite union of spherical Cantor sets. Such a set is reasonably tractable in terms of the calculation of its Hausdorff dimension and Beardon shows that this dimension is positive. His main result then follows from the fact that a non-elementary discrete group has a non-cyclic Schottky subgroup.

Meanwhile, Akaza [4] found a Schottky group having a limit set of Hausdorff dimension greater than 1. More importantly, in this paper and two others [5,6] Akaza indicated a connection between the convergence of the Poincaré series at the exponent α and the α-dimensional measure of $\Lambda(\Gamma)$. In 1971 Beardon [12] proved the following result which clarified this connection.

THEOREM 8.1. *Let Γ be a finitely generated Fuchsian group preserving the unit disk. If the series*

$$\sum_{\gamma \in \Gamma} (1 - |\gamma(0)|)^\alpha$$

converges then $\Lambda(\Gamma)$ has zero α-dimensional Hausdorff measure.

COROLLARY 8.2. *If Γ is a finitely generated Fuchsian group then*

$$d(\Lambda(\Gamma)) \le \delta(\Gamma).$$

In his proof Beardon shows that the conical limit set has zero α-dimensional measure and then uses Hedlund's result [17] that for a finitely generated Fuchsian group the limit set comprises conical limit points and (possibly) parabolic fixed points (necessarily countable in number). Beardon's method for the conical limit set works in all dimensions and we have

161

THEOREM 8.3. *If Γ is a discrete group preserving the unit ball of R^n and if the series*

$$\sum_{\gamma \in \Gamma} (1 - |\gamma(0)|)^\alpha$$

converges then the conical limit set has zero α-dimensional measure.

COROLLARY 8.4. *For the discrete group Γ*

$$d(C) \leq \delta(\Gamma).$$

Now using the Beardon Maskit generalization (Theorem 5.7) of Hedlund's result we have

COROLLARY 8.5. *If Γ is a geometrically finite discrete group preserving the unit ball in R^n then*

$$d(\Lambda(\Gamma)) \leq \delta(\Gamma).$$

We next give the proof of Theorem 8.3.

PROOF: Recalling the definition of the set $L(a : k, 1)$ and the definition of the Hausdorff α-dimensional measure, we see that if the series

$$\sum_{\gamma \in \Gamma} (1 - |\gamma(0)|)^\alpha$$

converges then the set $\cup_{k>0} L(a : k, 1)$ has zero α-dimensional measure. However, if $\{x_n\}$ is a countable dense subset of B then

$$C = \bigcap_{n \geq 1} \cup_{k>0} L(x_n : k, 1)$$

and the result follows. $\qquad\square$

Clearly, if Γ is a geometrically finite discrete group of the first kind then

$$d(C) = d(\Lambda(\Gamma)) = \delta(\Gamma) = n - 1$$

and one would hope in general that $d(C) = \delta(\Gamma)$ (there are several examples to show that in general $d(\Lambda(\Gamma) \neq \delta(\Gamma)$—see [28] for instance). Results of this type lie much deeper and it is beyond the scope of this paper to give the proofs. The critical tool which is required is the construction of an invariant measure supported on the limit set of a discrete group. This construction was first carried out by Patterson [27] for Fuchsian groups. He was able to prove that for a finitely generated Fuchsian group (either without cusps or with $\delta \geq 2/3$)

$$d(C) = d(\Lambda(\Gamma)) = \delta(\Gamma).$$

Sullivan [35] later constructed an invariant measure on the limit set of a discrete group in any dimension and proved the following results.

THEOREM 8.6. *For any Fuchsian group*

$$d(C) = \delta(\Gamma).$$

THEOREM 8.7. *If Γ is a geometrically finite discrete group preserving the unit ball in R^n then*

$$d(C) = d(\Lambda(\Gamma)) = \delta(\Gamma).$$

Results of this type are but one application of measures on the limit set. The reader is referred to the papers of Sullivan [35,38,39] for a full account of the amazing properties and applications of this remarkable measure.

One of the most famous problems in the theory of Kleinian groups is Ahlfor's question [2]. If Γ is finitely generated Kleinian group whose limit set is not the entire sphere then is the two-dimensional Lebesgue measure of $\Lambda(\Gamma)$ equal to zero?

Ahlfors settled this question affirmatively for geometrically finite groups and Thurston [40] has answered it affirmatively for large classes of non-geometrically finite groups.

In fact for geometrically finite groups it is a result of Tukia [42] that the Hausdorff dimension of the limit set is strictly less than $n - 1$ (this was previously established in the Fuchsian case by Beardon [12] and Patterson [27] and in the Kleinian case by Sullivan [39]).

In the opposite direction we recall Akaza's construction [4] yielding a Schottky group having a limit set of Hausdorff dimension greater than 1. In two later papers [7,8] he found a Kleinian group with a limit set of positive 3/2-dimensional Hausdorff measure. In the case of Fuchsian groups Beardon [12] constructed examples of finitely generated groups having limit sets of Hausdorff dimension arbitrarily close to 1. However, in terms of the Ahlfors question the best result of this type is due to Sullivan [37] who finds examples of finitely generated but geometrically infinite groups whose limit sets have zero planar measure and Hausdorff dimension 2.

Department of Mathematical Sciences, Northern Illinois University, De Kalb, Illinois 60115

REFERENCES

1. Agard, S., *A geometric proof of Mostow's rigidity theorem for groups of divergence type*, Acta Math. **151** (1983), 231–252.
2. Ahlfors, L. V., *Fundamental polyhedrons and limit point sets of Kleinian groups*, Proc. Nat. Acad. Sci. **55** (1966), 251–254.
3. __________, *Mobius transformations in several dimensions*, Lecture Notes, School of Mathematics, University of Minnesota (1981).
4. Akaza, T., *Poincaré theta series and singualr sets of Schottky groups*, Nagoya Math. J. **24** (1964), 43–65.
5. __________, *Singular sets of some Kleinian groups*, Nagoya Math. J..
6. __________, *Singular sets of some Kleinian groups (II)*, Nagoya Math. J. **29** (1967), 145–162.
7. __________, *(3/2)-dimensional measures of singular sets of some Kleinian groups*, J. Math. Soc. Japan **24** (1972), 448–464.
8. __________, *Remarks and corrections to the paper "(3/2)-dimensional measures of singular sets of some Kleinian groups"*, Science reports, Kanazawa University **14** (1974), 15–24.
9. Artin, E., *Ein mechanische system mit quasiergodischen bahnen*, Abh. Math. Sem. Univ. Hamburg **3** (1924), 170–175.
10. Beardon, A. F., *The Hausdorff dimension of singular sets of properly discontinuous groups in N-dimensional space*, Bull. Amer. Math. Soc. **71** (1965), 610–615.
11. __________, *The Hausdorff dimension of singular sets of properly discontinuous groups*, Amer. J. Math. **88** (1966), 722–736.
12. __________, *Inequalities for certain Fuchsian groups*, Acta Math. **127** (1971), 221–258.
13. __________, *The geometry of discrete groups*, Graduate Texts in Mathematics **91**, Springer Verlag, New York (1983).
14. Beardon, A. F. and Maskit, B., *Limit points of Kleinian groups and finite sided fundamental polyhedra*, Acta Math. **132** (1974), 1–12.
15. Beardon, A. F. and Nicholls, P. J., *Ford and Dirichlet regions for Fuchsian groups*, Can. J. Math. **34** (1982), 806–815.
16. Gottschalk, W. H. and Hedlund, G. A., "Topological Dynamics," AMS Colloquium Publications Vol. 36, American Math. Society, Providence, 1955.
17. Hedlund, G. A., *Fuchsian groups and transitive horocycles*, Duke Math. J. **2** (1936), 530–542.
18. Koebe, P., *Riemannsche mannigfaltigkeiten und nicht euklidische raumformen VI*, S.–B. Deutsche. Akad. Wiss. Berlin K1 Math. Phys. Tech. (1930), 504–541.
19. Lehner, "Discontinous groups and automorphic functions," Math. Survey 8 Amer. Math. Soc., Providence, R.I., 1964.
20. Lobell, F., *Uber die geodatischen linien der Clifford–Kleinschen Flachen*, Math. Zeit. **30** (1929), 572–607.
21. Myrberg, P. J., *Ein approximationsatz fur die Fuchssen gruppen*, Acta Math. **57** (1931), 389–409.
22. __________, *Die kapazitat der singularen Menge der linearen Gruppe*, Ann. Acad. Sci. Fenn. Ser A **10** (1941), 19 pp.
23. Nicholls, P. J., *The boundary behavior of automorphic forms*, Duke Math. J. **48** (1981), 807–812.
24. __________, *Ford and Dirichlet regions for discrete groups of hyperbolic motions*, Trans. Amer. Math. Soc. **282** (1984), 355–365.
25. __________, *Garnett points for Fuchsian groups*, Bull. London Math. Soc. **12** (1980), 216–218.
26. __________, *Discrete groups on the sphere at infinity*, Bull. London Math. Soc. **15** (1983), 488–492.

27. Patterson, S. J., *The limit set of a Fuchsian group*, Acta Math. **136** (1976), 241–273.

28. __________, *Some examples of Fuchsian groups*, Proc. London Math. Soc. **39** (1979), 276–298.

29. Pommerenke, Ch., *On the Green's function of Fuchsian groups*, Ann. Acad. Sci. Fenn. A1 Math. **2** (1976), 409–427.

30. Sario, L., *Uber Riemannsche Flachen mit hebbarem Rand*, Ann. Acad. Sci. Fenn. **50** (1948), 1–79.

31. Sheingorn, M., *Transitivity for the modular group*, Math. Proc. Camb. Phil. Soc. **88** (1980), 409–423.

32. __________, *Boundary behavior of automorphic forms and transitivity for the modular group*, Illinois J. Math. **24** (1980), 440–451.

33. Shimada, S., *On P. J. Myrberg's approximation theorem on Fuchsian groups*, Mem. Coll. Sci. Kyoto U. Ser. A **33** (1960), 231–241.

34. Sprindzuk, V. G., "Metric Theory of Diophantine Approximation," Wiley, New York, 1979.

35. Sullivan, D., *The density at infinity of a discrete group of hyperbolic motions*, Inst. Hautes Etudes Sci. Publ. Math. **50** (1979).

36. __________, *On the ergodic theory at infinity of an arbitrary discrete group of hyperbolic motions*, in "Riemann surfaces and related topics: Proceedings of the 1978 Stony Brook conference", Annals of Math. Studies No. 97, Princeton University Press, Princeton, New Jersey, 1981.

37. __________, *Growth of positive harmonic functions and Kleinian group limit sets of zero planar measure and Hausdorff dimension two*, Lecture Notes in Math. **894**, Springer Verlag, New York, 1981.

38. __________, *Discrete conformal groups and measurable dynamics*, Bull. Amer. Math. Soc. **6** (1982), 57 73.

39. __________, *Hausdorff measures old and new, and limit sets of geometrically finite Kleinian groups*, Acta Math. **153** (1984), 259–277.

40. Thurston, W., *The geometry and topology of 3-manifolds*, Math. Dept. Princeton University (to be published by Princeton University Press).

41. Tsuji, M., "Potential Theory in Modern Function Theory," Maruzen, Tokyo, 1959.

42. Tukia, P., *The Hausdorff dimension of the limit set of a geometrically finite Kleinian group*, Acta Math. **152** (1984), 127–140.

43. __________, *Rigidity theorems for Mobius groups*, (to appear).

44. Tuller, A., *The measure of transitive geodesics on certain three dimensional manifolds*, Duke Math. J. **4** (1938), 78–94.

A remark on a paper by Floyd

BY PEKKA TUKIA

Let G and H be two discrete groups of Möbius transformations of $\overline{R}^n = R^n \cup \{\infty\}$ and let $\Phi : G \to H$ be an isomorphism. The question to what extent Φ can be realized geometrically has been the subject of many studies. Many of these have been based on the observation that under certain circumstances there is always a map $f : L(G) \to L(H)$ of the limit sets inducing Φ. That is, we have

$$(1) \qquad\qquad fg(x) = \Phi(g)f(x)$$

for all $x \in L(G)$ and $g \in G$.

For instance, if G and H are geometrically finite, and if Φ carries parabolic elements of G bijectively onto parabolic elements of H, then one knows that there is a homeomorphism $f : L(G) \to L(H)$ satisfying (1), cf. [6, Theorem 3.3] (or [2] which also implies this as we will see below). The purpose of this note is to describe what happens if the condition on parabolic elements is omitted (which is actually a condition on parabolic elements of rank 1 since parabolic elements of rank $k > 1$ are preserved for algebraic reasons [6, Lemma 3.2]).

It turns out that there always is such a map of the limit sets but if the condition on parabolic elements is not satisfied, then this map is non-continuous at a dense set (but continuous outside parabolic fixed points). We give an example to describe this situation which seems to present some interesting and unusual features.

It is here that Floyd's paper [2] comes in. We need only to point out some consequences of it, as well as remark that Floyd's theorem which he states only for $n = 2$ is valid in all dimensions. The reason why Floyd excluded the case $n > 2$ is evidently that the proof makes use of the existence of cusp neighborhoods at parabolic fixed points and these results were not then available for $n > 2$.

Floyd constructed for every group G a group completion $\overline{G}$ which is a compact set which is almost, up to rank-one parabolic fixed points, homeomorphic to the limit set $L(G)$ if G is a geometrically finite Kleinian group. We will also show that if $n \leq 2$, then $\overline{G}$ can be realized as the limit set of another geometrically finite Kleinian group.

For the definitions of notions such as a geometrically finite Kleinian group, cusp neighborhoods, rank of a parabolic fixed point, etc., see [6]. We differ from [6] in that we let the term a *Kleinian group* denote any discrete group of Möbius transformations of $\overline{R}^n$, also such that are of the first kind. The n-dimensional hyperbolic space is $H^n = \{x \in R^n : x_n > 0\}$ and $\overline{H}^n = H^n \cup \overline{R}^{n-1}$.

FLOYD'S THEOREM. Let G be a finitely generated group and let Σ be a finite set of generators for G. Let $K(G, \Sigma)$ be the 1-complex whose set of vertices is G and such that $a, b \in G$ are joined by an edge if and only if $a = bg^{\pm 1}$ for some $g \in \Sigma$. Let $|g|$ be the word norm for $g \in G$. Let $f(r) = r^{-2}$. Set

$$d(a, b) = \min\{f(|a|), f(|b|)\}$$

when a and b are joined by an edge and extend this into a metric of $K(G, \Sigma)$ by taking shortest paths (each edge is assumed to be isometric to an interval). This makes $K(G, \Sigma)$ to a metric space which can be completed. Define

$$\overline{G} = \text{(completion of } K(G, \Sigma))\backslash K(G, \Sigma).$$

This is the *group completion* of G. It does not depend on the used set of generators Σ and is canonical in the sense that an isomorphism induces a homeomorphism of the group completions. See Floyd [2] for details.

For geometrically finite Kleinian groups, the group completion is closely related to the limit set, as was shown by Floyd. He proved

THEOREM 1. *Let G be a geometrically finite Kleinian group of $\overline{R}^n$. Then there is a continuous map $\varphi_G : \overline{G} \to L(G)$ which is 2-to-1 onto parabolic fixed points of rank one and injective everywhere else.*

PROOF: As we have remarked, Floyd's proof in [2] is valid in every dimension, provided that some results on parabolic cusps are. So we only give references to these results. What is essentially needed is that the hyperbolic convex hull of $L(G)$ minus some horoballs at parabolic fixed points is a closed set whose quotient by G is compact [2, p. 213]. However, this was proved in [6, Theorem 2.4] and Lemma B of [7] formulates it in exactly the form needed. It then follows, for instance, that all points of $L(G)$ other than parabolic fixed points are points of approximation as defined in [2, p. 212]. So the argument of [2, pp. 213–217] is valid. Note that the stabilizer of a parabolic fixed point contains a subgroup of finite index isomorphic to some Z^k [**6**, Theorem 2.1] and hence its completion is a point. This fact is needed on p. 216.

This has as an almost immediate corollary the result to which we have been aiming at:

COROLLARY. *Let G and H be geometrically finite Kleinian groups of $\overline{R}^n$ and let $\Phi : G \to H$ be an isomorphism. Then there is a map $f : L(G) \to L(H)$ inducing Φ such that f is continuous at all points $x \in L(G)$ such that x is not fixed by a parabolic $g \in G$ with $\Phi(g)$ loxodromic.*

Furthermore, if G is non-elementary, f is otherwise unique except that, if $x \in L(G)$ is fixed by a parabolic $g \in G$ such that $\Phi(g)$ is loxodromic, then $f(x)$ may be either one of the two points fixed by $\Phi(g)$.

PROOF: Consider the maps

$$L(G) \xleftarrow{\;\varphi_G\;} \overline{G} \xrightarrow{\;\overline{\Phi}\;} \overline{H} \xrightarrow{\;\varphi_H\;} L(H)$$

where $\overline{\Phi}$ is the map between the completions induced by Φ and φ_G and φ_H are the maps of the above theorem. Let f' be a right-inverse of φ_G which is compatible with the action of G on $\overline{G}$ and $L(G)$. By Floyd's theorem, it exists and is well-defined except if x is fixed by a parabolic $g \in G$ such that $\Phi(g)$ is loxodromic at which points we have two possible values for $f'(x)$.

Define now $f = \varphi_H \overline{\Phi} f'$. We claim that this satisfies the conditions of the corollary.

By compactness of $\overline{G}$ one easily sees that given any neighborhood U of the one- or two-point set $\varphi_G^{-1}(x)$, $x \in L(G)$, there is a neighborhood V of x such that $\varphi_G^{-1}(V) \subset U$. This implies that f has the kind of continuity property that we claimed. If $x \in L(G)$ is the attractive fixed point of some loxodromic $g \in G$, then $f(x)$ must be the attractive fixed point of $\Phi(g)$ if the groups are non-elementary. Since these points are dense in $L(G)$, f is unique at points of continuity. If x is fixed by a parabolic $g \in G$ such that $\Phi(g)$ is loxodromic, then $f(x)$ is fixed by $\Phi(g)$ and so we have the uniqueness property claimed.

Note that since the number of conjugacy classes of parabolic fixed points of a geometrically finite group is finite, it follows that f is unique up to a finite indeterminacy. We could recover the uniqueness by considering instead of point functions such functions which map a point $x \in L(G)$ to a one- or two-point set.

EXAMPLE: Let G and H be finitely generated Fuchsian groups of the first kind and let $\Phi : G \to H$ be an isomorphism. Use the upper-half-space model so that $\overline{R}$ is the limit set. Then Φ is induced by a homeomorphism of $\overline{R}$ if parabolic elements are carried bijectively onto parabolic elements.

However, even if this condition is not met we have by the corollary a map inducing Φ which is well-defined and continuous outside the set

$$X = \{x \in \overline{R}: x \text{ is fixed by some parabolic } g \in G$$
$$\text{such that } \Phi(g) \text{ is not parabolic}\}.$$

Since X is dense in $\overline{R}$ the map f seems rather wild. In order to understand it better, it is useful to consider the group completions $\overline{G}$ and $\overline{H}$ which are Cantor sets (see Theorem 2 below). The map $\varphi_G : \overline{G} \to L(G)$ identifies some points of $\overline{G}$ so that we obtain the connected set $\overline{R}$. To obtain f, we sever some or all of these identifications and form new identifications so that first $\overline{R}$ becomes totally disconnected and then again connected. If we would define f as a function whose point-images are either one- or two-point sets, we would get a more faithful representation of this process.

Intuitively, f cannot have much regularity. Indeed, by [5], f cannot be absolutely continuous and by [8] it cannot be differentiable outside the parabolic fixed points (although it is not difficult to see that if f is continuous at a parabolic fixed point, then it is also differentiable at this point).

If $n \leq 2$, then, given G, it is possible to construct canonically another geometrically finite group H such that $\overline{G} = L(H)$. We characterize the relation of G and H in

THEOREM 2. *Let G be a geometrically finite Kleinian group of $\overline{R}^n$, $n \leq 2$. Then there is another geometrically finite Möbius group H without parabolic elements of rank one and a continuous map $f : \overline{R}^n \to \overline{R}^n$ inducing an isomorphism $\Phi : H \to G$ such that $f^{-1}(x)$ is a point except if x is a parabolic fixed point of rank 1 in which case $f^{-1}(x)$ is a closed arc J. The arc J lies otherwise in the ordinary set of H but its endpoints are the fixed points of a loxodromic $h \in H$ such that $\Phi(h)$ is parabolic with the fixed point x.*

Furthermore, if $n = 1$, or if $n = 2$ and G is torsionless, then f can be extended to a continuous map $\overline{H}^{n+1} \to \overline{H}^{n+1}$ inducing Φ so that what has been said above remains true.

PROOF: If $n = 1$, to obtain H, we must basically only replace parabolic punctures in the quotient surface by removed disks. The construction of f is also straightforward (the methods we use below are valid slightly modified) and so we assume that $n = 2$. We can also assume that G is of the second kind since a geometrically finite group of the first kind can have only rank two cusps.

Let P_i, $i \in I$, be the parabolic fixed points of G of rank one and let U_i be a cusp neighborhood for P_i in $\overline{H}^3$ such that $U_i \cap U_j = \emptyset$ if $i \neq j$ and that $g(U_i) = U_j$ if $g(P_i) = P_j$, $g \in G$ (cf. [6, Section 2D]). Let

$$D = \overline{H}^3 \backslash (L(G) \cup (\cup \{U_i : i \in I\})).$$

If G is torsionless, then Thurston's characterization of hyperbolic 3-manifolds implies that D can be realized as a hyperbolic 3-manifold, that is D/G is homeomorphic to $(\overline{H}^3 \backslash L(H))/H$ for some Kleinian group H (which must be geometrically finite). The details of how to apply Thurston's theorem are given in a moment but at this point we note that then H has no rank-1 parabolics and hence $\overline{H}$, $L(H)$ and $\overline{G}$ are homeomorphic. Thus if we only want to find a geometrically finite Kleinian group H isomorphic to G such that $L(H)$ and $\overline{G}$ are homeomorphic, our theorem in the torsionless case is a simple application of Thurston's theorem.

However, to obtain all of our theorem a more complicated argument is necessary. Let G' be a torsionless normal subgroup of G of finite index. As we have already indicated, Thurston's characterization of hyperbolic 3-manifolds implies that D/G' can be realized as a hyperbolic manifold, see [3, Sections 4–6]. To apply it, note that once we have removed rank-1 cusps to form D/G', there are only rank-2 cusps to worry about. Hence D/G' is obtained from a compact manifold M' by the removal of certain number of incompressible tori on the boundary corresponding to conjugacy classes of stabilizers of rank-2 parabolic fixed points of G'. Let P' be the union of these tori.

It is easy to check that (M', P') satisfies the conditions of a pared 3-manifold, that is, (i) and (ii) of [3, pp. 58–59] are true. Since $\partial M \neq \emptyset$, M is Haken. Hence [3, Theorem B' p. 70 and the discussion on p. 62] imply that there is a geometrically finite Kleinian group H' and a homeomorphism $f : \overline{H}^3 \backslash L(H') \to D$ which induces an isomorphism $\Phi : H' \to G'$.

Here we can choose the map f to be quasiconformal since D/G' is a compact manifold plus some rank-two cusps. Hence

$$H = f^{-1}(G|D)f$$

is a quasiconformal group of $\overline{H}^3 \backslash L(H')$ and so it can be extended to a quasiconformal group of $\overline{H}^3$. It is known that quasiconformal groups of $\overline{R}^2$ can be conjugated by quasiconformal maps to Möbius groups (Sullivan, Tukia, see especially [9, Theorem F and remark F2]). Hence there is a quasiconformal map h of $\overline{R}^2$ such that $h(H|\overline{R}^2)h^{-1}$ is a Möbius group.

Thus h is compatible with the Möbius group H' (considered to act in $\overline{R}^2$), that is, hgh^{-1} is a Möbius transformation for any $g \in H'$. By [1, Section 11] or [4], h can be extended to a quasiconformal homeomorphism of $\overline{H}^3$ (still denoted by h) in such a way that $hH'h^{-1}$ is a Möbius group of $\overline{H}^3$. (Note that we *do not* claim that hHh^{-1} is a Möbius group.)

We replace H by hHh^{-1}. If G is torsionless, H is a Möbius group of $\overline{H}^3$, otherwise only $H|\overline{R}^2$ is Möbius. We show that it is the group we are seeking.

We start from the fact that there is a quasiconformal map $f : \overline{H}^3 \to D$ conjugating H to G. The components of $\overline{H}^3 \backslash D$ are the 1-cusps U_i which have the property that only a finite number of them have spherical diameter greater than a given $\varepsilon > 0$ [6, Lemma 2.3]. Furthermore, their closures are disjoint and clearly each $\text{int}(H^3 \backslash U_i)$ is finitely connected (cf. Väisälä [10, 17.5]). It follows that int D is finitely connected and then [10, 17.15 and 17.16] imply that we can extend f to a continuous map $\overline{H}^3 \to D$.

Pick now some U_i. Transform the situation [6, Section 2C] so that $P_i = \infty$, that R is G_i-invariant where G_i is the stabilizer of P_i and that

$$U_i = (H^3 \cup R^2) \backslash (R \times \overline{B}^2).$$

(Above $\overline{B}^2$ is the closed unit ball of $\overline{R}^2$.) Using this representation it is easy to see that there is a slightly larger open set $V_i \supset U_i$ such that $\overline{R}^2 \cap V_i = \overline{R}^2 \cap U_i$ and a continuous map $f_i : \overline{V}_i \backslash U_i \to \overline{V}_i$ inducing id $: G_i \to G_i$ such that

(i) $f_i|\partial V_i = \text{id}$ $(\partial = \text{boundary in } \overline{H}^3)$,
(ii) $f_i^{-1}(P_i) = \{(x, 0, 1)|x \in R\}$,
(iii) $f_i|(\overline{V}_i \backslash U_i) \backslash \{P_i\}$ is a homeomorphism onto $\overline{U}_i \backslash \{P_i\}$.

Find for each i such V_i and f_i. We can assume that $\overline{V}_i$'s are disjoint and that their spherical diameters tend to zero. Let $W_i = f^{-1}(\overline{V}_i \backslash U_i)$.

Since V_i is stabilized by a rank-1 parabolic of G, the stabilizer of W_i is generated by a loxodromic element g_i and one easily sees that $\overline{W}_i = W_i \cup \{\text{fixed points of } g_i\}$ and that each $f_i f | \overline{W}_i$ is continuous.

Redefine f so that f is unchanged outside W_i's and that f is replaced by $f_i f$ in each W_i. Then f is continuous in $\mathrm{cl}(\overline{H}^3 \setminus (\cup \{W_i : i \in I\}))$ and each $f|\overline{W}_i$ is continuous. Since the spherical diameters of $\overline{V}_i = f(\overline{W}_i)$ tend to zero, it follows that f is continuous.

It is easy to check that f has the required properties. Note that since H has no rank-1 parabolics, we can identify $\overline{G}$, $\overline{H}$ and $L(H)$ and under this identification, by Theorem 1, $f|L(H)$ is just the map φ_G. So $f^{-1}(x) \cap L(H)$, $x \in L(G)$, is just a one- or two-point set and otherwise the construction show that pre-images of points are as claimed. Note that (ii) implies that pre-images of rank-1 parabolic fixed points are arcs.

University of Helsinki, Helsinki, Finland, and MSRI, Berkeley, USA

REFERENCES

1. Douady, A. and Earle, C., *Conformally natural extension of homeomorphisms of the circle*, Acta Math. **157** (1986), 23–48.
2. Floyd, W.J., *Group completions and limit sets of Kleinian groups*, Inventiones Math. **57** (1980), 205–218.
3. Morgan, J.W., *Uniformization theorem for three-dimensional manifolds*, in "The Smith Conjecture," ed. by J.W. Morgan and H. Bass, Academic Press 1984, 37–125.
4. Reimann, H.M., *Invariant extension of quasiconformal deformations*, Ann. Acad. Sci. Fenn. Ser. AI **10** (1985), 477–492.
5. Sullivan, D., *Discrete conformal groups and measurable dynamics*, Bull. Amer. Math. Soc. **6** (1982), 57–73.
6. Tukia, P., *On isomorphisms of geometrically finite Möbius groups*, Inst. Hautes Étud. Sci. Publ. Math. **61** (1985), 171–214.
7. ________, *The Hausdorff dimension of the limit set of a geometrically finite Kleinian group*, Acta Math. **152** (1985), 127–140.
8. ________, *Differentiability and rigidity of Möbius groups*, Inventiones Math. **82** (1985), 557–578.
9. ________, *On quasiconformal groups*, J. Analyse Math. **46** (1986), 318–346.
10. Väisälä, J., "Lectures on n-dimensional quasiconformal mappings," Lecture Notes in Mathematics 229, Springer Verlag, 1971.

Purely elliptic Möbius groups

BY P.L. WATERMAN

Introduction.

Let $M(n)$ denote the full Möbius group of conformal and anti-conformal self maps of $\hat{R}^n = R^n \cup \{\infty\}$. Such maps are a composition of inversions in spheres and reflections in planes. They may be extended to act as Möbius transformations of $\hat{R}^{n+1}$ preserving hyperbolic $n + 1$-space:

$$H^{n+1} = \{x = (x_1, \ldots, x_{n+1}) \in R^{n+1} | x_{n+1} > 0\}$$

with the line element

$$ds^2 = \frac{dx_1^2 + \cdots + dx_{n+1}^2}{x_{n+1}^2}.$$

Viewed as such $M(n)$ is the full isometry group of H^{n+1}.

The elements of $M(n)$ are classified as follows [3]:

> $T \in M(n)$ is *elliptic* if it has a fixed point in H^{n+1}. Equivalently, T is elliptic if it is $M(n+1)$ conjugate to an orthogonal transformation of R^{n+1}.

> $T \in M(n)$ is *parabolic* if it has exactly one fixed point, necessarily in $\hat{R}^n$. Such maps are $M(n)$ conjugate to a screw motion or translation of R^n.

> $T \in M(n)$ is *loxodromic* if it has exactly two fixed points, both in $\hat{R}^n$. Such maps are $M(n)$ conjugate to a transformation of the form $x \to rAx$ with $r > 0$, $r \neq 1$ and $A \in 0(n)$, the orthogonal group of R^n.

Observe, by considering the normal forms, that a loxodromic element T is characterized by the property that there is an open ball $B \subset \hat{R}^{n+1}$ with $T(\overline{B}) \subset B$. Here, $\overline{B}$ denotes the closure of B.

We wish to describe subgroups of $M(n)$ all of whose elements are elliptic.

The following theorem is well known:

THEOREM A [**1**, P. 70]. *If Γ is a purely elliptic subgroup of $M(2)$ then the elements of Γ have a common fixed point in H^3 and Γ is $M(3)$ conjugate to a subgroup of $0(3)$.*

In this note we consider the situation in higher dimensions, proving in §2 that Theorem A extends to dimension 3. Finally, in §3 we give an example that shows the result to be false in general.

§1. Background results.

If Γ is a subgroup of $M(n)$ we define, following Greenberg [**3**], the *limit set* of Γ by

$$L(\Gamma) = \{x \in \hat{\mathsf{R}}^n | \exists T_m \in \Gamma \text{ with } T_m w \to x \text{ for some } w \in H^{n+1}\}$$

As in [**3**] we observe that $L(\Gamma)$ is closed, invariant under Γ and satisfies $L(\Gamma) = L(\overline{\Gamma})$. Also $L(T\Gamma T^{-1}) = TL(\Gamma)$.

The following is implicit in [**3**].

THEOREM B.

   (i) *If $L(\Gamma) = \emptyset$ then the elements of Γ have a common fixed point in H^{n+1} and Γ is $M(n+1)$ conjugate to a subgroup of $0(n+1)$. [**3**, Proposition 13].*

  (ii) *If $L(\Gamma)$ consists of exactly one point then Γ is $M(n)$ conjugate to a group of Euclidean isometries of R^n.*

 (iii) *If $L(\Gamma)$ contains at least two points then Γ contains a loxodromic element.*

PROOF:

   (i) See [**3**].

  (ii) Normalize by conjugating so that $L(\Gamma) = \{\infty\}$. Since $L(\Gamma)$ is invariant under Γ infinity is fixed by every element of Γ. On noting that loxodromic fixed points are limit points the result follows.

 (iii) Normalize so that $0, 1 \in L(\Gamma)$. Thus there exist $S, T \in \Gamma$ and disjoint open balls B_0, B_1, centered at $0, 1$ respectively, such that either

     a) $S(\overline{B}_0) \subset B_0$ in which case S is loxodromic.

or

     b) $T(\overline{B}_1) \subset B_1$ in which case T is loxodromic.

or

     c) $ST(\overline{B}_0) \subset B_0$ in which case ST is loxodromic.

COROLLARY B. *If Γ is a subgroup of $M(n)$ which contains only elliptic transformations then either*

(i) *the elements of Γ have a common fixed point in H^{n+1} and Γ is $M(n+1)$ conjugate to a subgroup of $0(n+1)$*

   or

(ii) *the elements of Γ have a common fixed point on the boundary of $H^{n+1}: \hat{\mathbb{R}}^n$ and Γ is $M(n)$ conjugate to a group of Euclidean isometries of $\mathbb{R}^n$.*

If all the elements of Γ are of finite order then we can say more using Selberg's lemma:

THEOREM C (SELBERG'S LEMMA [4]). *If G is a finitely generated subgroup of $GL(n,\mathbb{C})$ then it contains a torsion free normal subgroup of finite index.*

COROLLARY C. *If Γ is a subgroup of $M(n)$ such that each element is of finite order then the elements of Γ have a common fixed point in H^{n+1} and Γ is $M(n+1)$ conjugate to a subgroup of $0(n+1)$. If, further, Γ is finitely generated or discrete then Γ is finite.*

PROOF: Consider $\Gamma < M(n)$ with each element of Γ of finite order. Let Γ be the direct limit of finitely generated groups Γ_α whose elements are of finite order. Since $M(n)$ is isomorphic to a subgroup of the Lorentz group in $n+2$ variables Selberg's lemma applies and thus the Γ_α are finite. Thus $L(\Gamma_\alpha) = \emptyset$ and hence by Theorem B the elements of Γ_α have a common fixed point in H^{n+1} and each Γ_α is $M(n+1)$ conjugate to a subgroup of $0(n+1)$. Since the fixed point set of an orthogonal transformation is a plane it follows that the elements of Γ have a common fixed point in H^{n+1} and therefore that Γ is $M(n+1)$ conjugate to a subgroup of $0(n+1)$. Since $0(n+1)$ is compact the results follow.

§2. The situation in dimension three.

LEMMA 1. *If $S,T \in S0(3) - \{I\}$ and the axes of S,T,TS are distinct then they are not coplanar.*

PROOF: We normalize so that using the quaternionic description of rotations of $\mathbb{R}^3$:

$$Sw = e^{i\theta}we^{i\theta} \quad \text{where} \quad w = z + tj; z \in \mathbb{C}; \theta, t \in \mathbb{R}$$
$$Tw = awa^* \qquad\qquad a = \alpha + \beta j; \alpha, \beta \in \mathbb{C}, |a| = 1, a^* = \alpha + \overline{\beta}j$$

and assume the axes of S, T, TS lie in the i, j plane. Hence for $n = 0, 1$ there exist $y, t \in \mathbb{R}$ with

$$(\alpha + \beta j)(e^{2in\theta} yi + tj) = (yi + tj)(\overline{\alpha} - \overline{\beta} j)$$
$$\Rightarrow (\alpha e^{2in\theta} yi - \beta t) + (\alpha t - \beta e^{-2in\theta} yi)j = (\overline{\alpha} yi + \beta t) + (\alpha t - \overline{\beta} yi)j$$

This is impossible unless $\beta = 0$ in which case the axes coincide.

LEMMA 2. $T_1 x = Tx + b$ with $T \in SO(3) - \{I\}$ and $b \in \mathbb{R}^3 - \{0\}$ is elliptic iff b is perpendicular to the axis of T.

PROOF: Normalize so that $T = \begin{pmatrix} \cos\theta & \sin\theta & 0 \\ -\sin\theta & \cos\theta & 0 \\ 0 & 0 & 1 \end{pmatrix}$. On noting that T_1 is elliptic iff it has a fixed point in $\mathbb{R}^3$ the result follows.

THEOREM. If Γ is a purely elliptic subgroup of $M(3)$ then the elements of Γ have a common fixed point in H^4 and Γ is $M(4)$ conjugate to a subgroup of $0(4)$.

PROOF: Assume not, then by Corollary B we may assume that the elements of Γ are Euclidean isometries of $\mathbb{R}^3$ not having a common fixed point in $\mathbb{R}^3$. Also note that it suffices to consider the case in which the elements of Γ are orientation preserving.

If there exist $S, T_1 \in \Gamma$ with no common fixed point then we may normalize so that

$$Sw = e^{i\theta} w e^{i\theta} \text{ where} \qquad w = z + tj; z \in \mathbb{C}; \theta, t \in \mathbb{R}$$
$$T_1 w = awa^* + 1 = Tw + 1 \quad a = \alpha + \beta j; \alpha, \beta \in \mathbb{C}; |a| = 1, a^* = \alpha + \overline{\beta} j$$

In order that $T_1 S^m$ be elliptic the real axis must, by Lemma 2, be orthogonal to the axis of TS^m. By Lemma 1 this is impossible unless the axes of S, T_1 are parallel in which case $[S, T_1]$ is a translation. Thus any two elements of Γ have a common fixed point in $\mathbb{R}^3$. For this to be so the axes of the elements of Γ must be coplanar. By Lemma 1 the axes coincide and the theorem follows.

§3. The example.

We now give an example of a purely elliptic subgroup of $M(4)$ which is not conjugate to a subgroup of $0(5)$. Explicitly we construct a purely elliptic group of Euclidean isometries of $\mathbb{R}^4$ with infinity as its sole limit point showing that case (ii) of Corollary B can occur if $n \geq 4$.

Consider

$$S_1 = \begin{pmatrix} 1-\sqrt{2} & \sqrt{2\sqrt{2}-2} \\ -\sqrt{2\sqrt{2}-2} & 1-\sqrt{2} \end{pmatrix} \quad T_1 = \begin{pmatrix} 1-\sqrt{2}+i\sqrt{2\sqrt{2}-2} & 0 \\ 0 & 1-\sqrt{2}-i\sqrt{2\sqrt{2}-2} \end{pmatrix}$$

$$S_2 = \begin{pmatrix} 1+\sqrt{2} & \sqrt{2\sqrt{2}+2} \\ \sqrt{2\sqrt{2}+2} & 1+\sqrt{2} \end{pmatrix} \quad T_2 = \begin{pmatrix} 1+\sqrt{2}+\sqrt{2\sqrt{2}+2} & 0 \\ 0 & 1+\sqrt{2}-\sqrt{2\sqrt{2}+2} \end{pmatrix}$$

and note that $\langle S_2, T_2 \rangle$ is a Schottky group and therefore free on two generators. We claim that $\langle S_1, T_1 \rangle$, a subgroup of $SU(2, \mathbb{C})$, is also free on two generators. Consider the field $K = \mathbb{Q}(i, \sqrt{2\sqrt{2}-2})$ and let σ be the Galois automorphism given by

$$\sigma(\sqrt{2\sqrt{2}+2}) = i\sqrt{2\sqrt{2}-2}, \quad \sigma(i) = i.$$

σ extends to an automorphism of $SL(2, K)$ with $\sigma(S_2) = S_1$ and $\sigma(T_2) = T_1$ and hence $\langle S_1, T_1 \rangle$ is isomorphic $\langle S_2, T_2 \rangle$.

Now consider the embedding of $SU(2, \mathbb{C})$ in $0(4)$ given by

$$X = \begin{pmatrix} x+yi & u+vi \\ -u+vi & x-yi \end{pmatrix} \rightarrow \begin{pmatrix} \begin{pmatrix} x & y \\ -y & x \end{pmatrix} & \begin{pmatrix} u & v \\ -v & u \end{pmatrix} \\ \begin{pmatrix} -u & v \\ -v & -u \end{pmatrix} & \begin{pmatrix} x & -y \\ y & x \end{pmatrix} \end{pmatrix} = \tilde{X}.$$

Note that $\det(\tilde{X} - I) = 0$ iff $X = I$ so that if $X \neq I$ then $\tilde{X}$ has no fixed points other than $0, \infty$. This follows easily by direct calculation or via results on partitioned matrices.

If we let $\Gamma_1 = \langle \tilde{S}_1, \tilde{T}_1 \rangle$ then it is, by the above, a free group acting freely on $\mathbb{R}^4 - \{0\}$. (See [2] for the higher dimensional case.) Hence if $W_1 \in \Gamma_1 - \{I\}$ and $b \in \mathbb{R}^4$ then $Wx = W_1 x + b$ is elliptic since $W_1 - I$ is invertible.

Finally, if $\Gamma = \langle S, T \rangle$ where

$$Sx = \tilde{S}_1 x$$

$$Tx = \tilde{T}_1 x + \begin{pmatrix} 1 \\ 0 \\ 0 \\ 0 \end{pmatrix} \quad \text{and } x \in \mathbb{R}^4.$$

then Γ is a purely elliptic group of Möbius transformations not conjugate to a subgroup of $0(4)$ (S, T have no common fixed point in $\mathbb{R}^4$).

Acknowledgements. In conclusion I should like to thank Alan Beardon, Bill Goldman, Leon Greenberg and Colin Maclachlan for their helpful comments.

Dept. of Mathematical Sciences, Northern Illinois University, DeKalb, Illinois 60115

178

REFERENCES

1. Beardon, A.F., "The Geometry of Discrete Groups," Springer–Verlag, New York, 1983.
2. Deligne, P. and Sullivan, D., *Division algebras and the Hausdorff–Banach–Tarski paradox*, L'Enseignement Math. **29** (1983), 145–150.
3. Greenberg, L., *Discrete subgroups of the Lorentz group*, Math. Scand. **10** (1962), 85–107.
4. Selberg, A., *On discontinuous groups in higher-dimensional symmetric spaces*, in "Contributions to Function Theory," pp. 147–164, Tata Institute, Bombay, 1960.

Conformally natural reflections in
Jordan curves with applications to
Teichmüller spaces

BY CLIFFORD J. EARLE AND SUBHASHIS NAG

§1. Introduction.

In his fundamental paper [1] Ahlfors initiated the study of quasiconformal reflections. Using the results of Beurling and Ahlfors [6] he showed that every quasicircle that passes through ∞ permits a quasiconformal reflection that satisfies a global Lipschitz condition (with exponent one) in the plane. Using that result he proved by a direct construction that the Bers embedding of the universal Teichmüller space has an open image. Lipschitz continuous quasiconformal reflections also play a crucial role in Bers's subsequent proof (see [4] and [5]) that for any Teichmüller space the Bers embedding not only has an open image but also has local cross sections. That result is one of the cornerstones of Teichmüller theory.

The quasiconformal extensions defined by Beurling and Ahlfors [6] do not behave well with respect to composition with Möbius transformations. That is why Ahlfors [1] studies only the universal Teichmüller space. In [7] a "conformally natural" extension operator was introduced. Let Δ be the open unit disk and let S^1 be its boundary. Given any homeomorphism $\varphi : S^1 \to S^1$, that operator produces a homeomorphism $ex(\varphi) : Cl\Delta \to Cl\Delta$ in a conformally natural way: if g and h are Poincaré isometries of Δ, then

$$(1.1) \qquad\qquad ex(g \circ \varphi \circ h) = g \circ ex(\varphi) \circ h.$$

In this paper we apply the methods of Ahlfors [1] to general Teichmüller spaces by using the conformally natural extension [7] in place of the Beurling-Ahlfors extension. In this way we obtain the first direct construction of "explicit" local cross sections of the Bers embedding. (Bers [5] relies on the implicit function theorem.)

Our paper, which is in essence a paraphrase of portions of Ahlfors [1], is organized as follows. In §§2, 3, and 4 we study the conformally natural

---

This work was supported in part by National Science Foundation grant MCS-8301564.

reflections obtained from the extension operator [7]. These may have independent interest, quite apart from their use here to construct local cross sections. In §§5 and 6, using these reflections, we carry out the Ahlfors construction of explicit quasiconformal extensions of locally injective functions with small Schwarzian derivatives. Finally, in §7 we discuss the Bers embedding and use our earlier results to obtain the desired local cross sections.

§2. Reflections in a Jordan Curve.

Let D be a Jordan region in the Riemann sphere $\mathbb{C} \cup \{\infty\} = S^2$, and let $D^* (= S^2 \backslash \mathcal{C}\ell D)$ be the complementary region. Recall that Δ is the open unit disk and S^1 is its boundary. Choose conformal maps f and f^* of Δ onto D and D^* respectively. Since f and f^* both define homeomorphisms of S^1 onto the common boundary of D and D^*, we obtain a sense-reversing homeomorphism $f^{-1} \circ f^* : S^1 \rightarrow S^1$. This extends, by the methods of [7], to the homeomorphism $ex(f^{-1} \circ f^*) : \mathcal{C}\ell\Delta \rightarrow \mathcal{C}\ell\Delta$. The conformally natural reflection $j(D) : S^2 \rightarrow S^2$ is defined by setting

$$
\begin{aligned}
(2.1) \quad j(D) &= f^* \circ (f \circ ex(f^{-1} \circ f^*))^{-1} && \text{in } \mathcal{C}\ell D, \\
j(D) &= f \circ ex(f^{-1} \circ f^*) \circ (f^*)^{-1} && \text{in } \mathcal{C}\ell D^*.
\end{aligned}
$$

Obviously $j(D)$ is a sense-reversing homeomorphism of order two that fixes ∂D pointwise and interchanges D and D^*. Since $ex(f^{-1} \circ f^*)$ is a real analytic diffeomorphism inside Δ (see [7]), $j(D)$ is a real analytic diffeomorphism of $D \cup D^*$ onto itself. In general $j(D)$ is not smooth at the boundary points of D.

LEMMA 1. *The reflection $j(D)$ depends only on D, not on the choice of the conformal maps f and f^*.*

PROOF: The most general conformal maps of Δ onto D and D^* are $f \circ g$ and $f^* \circ g^*$, where g and g^* are conformal automorphisms of Δ. According to (2.1) these conformal maps produce a reflection j that satisfies

$$
j = (f^* \circ g^*) \circ [(f \circ g) \circ ex(g^{-1} \circ f^{-1} \circ f^* \circ g^*)]^{-1}
$$

in D. By (1.1),

$$
j = f^* \circ g^* \circ [(f \circ g) \circ g^{-1} \circ ex(f^{-1} \circ f^*) \circ g^*]^{-1} = j(D),
$$

in D and hence everywhere. $\qquad\square$

Lemma 1 implies the following conformal naturality property of the reflection $j(D)$.

THEOREM 1. *Let $g : S^2 \to S^2$ be a homeomorphism whose restriction to $D \cup D^*$ is conformal. Then*

$$(2.2) \qquad\qquad g{\circ}j(D){\circ}g^{-1} \;=\; j(g(D)).$$

PROOF: Choose conformal maps f and f^* of Δ onto D and D^*. Formula (2.2) follows directly from definition (2.1) if we compute $j(D)$ using f and f^* and compute $j(g(D))$ using $g{\circ}f$ and $g{\circ}f^*$. $\qquad\square$

REMARKS: (1) The most important application of Theorem 1 is to the case where g is a Möbius transformation. For instance (2.2) implies that $j(D)$ commutes with every Möbius transformation that maps D onto itself. In particular, if ∂D is a line or circle, $j(D)$ is the standard reflection or inversion in ∂D, and $j(D) \;=\; j(D^*)$.

(2) It would be nice always to have $j(D) \;=\; j(D^*)$. Unfortunately that is false whenever

$$ex((f^*)^{-1}{\circ}f) \;\neq\; (ex(f^{-1}{\circ}f^*))^{-1}.$$

(3) Since (1.1) holds for all Poincaré isometries of Δ, even the sense-reversing ones, Lemma 1 and Theorem 1 remain true as stated even if the conformal maps f, f^*, and g are allowed to be sense-reversing. However, for our purposes in this paper we have found it convenient to reserve the terms "conformal map" and "quasiconformal map" for sense-preserving maps unless the contrary is explicitly stated.

§3. Continuous Dependence on D.

By definition, the Fréchet distance $\rho(C, C')$ between the Jordan curves C and C' is the infimum of the numbers $\epsilon \;>\; 0$ such that there is a homeomorphism $f : C \to C'$ satisfying $d(z, f(z)) \;\leq\; \epsilon$ for all $z \in C$. Here d is the spherical distance on S^2.

Let (C_n) be a sequence of Jordan curves such that $\rho(C_n, C) \to 0$ for some Jordan curve C. If D is a component of $S^2 \backslash C$ and $z_0 \in D$ then, for n sufficiently large, z_0 belongs to exactly one component D_n of $S^2 \backslash C_n$. We say that the regions D_n converge as Jordan regions to D. At the same time, one can prove that the complementary regions D_n^* converge as Jordan regions to D^*.

THEOREM 2. *Let the regions D_n converge as Jordan regions to D. Then $j(D_n) \to j(D)$ uniformly in S^2. In addition all the derivatives of $j(D_n)$*

converge to those of $j(D)$ uniformly on compact subsets of $(D \cup D^*) \backslash \{\infty, j(D)(\infty)\}$.

PROOF: By a theorem of Radó (see [9, pp. 59-62]) there are conformal maps $f_n : \Delta \to D_n$, $f_n^* : \Delta \to D_n^*$, $f : \Delta \to D$, and $f^* : \Delta \to D^*$ such that $f_n \to f$ and $f_n^* \to f^*$ uniformly in $Cl\Delta$. Therefore $f_n^{-1} \circ f_n^* \to f^{-1} \circ f^*$ uniformly on S^1. By Proposition 2 of [7],

$$h_n = ex(f_n^{-1} \circ f_n^*) \to ex(f^{-1} \circ f^*) = h$$

uniformly on $Cl\Delta$. Define homeomorphisms $w_n : S^2 \to S^2$ by putting $w_n = f_n^* \circ j(\Delta)$ in Δ^* $(= S^2 \backslash Cl\Delta)$ and $w_n = f_n \circ h_n$ in $Cl\Delta$. Similarly, put $w = f^* \circ j(\Delta)$ in Δ^* and $w = f \circ h$ in $Cl\Delta$. Then $w_n \to w$ uniformly on S^2, so

$$j(D_n) = w_n \circ j(\Delta) \circ w_n^{-1} \to w \circ j(\Delta) \circ w^{-1} = j(D)$$

uniformly on S^2.

To verify convergence of the derivatives we observe that $h_n \to h$ and $h_n^{-1} \to h^{-1}$ in $C^\infty(\Delta, \mathbb{C})$, by Proposition 2 of [7]. The conclusion of Theorem 2 now follows from definition (2.1) and the known convergence properties of the conformal maps f_n, f_n^*, and their inverses. $\qquad\square$

§4. Reflections in a Quasicircle.

By definition, a Jordan curve is a K-quasicircle if it is the image of a line or circle in $S^2 = \mathbb{C} \cup \{\infty\}$ under a K-quasiconformal homeomorphism of S^2 onto itself. A quasicircle is a Jordan curve that is a K-quasicircle for some K. The following lemma is well known (see Ahlfors [1]). We include its proof for the reader's convenience.

LEMMA 2. *Let D and D^* be complementary Jordan regions bounded by a K-quasicircle. Any conformal map $f^* : \Delta \to D^*$ has a K^2-quasiconformal extension $g : S^2 \to S^2$. If in addition $f : \Delta \to D$ is conformal, then the homeomorphism $f^{-1} \circ f^* : S^1 \to S^1$ has a sense-reversing K^2-quasiconformal extension mapping Δ onto itself.*

PROOF: By hypothesis there is a K-quasiconformal map $w : S^2 \to S^2$ such that $w(\Delta) = D$. The map $j = w \circ j(\Delta) \circ w^{-1}$ is a sense-reversing K^2-quasiconformal map that fixes ∂D pointwise and interchanges D and D^*.

Given the conformal map $f^* : \Delta \to D^*$ we define the required extension $g : S^2 \to S^2$ by putting $g = f^*$ in $Cl\Delta$ and $g = j \circ f^* \circ j(\Delta)$ in Δ^*. Similarly, the required extension of $f^{-1} \circ f^*$ to Δ is $f^{-1} \circ j \circ f^*$. $\qquad\square$

By definition, a K-quasiconformal reflection is a sense-reversing K-quasiconformal map $j : S^2 \to S^2$ of order two that interchanges two complementary Jordan regions D and D^* and fixes their common boundary pointwise. We shall need the following consequence of Lemma 1 in [**1**].

LEMMA 3. *Let j be a K-quasiconformal reflection interchanging the complementary Jordan regions D and D^* and let $\lambda(z)|dz|$ be the Poincaré metric in $D \cup D^*$. There is a number C, depending only on K, such that*

$$(4.1) \qquad C^{-1} \leq |z - z^*|^2 \lambda(z)\lambda(z^*) \leq C$$

for any finite points $z \in D$ and $z^* = j(z) \in D^*$.

PROOF: First assume $\infty \in \partial D$. Lemma 1 of [**1**] gives a number C_1, depending only on K, such that

$$(1 \leq) \; \frac{|z - z^*|^2}{\delta(z)\delta(z^*)} \leq C_1$$

for all $z \in D$. Here $\delta(\varsigma)$ is the Euclidean distance from ς to ∂D. Since

$$(4\delta(\varsigma))^{-1} \leq \lambda(\varsigma) \leq \delta(\varsigma)^{-1}$$

for all $\varsigma \in D \cup D^*$, we obtain (4.1) with C equal to the maximum of C_1 and 16.

For the general case choose a Möbius transformation g that maps a point of ∂D to ∞. Let $\lambda_1(w)|dw|$ be the Poincaré metric in $g(D \cup D^*)$, so that

$$\lambda_1(g(\varsigma))|g'(\varsigma)| = \lambda(\varsigma)$$

for all finite points $\varsigma \in D \cup D^*$. Let $j_1 = g \circ j \circ g^{-1}$. By what we have already proved

$$C^{-1} \leq |w - j_1(w)|^2 \lambda_1(w)\lambda(j_1(w)) \leq C$$

for all $w \in g(D)$. Taking $w = g(z)$, where $z \in D$ and $z^* = j(z) \in D^*$ are finite, we obtain

$$(4.2) \qquad C^{-1} \leq |g(z) - g(z^*)|^2 \lambda_1(g(z))\lambda_1(g(z^*)) \leq C.$$

Now

$$|g(z) - g(z^*)|^2 \lambda_1(g(z))\lambda_1(g(z^*)) = |z - z^*|^2 |g'(z)||g'(z^*)|\lambda_1(g(z))\lambda_1(g(z^*))$$
$$= |z - z^*|^2 \lambda(z)\lambda(z^*),$$

so (4.2) implies the required inequality (4.1). $\square$

We are now ready to prove the principal result of this section.

THEOREM 3. *Let ∂D be a K-quasicircle, and let $\lambda(z)|dz|$ be the Poincaré metric in $D \cup D^*$. There are numbers C_1 and C_2, depending only on K, such that*

$$(4.3) \qquad j(D): S^2 \to S^2 \text{ is a } C_1 - \text{quasiconformal reflection,}$$

$$(4.4) \qquad C_2^{-1} \leq |z - z^*|^2 \lambda(z^*)^2 |j(D)_{\bar{z}}(z)| \leq C_2$$

for any finite points $z \in D$ and $z^* = j(D)(z) \in D^*$.

PROOF: Let $f : \Delta \to D$ and $f^* : \Delta \to D^*$ be conformal maps. Since ∂D is a K-quasicircle, $f^{-1} \circ f^* : S^1 \to S^1$ has a sense-reversing K^2-quasiconformal extension to Δ, by Lemma 2. Let $h = ex(f^{-1} \circ f^*)$ be the conformally natural extension, and let

$$\sigma(z,w) = \tanh^{-1}\left|\frac{z - w}{1 - \bar{z}w}\right|, \qquad z, w \in \Delta,$$

be the Poincaré distance in Δ. The proof of Theorem 2 in [7] shows there is a number C^*, depending only on K, such that

$$(4.5) \qquad (C^*)^{-1}\sigma(z,w) \leq \sigma(h(z),h(w)) \leq C^*\sigma(z,w)$$

for all $z, w \in \Delta$. Since f and f^* are Poincaré isometries of Δ onto D and D^*, (4.5) and definition (2.1) imply that $j(D) : D \cup D^* \to D \cup D^*$ changes Poincaré distances by at most a factor C^*. That implies (4.3), with $C_1 = (C^*)^2$. It also implies that

$$(C^*)^{-1}\lambda(z) \leq \lambda(z^*)|j(D)_{\bar{z}}(z)| \leq C^*\lambda(z)$$

if $z \in D$ and $z^* = j(D)(z) \in D^*$ are both finite. That inequality, together with (4.3) and Lemma 2, gives (4.4) with a constant C_2 that depends only on K. $\square$

§5. A Map from Quadratic to Beltrami Differentials.

Let D and D^* be complementary Jordan regions bounded by a K-quasicircle, and let $\lambda(\varsigma)|d\varsigma|$ be the Poincaré metric in $D \cup D^*$. As usual we denote by $B(D^*)$ the Bers space of holomorphic functions $\varphi(\varsigma)$ in D^* such that

$$\|\varphi\| = \sup\{\lambda(\varsigma)^{-2}|\varphi(\varsigma)|;\ \varsigma \in D^*\} < \infty.$$

(If $\infty \in D^*$ and $\varphi \in B(D^*)$, then $\varphi(\varsigma)$ automatically has a zero of order ≥ 4 at ∞.)

Put $\epsilon(K) = C_2^{-1}$, where the constant C_2 is given by Theorem 3, and write $z^* = j(D)(z)$ if $z \in D$. The Ahlfors map $\varphi \mapsto \alpha(\varphi)$ is defined on the open ball

$$B(D^*)_{\epsilon(K)} = \{\varphi \in B(D^*);\ \|\varphi\| < \epsilon(K)\}$$

by putting

$$(5.1) \qquad \alpha(\varphi)(z) = \frac{\varphi(z^*)(z - z^*)^2 j(D)_{\bar{z}}(z)}{2 + \varphi(z^*)(z - z^*)^2 j(D)_z(z)}, \qquad z \in D.$$

LEMMA 4. *The Ahlfors map is a holomorphic map of $B(D^*)_{\epsilon(K)}$ into the open unit ball of $L^\infty(D, \mathbb{C})$.*

PROOF: Inequality (4.4) of Theorem 3 implies that

$$|\varphi(z^*)(z - z^*)^2 j(D)_{\bar{z}}(z)| \leq C_2\|\varphi\|$$

for all $z \in D$ and $\varphi \in B(D^*)$. Since $|j(D)_z| < |j(D)_{\bar{z}}|$ everywhere in D, the L^∞ norm $\|\alpha(\varphi)\|_\infty$ of $\alpha(\varphi)$ satisfies

$$(5.2) \qquad \|\alpha(\varphi)\|_\infty \leq \frac{C_2\|\varphi\|}{2 - C_2\|\varphi\|} \leq \frac{\|\varphi\|}{\epsilon(K)} < 1$$

if $\|\varphi\| < \epsilon(K)$. The holomorphic dependence on φ is obvious. $\square$

There is an Ahlfors map $\alpha:\ B(D^*)_{\epsilon(K)} \to L^\infty(D, \mathbb{C})$ whenever D is a Jordan region bounded by a K-quasicircle. We shall sometimes write $\alpha = \alpha_D$ if the dependence on D is important.

The spaces $B(D^*)$ and $L^\infty(D, \mathbb{C})$ transform under conformal maps like quadratic and Beltrami differentials respectively. In other words, if g is a conformal map of D_1^* onto D_2^* or D_1 onto D_2, there is an induced linear isometry $g^* : B(D_2^*) \rightarrow B(D_1^*)$ or $g^* : L^\infty(D_2, \mathbb{C}) \rightarrow L_\infty(D_1, \mathbb{C})$ given by

$$(5.3) \qquad g^*(\varphi) = (\varphi \circ g)(g')^2 \qquad \text{if } \varphi \in B(D_2^*),$$

$$(5.4) \qquad g^*(\mu) = (\mu \circ g)\overline{g}'/g' \qquad \text{if } \mu \in L^\infty(D_2, \mathbb{C}).$$

The conformal naturality property (2.2) of the reflections $j(D)$ leads to the following important invariance property of the Ahlfors maps.

LEMMA 5. *Let D be a Jordan region bounded by a K-quasicircle. If g is any Möbius transformation, then*

$$g^*(\alpha_{g(D)}(\varphi)) = \alpha_D(g^*(\varphi)) \qquad \text{for all } \varphi \in B(g(D^*))_{\epsilon(K)}.$$

PROOF: If $z \in D$, $\varphi \in B(g(D^*))_{\epsilon(K)}$, and $\psi = g^*(\varphi)$, we must show that

$$(5.5) \qquad \alpha_{g(D)}(\varphi)(g(z))\overline{g'(z)}/g'(z) = \alpha_D(\psi)(z).$$

Formula (5.1) defines $\alpha_{g(D)}(\varphi)(g(z))$ as a fraction whose numerator and denominator we denote temporarily by

$$A(g(z)) = \varphi(g(z)^*)(g(z) - g(z)^*)^2 j(g(D))_{\overline{z}}(g(z)),$$

$$B(g(z)) = 2 + \varphi(g(z)^*)(g(z) - g(z)^*)^2 j(g(D))_z(g(z)).$$

Here $g(z)^* = j(g(D))(g(z))$. Formula (2.2) gives $g(z)^* = g(z^*)$ if $z^* = j(D)(z)$. Thus

$$A(g(z)) = \varphi(g(z^*))(g(z) - g(z^*))^2 j(g(D))_{\overline{z}}(g(z))$$
$$= \varphi(g(z^*))(z - z^*)^2 g'(z)g'(z^*) j(g(D))_{\overline{z}}(g(z)).$$

Writing (2.2) in the form

$$(5.6) \qquad j(g(D)) \circ g = g \circ j(D)$$

and differentiating both sides of (5.6) with respect to $\overline{z}$, we find that

$$A(g(z))\overline{g'(z)}/g'(z) = \varphi(g(z^*))g'(z^*)^2(z - z^*)^2 j(D)_{\overline{z}}(z)$$
$$= \psi(z^*)(z - z^*)^2 j(D)_{\overline{z}}(z).$$

A similar calculation, which involves differentiating both sides of (5.6) with respect to z, shows that

$$B(g(z)) \;=\; 2 \;+\; \psi(z^*)(z - z^*)^2 j(D)_z(z).$$

That proves (5.5). $\qquad\qquad\qquad\qquad\qquad\qquad\qquad\qquad\qquad\qquad\Box$

REMARKS: (1) We shall use the Ahlfors map in §7 to obtain local cross sections of the Bers embedding. This works in arbitrary Teichmüller spaces, not just the universal Teichmüller space, because of the invariance property stated in Lemma 5.

(2) Inequality (4.4) in Theorem 3 implies that if ∂D is a quasicircle, then

$$(5.7) \qquad\qquad \sup\{|\varphi(z^*)(z - z^*)^2 j(D)_{\bar{z}}(z)|; \; z \in D\}$$

defines a norm on $B(D^*)$ that is equivalent to the standard norm. One can use (5.7) to define a norm on the holomorphic functions φ in any Jordan region D^*. We suspect that there are no nontrivial functions of finite norm unless D is bounded by a quasicircle, but we have no proof.

§6. A Generalized Ahlfors-Weill Theorem.

Let D and D^* be complementary Jordan regions bounded by a K-quasicircle, and let $\varphi \in B(D^*)$ satisfy $\|\varphi\| < \epsilon(K)$. Choose linearly independent solutions v_1, v_2 of the differential equation

$$(6.1) \qquad\qquad\qquad\qquad 2v'' + \varphi v = 0$$

in D^*. The functions v_1 and v_2 are holomorphic in D^* except for possible simple poles at ∞. We normalize them by setting the Wronskian

$$(6.2) \qquad\qquad\qquad\qquad v_1 v_2' - v_2 v_1' \equiv -1$$

in D^*.

Following Ahlfors [1] we define a map w from $D \cup D^*$ into the Riemann sphere by putting

$$(6.3\text{a}) \qquad\qquad w(z) \;=\; v_1(z)/v_2(z), \qquad\qquad\qquad z \in D^*,$$

$$(6.3\text{b}) \qquad\qquad w(z) \;=\; \frac{v_1(z^*) + (z - z^*)v_1'(z^*)}{v_2(z^*) + (z - z^*)v_2'(z^*)}, \qquad z \in D.$$

Here as usual $z^* = j(D)(z)$ if $z \in D$. If z and z^* are finite, the numerator and denominator in either (6.3a) or (6.3b) are finite numbers that (as Ahlfors observed) do not vanish simultaneously because of (6.2). Thus $w(z)$ is a well defined point on the Riemann sphere.

To interpret formulas (6.3) when z, z^*, or w equals ∞, we move ∞ to a finite point by a Möbius transformation. That preserves the form of formulas (6.3). In fact if $w(z) = \infty$ we can transform w to $1/w$ (and $w(z)$ to zero) simply by replacing the solutions v_1 and v_2 of equation (6.1) by the solutions iv_2 and iv_1 respectively. If z or z^* is ∞, we make a change of variables

$$z = g(\varsigma) = \frac{a\varsigma + b}{c\varsigma + d}, \qquad ad - bc = 1,$$

and we put $D_1 = g^{-1}(D)$ and $D_1^* = g^{-1}(D^*)$. Then

$$\psi = g^*(\varphi) = (\varphi \circ g)(g')^2 \in B(D_1^*).$$

Moreover, the functions

$$u_j(\varsigma) = v_j(g(\varsigma))g'(\varsigma)^{-1/2} = (c\varsigma + d)v_j(g(\varsigma)), \qquad j = 1,2,$$

solve the equation $2u'' + \psi u = 0$ in D_1^* and satisfy $u_1 u_2' - u_2 u_1' \equiv -1$. Direct computation (using (2.2)) shows that

$$w(g(\varsigma)) = u_1(\varsigma)/u_2(\varsigma), \qquad\qquad \varsigma \in D_1^*,$$

$$w(g(\varsigma)) = \frac{u_1(\varsigma^*) + (\varsigma - \varsigma^*)u_1'(\varsigma^*)}{u_2(\varsigma^*) + (\varsigma - \varsigma^*)u_2'(\varsigma^*)}, \qquad \varsigma \in D_1,$$

where $\varsigma^* = j(D_1)(\varsigma)$. These formulas have the desired form (6.3).

Differentiation of formulas (6.3) gives

(6.4a)
$$w_{\bar{z}}(z) = 0, \qquad\qquad z \in D^*$$

(6.5a)
$$w_z(z) = w'(z) = \frac{1}{v_2(z)^2}, \qquad\qquad z \in D^*,$$

(6.4b)
$$w_{\bar{z}}(z) = \frac{(z - z^*)^2 \varphi(z^*) j(D)_{\bar{z}}(z)}{2(v_2(z^*) + (z - z^*)v_2'(z^*))^2}, \qquad z \in D,$$

(6.5b)
$$w_z(z) = \frac{2 + (z - z^*)^2 \varphi(z^*) j(D)_z(z)}{2(v_2(z^*) + (z - z^*)v_2'(z^*))^2}, \qquad z \in D.$$

After these preliminaries we are ready to prove our generalized Ahlfors-Weill theorem.

THEOREM 4. *If ∂D is a K-quasicircle, $\varphi \in B(D^*)$, and $\|\varphi\| < \epsilon(K^2)$, the mapping $w : D \cup D^* \to S^2$ defined by (6.3) extends to a quasiconformal homeomorphism of S^2 onto itself.*

PROOF: We paraphrase the proof of Theorem 2 in Ahlfors [1]. (See also [2, pp. 128-133].) First we assume that ∂D is a real analytic curve and a K^2-quasicircle. We assume also that φ is holomorphic in a neighborhood of $C\ell\Delta^*$ (with a zero of order ≥ 4 at ∞ if $\infty \in C\ell D^*$). Under those circumstances we claim that $w(z)$ extends to a C^1 local diffeomorphism of S^2, hence a diffeomorphism. In fact the assumptions on φ and the estimate

$$|z - z^*|^2 |j(D)_z(z)| < |z - z^*|^2 |j(D)_{\bar z}(z)| \leq C_2 \lambda(z^*)^{-2}$$

from Theorem 3 imply that (6.3a), (6.4a), and (6.5a) hold in $C\ell(D^*)$, (6.3b), (6.4b), and (6.5b) hold in $C\ell(D)$, and the two sets of formulas agree on ∂D. Thus $w \in C^1$. To check that w is a local diffeomorphism we first change variables as above to make z, z^*, and $w(z)$ finite, so that the denominators in (6.3), (6.4), and (6.5) are nonzero. Formula (6.5a) shows that $w(z)$ has a positive Jacobian in $C\ell D^*$. In D, (6.4b) and (6.5b) imply that

$$|w_z(z)| - |w_{\bar z}(z)| > \frac{1 - |(z - z^*)^2 \varphi(z^*) j(D)_{\bar z}(z)|}{|v_2(z^*) + (z - z^*)v_2'(z^*)|^2}.$$

By Theorem 3 in §4, that expression is positive in D if ∂D is a K^2-quasicircle and $\|\varphi\| < \epsilon(K^2)$.

We have proved that $w(z)$ is a C^1 local diffeomorphism of S^2, hence a diffeomorphism $w : S^2 \to S^2$, under the regularity assumptions stated above. Since the complex dilatation $\mu = w_{\bar z}/w_z$ is zero in $C\ell D^*$ and equals $\alpha(\varphi)$ in D, inequality (5.2), with K^2 in the role of K, implies further that w is a K'-quasiconformal map, with

$$K' = \frac{\epsilon(K^2) + \|\varphi\|}{\epsilon(K^2) - \|\varphi\|}.$$

For the general case we use the approximation argument from Ahlfors [1, p. 300]. Let $\varphi \in B(D^*)$ be given, and assume $\|\varphi\| < \epsilon(K^2)$. Fix functions v_1 and v_2 satisfying (6.1) and (6.2). Since ∂D is a K-quasicircle Lemma 2 gives us a K^2-quasiconformal map $g : S^2 \to S^2$ that maps Δ conformally onto D^*. Let (r_n) be a sequence in $(0,1)$ that converges to 1, and let $D_n^* = g(\{\varsigma \in \Delta; |\varsigma| < r_n\})$. The boundary of D_n^* is a real analytic K^2-quasicircle to which our previous arguments apply. Since the $B(D_n^*)$ norm of φ is less than its $B(D^*)$ norm, we obtain a sequence of K'-quasiconformal maps $w_n : S^2 \to S^2$ that satisfy

$$w_n(z) = v_1(z)/v_2(z), \qquad\qquad z \in D_n^*,$$

$$w_n(z) = \frac{v_1(z_n^*) + (z - z_n^*)v_1'(z_n^*)}{v_2(z_n^*) + (z - z_n^*)v_2'(z_n^*)}, \qquad z \in D_n.$$

Here $z_n^* = j(D_n)(z)$ if $z \in D_n$.

Since the maps w_n converge in D^* to the nonconstant map v_1/v_2, we may assume that w_n converges uniformly in S^2 to a K'-quasiconformal map w. To complete the proof of Theorem 4 we must show that the limit function $w(z)$ satisfies (6.3b) in D. It suffices to prove that $j(D_n)$ converges to $j(D)$ in D, and that follows readily from Theorem 2. In fact the maps $z \mapsto g(r_n g^{-1}(z))$ converge to the identity uniformly on S^2 as $n \to \infty$. Therefore the regions D_n converge as Jordan regions to D in the sense of §3, and $j(D_n) \to j(D)$. $\qquad\square$

REMARKS: (1) Theorem 4 reduces to the classical Ahlfors-Weill theorem (see [**3**] or [**1**, Lemma 3]) if ∂D is S^1 or $\mathbb{R} \cup \{\infty\}$.

(2) The only essential difference between our proof of Theorem 4 and Ahlfors's original proof is that we can use Theorem 2 to determine the limit function $w(z)$ in the approximation argument above. Without that continuity theorem, we could assert by compactness that (6.3b) holds in D for some quasiconformal reflection $z \mapsto z^*$, but we would have little control over its properties. In particular we could not guarantee the invariance properties that we shall need in the next section.

(3) If $w(z)$ is defined by (6.3), the Schwarzian derivative of w in D^* is the function $\varphi(z)$. We shall have more to say about Schwarzian derivatives in the next section.

§7. Holomorphic Local Cross Sections in Teichmüller Space.

For the remainder of this paper D and D^* will be complementary Jordan regions bounded by a quasicircle. We shall assume that $\infty \in D^*$. Let G be any quasi-Fuchsian group that leaves D and D^* invariant; G may be finitely or infinitely generated or trivial. As usual $B(D^*, G)$ is defined to be the closed subspace of $B(D^*)$ consisting of the functions φ such that $\varphi = g^*(\varphi)$ for all g in G. Similarly, $L^\infty(D, G)$ is the closed subspace of $L^\infty(D, \mathbb{C})$ consisting of the functions μ such that $\mu = g^*(\mu)$ for all $g \in G$. (The maps g^* are given by (5.3) and (5.4) respectively.)

As usual we let $M(D,G)$ be the open unit ball in $L^\infty(D,G)$. For each μ in $M(D,G)$ let $w^\mu(z)$ be the unique quasiconformal map of the sphere onto itself that solves the Beltrami equation

$$(7.1) \qquad\qquad w_{\bar{z}} = \mu w_z$$

in D and is conformal in D^* with the behavior

$$(7.2) \qquad\qquad w^\mu(z) = z + O(z^{-1})$$

at infinity.

We form the Schwarzian derivative

$$S(w^\mu) = \frac{(w^\mu)'''}{(w^\mu)'} - \frac{3}{2}\left(\frac{(w^\mu)''}{(w^\mu)'}\right)^2$$

of the conformal map w^μ in D^*. Bers [5] proved that $S(w^\mu) \in B(D^*,G)$ if $\mu \in M(G)$ and that the map $\Phi : M(D,G) \to B(D^*,G)$ defined by putting

$$(7.3) \qquad\qquad \Phi(\mu) = S(w^\mu)$$

is holomorphic. The image $\Phi(M(D,G))$ is the Teichmüller space $T(D,G)$ of the quasi-Fuchsian group G. The map (7.3) is called the Bers map, and the representation it induces of Teichmüller space as a subset of $B(D^*,G)$ is called the Bers embedding. Bers proved in [5] that Φ is a split submersion and therefore has local holomorphic cross sections. We shall now give an independent proof of that result by exhibiting such sections directly. We first prove:

THEOREM 5. *The Ahlfors map $\alpha(\varphi)$ defined by (5.1) is a holomorphic right inverse of $\Phi : M(D,G) \to B(D^*,G)$ in a neighborhood of zero in $B(D^*,G)$. In particular $T(D,G)$ contains a neighborhood of zero.*

PROOF: Let ∂D be a K-quasicircle. By Lemma 4 in §5, there is an $\epsilon(K) > 0$ such that α is a holomorphic map of

$$B(D^*,G)_{\epsilon(K)} = \{\varphi \in B(D^*,G); \|\varphi\| < \epsilon(K)\}$$

into the open unit ball of $L^\infty(D,\mathbb{C})$. By Lemma 5, α maps $B(D^*,G)_{\epsilon(K)}$ into $M(D,G)$. We must prove that $\Phi(\alpha(\varphi)) = \varphi$ for all φ in a neighborhood of zero.

192

Given $\varphi \in B(D^*, G)$ choose solutions v_1 and v_2 of equation (6.1) in D^* with behavior

$$v_1(z) = z + O(z^{-1}), \qquad v_2 = 1 + O(z^{-2})$$

at infinity. Then v_1 and v_2 also satisfy (6.2) in D^*. By Theorem 4, if $\|\varphi\| < \epsilon(K^2)$ the map $w : D \cup D^* \to S^2$ defined by (6.3) has a quasiconformal extension $w : S^2 \to S^2$. That extension is conformal in D^* with behavior (7.2) at infinity. In D it solves equation (7.1) with $\mu = \alpha(\varphi)$, so $w = w^\mu$ with $\mu = \alpha(\varphi)$. Since $w = v_1/v_2$ in D^*,

$$\Phi(\alpha(\varphi)) = S(v_1/v_2) = \varphi. \qquad \square$$

A standard technique (moving the origin by a right translation) leads easily from Theorem 5 to our goal, the well known

THEOREM 6 (SEE BERS [4] AND [5]). *$T(D,G)$ is open in $B(D^*,G)$. Moreover, if $\mu \in M(D,G)$ and $\varphi = \Phi(\mu)$, then $\Phi : M(D,G) \to B(D^*,G)$ has a holomorphic right inverse β defined in a neighborhood of φ and satisfying $\beta(\varphi) = \mu$.*

PROOF: Let μ in $M(D,G)$ and $\varphi = \Phi(\mu)$ in $B(D^*,G)$ be given. Form $D^\mu = w^\mu(D)$, $(D^\mu)^* = w^\mu(D^*)$, and the quasi-Fuchsian group $G^\mu = w^\mu \circ G \circ (w^\mu)^{-1}$. The Bers map $\Phi^\mu : M(D^\mu, G^\mu) \to B((D^\mu)^*, G^\mu)$ is given by

$$\Phi^\mu(\nu) = S(w^\nu) \qquad \text{if } \nu \in M(D^\mu, G^\mu).$$

By Theorem 5, Φ^μ has a holomorphic right inverse in a neighborhood of zero given by the Ahlfors map $\alpha(= \alpha_{D^\mu})$. Observe that $\alpha(0) = 0$.

Let $f : M(D^\mu, G^\mu) \to M(D,G)$ be "right translation by μ," so that $\sigma = f(\nu)$ means

$$(7.4) \qquad\qquad w^\sigma = w^\nu \circ w^\mu.$$

It is well known that f is a biholomorphic map. Formula (7.4) holds in all of S^2, hence in D^*, and the composition law for Schwarzian derivatives gives

$$\Phi(f(\nu)) = S(w^\nu \circ w^\mu)$$
$$= (S(w^\nu) \circ w^\mu)[(w^\mu)']^2 + S(w^\mu)$$
$$= (w^\mu)^*(\Phi^\mu(\nu)) + \varphi.$$

The required local right inverse can therefore be explicitly defined as

$$\beta \;=\; f \circ \alpha \circ h^{-1},$$

where $h : B((D^\mu)^*, G^\mu) \;\to\; B(D^*, G)$ is the biholomorphic map

$$h(\psi) \;=\; (w^\mu)^*(\psi) \;+\; \varphi. \qquad \square$$

REMARKS: (1) By the implicit function theorem, Theorem 6 is equivalent to Bers's theorem [5] that the Bers map Φ is a split submersion.

(2) Differentiation of the formula $\Phi \circ \alpha \;=\; id$ produces identities closely related to formulas of Bers [5] and Gardiner [8]. These identities are discussed in [10].

Clifford J. Earle, Cornell University, Ithaca, New York
and M.S.R.I., Berkeley, California

Subhashis Nag, Indian Statistical Institute, Calcutta, India
and M.S.R.I., Berkeley, California

REFERENCES

1. Ahlfors, L.V., *Quasiconformal reflections*, Acta Math. **109** (1963), 291–301.
2. __________, "Lectures on Quasiconformal Mappings," D. Van Nostrand, New York, 1966.
3. Ahlfors, L., and Weill, G., *A uniqueness theorem for Beltrami equations*, Proc. Amer. Math. Soc. **13** (1962), 975–978.
4. Bers, L., *Automorphic forms and general Teichmüller spaces*, in "Proceedings of the Conference on Complex Analysis (Minneapolis 1964)", pp. 109–113, Springer, Berlin, 1965.
5. __________, *A non-standard integral equation with applications to quasiconformal mappings*, Acta Math. **116** (1966), 113–134.
6. Beurling, A., and Ahlfors, L., *The boundary correspondence under quasiconformal mappings*, Acta Math. **96** (1956), 125–142.
7. Douady, A., and Earle, C.J., *Conformally natural extension of homeomorphisms of the circle*, Acta Math. **157** (1986), 23–48.
8. Gardiner, F., *An analysis of the group operation in universal Teichmüller space*, Trans. Amer. Math. Soc. **132** (1968), 471–486.
9. Goluzin, G.M., "Geometric Theory of Functions of a Complex Variable," Translations of Mathematical Monographs, Vol. 26, Amer. Math. Soc., Providence, 1969.
10. Nag, S., *Conformally natural Ahlfors-Weill sections and Bers's reproducing formulas in Teichmüller theory*, Bull. Austral. Math. Soc. **36** (1987), 187–196.

A theorem of Bers and Greenberg for infinite dimensional Teichmüller spaces

BY FREDERICK P. GARDINER

Introduction.

This paper concerns a theorem of Bers and Greenberg [**BG**] about isomorphisms between Teichmüller spaces. For a Fuchsian group acting on the upper half plane H, let H_Γ be H with all of the fixed points of elliptic elements of Γ removed. When Γ is finitely generated and of the first kind, the theorem says that the Teichmüller space, $T(\Gamma)$, of the Fuchsian group Γ depends only on the topological type of the surface H_Γ/Γ and not on the orders of the elliptic elements of Γ.

The heart of the theorem is the fact that a trivial mapping for the group Γ projects to a self-mapping of the surface H_Γ/Γ which is homotopic to the identity. When lifted to H_Γ, such a homotopy does not pass through the elliptic fixed points of Γ. Bers and Greenberg proved the theorem by showing the natural mapping from $T(H_\Gamma/\Gamma)$ to $T(\Gamma)$ is a unramified covering and then appealing to a theorem of Fricke, [**FK**], also proved by Keen in [**K**], which says that the Teichmüller space of a finitely generated Fuchsian group with elliptic elements is homeomorphic to Euclidean n-space. Marden [**M**] proved the theorem, shortly after it was announced by Bers and Greenberg, by showing directly the existence of the required homotopy. A simple new construction of the required homotopy has been given by Earle and McMullen and appears in this proceedings, (see [**EM**]).

A few years later Kra observed that the theorem, at least in the finite dimensional cases, is a straightforward consequence of Teichmüller's theorem, (see [**EK**]). In fact, the natural mapping from $T(H_\Gamma/\Gamma)$ to $T(\Gamma)$ is an isometry because it preserves Teichmüller-Beltrami differentials. The validity of this observation depends on knowing that Teichmüller-Beltrami differentials for a Fuchsian group with elliptic elements are uniquely extremal. This was a known fact and follows from the Teichmüller inequality, [**RS**]. The inequality can be obtained from the analogous inequality for Fuchsian groups without elliptic elements and from the fact that every finitely generated Fuchsian group contains a torsion-free subgroup of finite index.

Research partially supported by the Research Foundation of CUNY, New York, New York.

The objective of this paper is to prove the theorem, and a slight generalization of it, in both the finite and infinite dimensional cases using a single method. The method is based on a theorem of Reich and Strebel [RS] which asserts that Hamilton's necessary condition for extremality is also sufficient. In the finite dimensional case, our proof reduces to the proof indicated in [EK].

Sections 1 and 2 give the definition of Teichmüller space for a Riemann surface relative to a closed subset of its ideal boundary and the definition of Teichmüller space of a Fuchsian group relative to a closed invariant set containing its limit set.

Section 3 gives the natural mapping from the Teichmüller space of H_Γ/Γ relative to a particular closed Γ-invariant boundary subset to the Teichmüller space of the Fuchsian group Γ relative to the appropriate closed set. Moreover, it is observed that this mapping is functorial.

Section 4 concerns the definition of Teichmüller's metric for Teichmüller spaces of Fuchsian groups and Riemann surfaces.

Section 5 gives the theorem of Bers and Greenberg and shows how it is a consequence of the necessity and sufficienty of Hamilton's condition [RS], which implies that the natural mapping of section 3 is an isometry.

I would like to thank Irwin Kra and the referee for pointing out an essential error in my original argument.

§1. The Teichmüller space of a Fuchsian group relative to a closed invariant set containing the limit set.

Let Γ be a Fuchsian group acting on the upper half plane H and let C be a closed Γ-invariant set which is contained in $\hat{R} = R \cup \{\infty\}$ and which contains the limit set of Γ. Assume the closed set C contains at least three points. We will define $T(\Gamma, C)$, the Teichmüller space of Γ relative to C.

Let $M(\Gamma)$ be the space of Beltrami coefficients μ for Γ. These coefficients are complex-valued L_∞-functions defined on H satisfying

$$(1) \qquad \begin{array}{ll} \text{a)} & \mu(A(z))\overline{A'(z)} = \mu(z)A'(z) \text{ for all } A \text{ in } \Gamma \text{ and} \\[2mm] \text{b)} & \|\mu\|_\infty < 1. \end{array}$$

By the existence theorem for the Beltrami equation [AB], every such coefficient μ can be realized in the form

$$(2) \qquad w_{\bar{z}}(z) = \mu(z)w_z(z)$$

where w is a homeomorphism of $H \cup \hat{R}$ onto $H \cup \hat{R}$. The partial derivatives $w_{\bar{z}}$ and w_z are taken in the generalized sense and they are locally L_2-functions. Moreover, if one normalizes w to fix three points of the closed set C, then the function w satisfying (2) is unique.

Consider the group, $D_0(\Gamma, C)$, of quasiconformal homeomorphisms w of $H \cup \hat{R}$ onto $H \cup \hat{R}$ which satisfy

(3)
$$
\begin{aligned}
&\text{a)} \quad w {\circ} A = A {\circ} w \text{ for all } A \text{ in } \Gamma \text{ and}\\
&\text{b)} \quad w(x) = x \text{ for all } x \text{ in } C.
\end{aligned}
$$

The verificattion that $D_0(\Gamma, C)$ is a group is obvious. We will call $D_0(\Gamma, C)$ the group of trivial mappings for Γ and C. This group induces an equivalence relation on $M(\Gamma)$. For two elements μ and ν of $M(\Gamma)$ we form normalized solutions w^μ and w^ν of the equations:

$$w^\mu_{\bar{z}}(z) = \mu(z) w^\mu_z(z),$$

$$w^\nu_{\bar{z}}(z) = \nu(z) w^\nu_z(z).$$

(The mappings w^μ and w^ν are normalized to fix the same three points of Λ). We say that $\mu \sim \nu$ if there is an element w in $D_0(\Gamma, C)$ such that $w^\mu {\circ} w = w^\nu$.

DEFINITION. *The Teichmüller space $T(\Gamma, C)$ is the set of equivalence classes of elements of $M(\Gamma)$ with the equivalence relation induced by $D_0(\Gamma, C)$.*

REMARK: The definition given here is non-standard because the equivalence relation induced by $D_0(\Gamma, C)$ does not involve any homotopy condition. It is a fact, proved by Ahlfors for torsion free Fuchsian groups of the first kind [A] and by Marden for general Fuchsian groups [M], that if w satisfies conditions 3 a) then there is a homotopy of w restricted to the upper half-plane with the elliptic fixed points removed connecting w to the identity.

§2. The Teichmüller space of a Riemann surface relative to a distinguished closed subset of its ideal boundary.

In this section we will be concerned with a Riemann surface R realized as the quotient of the non-elliptic fixed points of the upper half plane factored by a Fuchsian group. In particular, let H_Γ be the upper half plane with all fixed points of elliptic elements of Γ removed. Let $R = H_\Gamma/\Gamma$. If Γ is of the second kind, assume C is a Γ-invariant closed subset of the extended real axis which contains the limit set Λ of Γ. Then $\sigma = (C - \Lambda)/\Gamma$ is a closed subset of the ideal boundary of R.

In a manner analogous to section 1 we will define the Teichmüller space $T(R,\sigma)$. Let $M(R)$ be the space of Beltrami coefficients μ on R. To each local parameter z, there is assigned a complex-valued L_∞-function μ^z, such that, for any two local parameters z_1 and z_2 with overlapping domains of definition and for the corresponding functions $\mu_1 = \mu^{z_1}$ and $\mu_2 = \mu^{z_2}$, one has

$$(4) \qquad \text{a)} \quad \mu_1(z_1(z_2)) \frac{\overline{dz_1}}{dz_2} \left(\frac{dz_1}{dz_2} \right)^{-1} = \mu_2(z_2).$$

Observe that $|\mu(z)|$ is a well-defined function on R and, thus $\|\mu\|_\infty = \operatorname{ess\,sup}\limits_{z \in R} |\mu(z)|$ is defined. We assume that the Beltrami coefficients μ in $M(R)$ also satisfy

$$(4) \qquad \text{b)} \quad \|\mu\|_\infty < 1.$$

Let $\pi : H_\Gamma \to H_\Gamma/\Gamma = R$ be the natural projection. Observe that π induces an isometric isomorphism $\pi^* : M(R) \to M(\Gamma)$ by letting

$$\pi^*(\mu) = \mu(\pi(z))\overline{\pi'(z)}/\pi'(z).$$

The conditions 1 a) and 1 b) correspond precisely to the conditions 4 a) and 4 b).

Consider the group $D_0(R,\sigma)$ of quasiconformal homeomorphisms w of R onto R which extend uniquely to the ideal boundary of R and which satisfy

(5)

a) w is homotopic to identity by a homotopy w_t which extends continuously to the ideal boundary of R

     $(w_0(p) = p$ for p in R and $w_1(p) = w(p)$ for p in $R)$ and

b) $w_t(x) = x$ for all x in σ and all t with $0 \leq t \leq 1$.

By the existence theorem for solutions to the Beltrami equation,

$$(6) \qquad\qquad w_{\bar{z}} = \mu(z)w_z,$$

for any element μ in $M(R)$, there exists a quasiconformal mapping $w = w^\mu$ which satisfies (6) and which maps the Riemann surface R onto a Riemann surface R_1.

The group $D_0(R,\sigma)$ induces an equivalence relation on $M(R)$. Two Beltrami coefficients μ and ν are equivalent if there exists a conformal map α and an element w_0 of $D_0(R,\sigma)$ which makes the following diagram commute:

$$(7) \qquad \begin{array}{ccc} R & \xrightarrow{\;w^\mu\;} & R_1 \\ {\scriptstyle w_0}\downarrow & & \downarrow{\scriptstyle \alpha} \\ R & \xrightarrow[\;w^\nu\;]{} & R_2. \end{array}$$

DEFINITION. *The Teichmüller space $T(R,\sigma)$ is the set of equivalence classes of elements of $M(R)$ with the equivalence relation induced by $D_0(R,\sigma)$.*

REMARK: For the case where σ is the empty set, $T(R,\sigma)$ is called reduced Teichmüller space. For the case where σ is the whole ideal boudnary of R, $T(R,\sigma)$ is called unreduced Teichmüller space. When there is no ideal boundary, that is, when R has a universal covering group which is of the first kind, there is no distinction to be made. In either of the two cases, that is, for both reduced or unreduced Teichmüller space, this definition is the standard one. It depends on the group of trivial mappings $D_0(R,\sigma)$, whose elements are defined so as to satisfy a homotopy condition.

§3. The mapping from Teichmüller space of a Riemann surface to Teichmüller space of a Fuchsian group.

As before, assume that Γ is a Fuchsian group acting on the upper half plane and that Λ is the limit set of Γ. Let C be a Γ-invariant subset of the extended real axis which contains Λ.

Let R be the Riemann surface H_Γ/Γ where H_Γ is the upper half plane with all of the elliptic fixed points of elements of Γ removed. Let $\sigma = (C - \Lambda)/\Gamma$. Thus σ is a closed subset of the ideal boundary of R. We have already remarked that the covering $\pi : H_\Gamma \to R$ induces an isometry of $M(R)$ onto $M(\Gamma)$ by formula $\pi^*(\mu) = \mu(\pi(z))\overline{\pi'(z)}/\pi'(z)$.

PROPOSITION 1. *The mapping $\pi^* : M(R) \to M(\Gamma)$ induces a well-defined mapping Φ from $T(R,\sigma)$ onto $T(\Gamma,C)$.*

PROOF: Since π^* is a surjection, so is Φ. To show that Φ is well-defined, we must show that if μ is equivalent to ν under the equivalence relation induced by $D_0(R,\sigma)$ then $\pi^*(\mu)$ is equivalent to $\pi^*(\nu)$ under the equivalence relation induced by $D_0(\Gamma,C)$.

200

Let $f : R \to R$ be an element of $D_0(R,\sigma)$. Consider the following diagram of covering spaces and liftings of the mapping f:

$$(8) \qquad
\begin{array}{ccc}
H & \xrightarrow{\;\tilde{w}\;} & H \\
\tilde{\pi}\downarrow\;\; \searrow^{\pi_1} \quad H_\Gamma \xrightarrow{w} H_\Gamma \quad {}^{\pi_1}\nearrow & & \downarrow\tilde{\pi} \\
\nearrow^{\pi} \qquad\qquad \pi\searrow & & \\
R & \xrightarrow[f]{} & R.
\end{array}$$

A lifting $\tilde{w}$ of f exists because H is the universal covering of R. To define $w(z)$, pick a point p in H such that $\pi_1(p) = z$. Then let $w(z) = \pi_1 \circ \tilde{w}(p)$. It is an exercise to show that this definition is independent of which point p one chooses in $\pi_1^{-1}(z)$.

LEMMA. *If the quasiconformal mapping f in (8) is in $D_0(R,\sigma)$ then it is a lifting w in $D_0(\Gamma, C)$.*

PROOF: Let $\tilde{\Gamma}$ be the covering group for $\tilde{\pi}$. Since homotopic mappings lift to the universal covering surfaces, the homotopy connecting f to the identity on R lifts to a homotopy $\tilde{w}_t$ which connects $\tilde{w}$ to the identity. Clearly, for any A in $\tilde{\Gamma}$, $\tilde{w} \circ A$ is another lifting of f and therefore $\tilde{w} \circ A = B \circ \tilde{w}$ for some B in $\tilde{\Gamma}$. The homotopy $\tilde{w}_t$ determines a continuous curve of homomorphisms $\chi_t : \tilde{\Gamma} \to \tilde{\Gamma}$ determined by

$$\tilde{w}_t \circ A = \chi_t(A) \circ \tilde{w}_t.$$

Clearly $\chi_0(A) = A$ and $\chi_t(A)$ is a continuous mapping into the discrete group $\tilde{\Gamma}$. Thus $\chi_t(A) = A$ for $0 \le t \le 1$. We conclude that $\tilde{w} \circ A \circ \tilde{w}^{-1} = A$ for all A in $\tilde{\Gamma}$.

The fact that $\tilde{w} \circ A = A \circ \tilde{w}$ for all A in $\tilde{\Gamma}$ implies that $w \circ B = B \circ w$ for all B in Γ. To see this, let B be in Γ and let A in $\tilde{\Gamma}$ satisfy $\pi_1 \circ A = B \circ \pi_1$. Then

$$w \circ B \circ \pi_1 = w \circ \pi_1 \circ A = \pi_1 \circ \tilde{w} \circ A$$

$$= \pi_1 \circ A \circ \tilde{w} = B \circ \pi_1 \circ \tilde{w} = B \circ w \circ \pi_1.$$

Thus, we see that w satisfies property 3 a) of trivial mappings in $D_0(\Gamma, C)$.

Since $w \circ B = B \circ w$ for all B in Γ, it follows that w fixes all fixed points of elements of Γ and, by continuity, w fixes all points in the limit set of Γ. Let w_t be a lifting of the homotopy f_t which connects f to the identity. Let J

be a component of $\hat{\mathbb{R}} - \Lambda$. J is an open interval and $w(J)$ must be the same interval since w fixes the endpoints of J, which are in the limit set Λ. For p in $C \cap J$ we know that $w(p) = B(p)$ where B is in the subgroup of Γ which fixes J. If B is not the identity then the continuous curve $w_t(p)$ which joins p to $B(p)$ would project to a continuous curve $f_t(\pi(p))$ which winds around the boundary component determined by J. But $f_t(\pi(p)) = \pi(p)$ for each t, $0 \le t \le 1$ and $\pi(p)$ in σ. Thus B is the identity and $w(x) = x$ for all x in C.

This concludes the proof of the lemma. Proposition 1 is an easy consequence of the lemma.

The next task is to observe that the mapping Φ in proposition 1 is functorial. To see this, let Γ_1 and Γ_2 be Fuchsian groups acting on H and let h be a quasiconformal homeomorphism of H onto H which conjugates Γ_1 into Γ_2, that is, $h \circ \Gamma_1 \circ h^{-1} = \Gamma_2$. Moreover, assume C_1 and C_2 are closed invariant sets of the extended real axis containing the limit sets Λ_1 and Λ_2 of Γ_1 and Γ_2, respectively, with $h(C_1) = C_2$.

Of course, there are corresponding subsets $\sigma_j = (C_j - \Lambda_j)/\Gamma_j$, for $j = 1$ and 2, of the ideal boundaries of the Riemann surfaces $R_j = H_{\Gamma_j}/\Gamma_j$.

The mapping h obviously induces an isomorphism $\tilde{S}$ of $T(\Gamma_2, \sigma_2)$ onto $T(\Gamma_1, \sigma_1)$ by associating to a Beltrami coefficient ν in $M(\Gamma_2)$ the Beltrami coefficient of $w_\nu \circ h$ in $M(\Gamma_1)$. It also induces an isometry S from $T(R_2, \sigma_2)$ onto $T(R_1, \sigma_1)$ in a similar manner. We have the following diagram:

$$
(9) \qquad
\begin{array}{ccc}
T(\Gamma_2, C_2) & \xrightarrow{\;\;\tilde{S}\;\;} & T(\Gamma_1, C_1) \\[4pt]
\Phi_2 \uparrow & & \uparrow \Phi_1 \\[4pt]
T(R_2, \sigma_2) & \xrightarrow[\;\;S\;\;]{} & T(R_1, \sigma_1),
\end{array}
$$

and the following proposition is obvious.

PROPOSITION 2. *The aforementioned diagram (9) is commutative, where Φ_2 and Φ_1 are the mappings coming from proposition 1 and S and $\tilde{S}$ are the isomorphisms induced by h.*

§4. Teichmüller's metric.

At first, we consider the Teichmüller space $T(\Gamma, C)$ for the Fuchsian group Γ relative to the Γ-invariant closed set C. Let Ψ be the natural projection from $M(\Gamma)$ onto $T(\Gamma, C)$ which identifies points of $M(\Gamma)$ that are equivalent with respect to the equivalence relation induced by $D_0(\Gamma, C)$. For a quasi-conformal mapping f, let $K(f)$ denote its maximal dilation. Teichmüller's metric is defined by

$$(10) \qquad d_\Gamma(\Psi(\mu), \Psi(\nu)) = \frac{1}{2} \inf \log K(w^\mu \circ w \circ (w^\nu)^{-1})$$

where the infimum is taken over all quasiconformal homeomorphisms w in $D_0(\Gamma, C)$. To shorten notation we often write $d_\Gamma(\mu, \nu)$ instead of (10).

It is an exercise to show that (10) is a nondegenerate metric on $T(\Gamma, C)$. Symmetry follows from the fact that $K(f) = K(f^{-1})$ and that $D_0(\Gamma, C)$ is a group. Transitivity follows from the law $K(f \circ g) \le K(f)K(g)$. Finally, the metric is nondegenerate because for normalized quasiconformal mappings f_n with $K(f_n) \to 1$ the sequence f_n converges normally to a holomorphic function.

For two quasiconformally conjugate Fuchsian groups Γ_1 and Γ_2, $(w \circ \Gamma_1 \circ w^{-1} = \Gamma_2)$ with invariant closed sets C_1 and C_2 for which $w(C_1) = C_2$, it is clear that the mapping $\tilde{S}$ of proposition 2 is an isometry with respect to the metrics d_{Γ_1} and d_{Γ_2}.

The Teichmüller metric d for the Teichmüller space $T(R, \sigma)$ is defined in a parallel manner and it is obvious that the mapping S of proposition 2 is an isometry from $T(R_2, \sigma_2)$ onto $T(R_1, \sigma_1)$ with respect to the respective Teichmüller metrics d_2 and d_1.

LEMMA 2. *Suppose that for arbitrary Γ, C, R, and Φ satisfying the conditions of proposition 1, one has $d_\Gamma(\Phi(\mu), 0) = d(\mu, 0)$ for every μ in $M(R)$. Then the mapping $\Phi : T(R, \sigma) \to T(\Gamma, C)$ is an isometry.*

PROOF: We use the functorial property of the mapping Φ as expressed in proposition 2. Referring to diagram (1), let d_1 and d_2 be the Teichmüller metrics for $T(R_1, \sigma_1)$ and $T(R_2, \sigma_2)$ and let d_{Γ_1} and d_{Γ_2} be the Teichmüller metrics for $T(\Gamma_1, C_1)$ and $T(\Gamma_2, C_2)$, respectively. Note that if S is induced by the mapping h where $w \circ h = $ identity and w has Beltrami coefficient μ, then $S[\mu] = [0]$. We have

$$d_2(\mu, \nu) = d_1(0, S(\nu)).$$

But by hypothesis, $d_1(0, S(\nu)) = d_{\Gamma_1}(0, \Phi_1 \circ S(\nu))$. By proposition 2, $d_{\Gamma_1}(0, \Phi_1 \circ S(\nu)) = d_{\Gamma_1}(0, \tilde{S} \circ \Phi_2(\nu)) = d_{\Gamma_2}(\tilde{S}^{-1}(0), \Phi_2(\nu)) = d_{\Gamma_2}(\Phi_2(\mu), \Phi_2(\nu))$. We conclude that $d_2(\mu, \nu) = d_{\Gamma_2}(\Phi_2(\mu), \Phi_2(\nu))$ for arbitrary points $[\mu]$ and $[\nu]$ in the Teichmüller space $T(R_2, \sigma_2)$.

§5. The theorem of Bers and Greenberg.

THEOREM [**BG**]. *The mapping* $\Phi : T(R,\sigma) \to T(\Gamma, C)$ *is an isomorphism. In fact, it is an isometry with respect to Teichmüller's metrics on the two Teichmüller spaces. In particular, when* $T(R,\sigma)$ *is finite dimensional, the Teichmüller space* $T(\Gamma, C)$ *depends only on the topology of the pair* (R,σ) *(where* $R - H_\Gamma / \Gamma$ *and* $\sigma = (C - \Lambda)/\Gamma$ *) and not on the orders of the elliptic fixed points of elements of* Γ.

PROOF: By Lemma 2, all we need to show is that $d(0,\mu) = d_\Gamma(0,\Phi(\mu))$ for arbitrary μ in $M(R)$. We shall use the necessity and sufficiency of Hamilton's condition for extremality. Let $\pi : H_\Gamma \to H_\Gamma / \Gamma = R$ be the covering mapping and $\pi^* : M(R) \to M(\Gamma)$ the induced mapping on Beltrami coefficients. Assume μ is extremal in its class for $T(R,\sigma)$. From the necessity of Hamilton's condition this means that $d(0,\mu) = \dfrac{1}{2} \log \dfrac{1+k}{1-k}$ where $k = \|\mu\|_\infty$ and

$$(11) \qquad k = \sup Re \iint_R \mu\phi\, dx dy$$

where the supremum is taken over all holomorphic quadratic differentials ϕ on R satisfying

$$(12) \qquad \|\phi\| = \iint_R |\phi(z)|\, dx dy = 1 \text{ and}$$

$$(13) \qquad \phi \text{ is real with respect to real boundary uniformizers}$$

along any part of the ideal boundary in the complement of σ.

We denote the space of holomorphic quadratic differentials ϕ satisfying (13) and of finite norm by $A(R,\sigma)$.

On the other hand, the supremum in (11) is the same as

$$(14) \qquad \sup Re \iint_{H/\Gamma} \pi^*(\mu)q\, dx dy$$

where the supremum is taken over all holomorphic quadratic differential forms q for Γ which satisfy

$$(15) \qquad \|q\| = \iint_{H/\Gamma} |q(z)| = 1 \text{ and}$$

$$(16) \qquad q \text{ is real-valued on } \hat{R} - C.$$

We denote the space of holomorphic quadratic differential forms q satisfying (16) and of finite norm by $A(\Gamma, C)$. To justify the statement that the supremum in (11) is the same as the supremum in (14) we must observe that the mapping $\pi_* : A(R, \sigma) \to A(\Gamma, C)$ given by $\pi_*(\phi) = \phi(\pi(z))\pi'(z)^2 = q(z)$ is an isometric isomorphism of Banach spaces. We leave this verification to the reader, (see [**Kr**]).

Now, from the sufficiency of Hamilton's condition, this implies that $\pi^*(\mu)$ is extremal and thus $d_\Gamma(0, \pi^*(\mu)) = \dfrac{1}{2} \log \dfrac{1+k}{1-k}$.

Brooklyn College, CUNY

205

REFERENCES

[A] Ahlfors, L.V., "Lectures on Quasiconformal Mapping," Van Nostrand, New York, 1966.

[AB] Ahlfors, L.V., and Bers, L., *Riemann's mapping theorem for variable metrics*, Ann. of Math. **72** (1960), 385–404.

[BG] Bers, L., and Greenberg, L., *Isomorphisms between Teichmüller Spaces*, Advances in the Theory of Riemann Surfaces, Study 66, P.U.P., 1969, 53–79.

[EK] Earle, C., and Kra, I., *On holomorphic mappings between Teichmüller spaces*, Contributions to Analysis, (L.V. Ahlfors et al. eds.), Academic Press, New York and London, 1974, 107–124.

[EM] Earle, C., and McMullen, C., *Quasiconformal isotopies*, this volume.

[FK] Fricke, R., and Klein, F., "Vorlesungen uber die Theorie der Automorphen Funktionen," Vol. 2, B.G. Teubner, 1926.

[G] Gardiner, F.P., *The Teichmüller–Kobayashi metric for infinite dimensional Teichmüller spaces*, Springer Lecture Notes in Mathematics **971**; Kleinian Groups and Related Topics, ed. D.M. Gallo and R.M. Porter, 1981, 48–67.

[K] Keen, L., *Intrinsic moduli on Riemann surfaces*, Ann. of Math. **84** (1966), 404–420.

[Kr] Kra, I., "Automorphic Forms and Kleinian Groups," W.A. Benjamin, Reading, Massachusetts, 1972.

[M] Marden, A., *On homotopic mappings of Riemann surfaces*, Ann. of Math. **90** (1969), 1–8.

[RS] Reich, E., and Strebel, K., *Extremal quasiconformal mappings with given boundary values*, Contributions to Analysis (L.V. Ahlfors et al. eds.), Academic Press, New York and London, 1974, 375–392.

A finiteness theorem for holomorphic families
of Riemann surfaces

Dedicated to Professor Tadashi Kuroda
on his sixtieth birthday

BY YOICHI IMAYOSHI AND HIROSHIGE SHIGA

§0. Introduction.

In this paper we will give an analytic proof of the following finiteness theorem for holomorphic families of Riemann surfaces:

FINITENESS THEOREM OF FAMILIES. *Let B be a Riemann surface of finite type. Then, there are only finitely many non-isomorphic and locally non-trivial holomorphic families of Riemann surfaces of fixed finite type (g, n) with $2g - 2 + n > 0$ over B.*

This fact was first conjectured by Shafarevich [18] and it is called *Shafarevich conjecture* in the function field case. Parshin [16] proved it for a closed Riemann surface B and for holomorphic families of type $(g, 0)$ with $g \geq 2$. Arkelov [3] proved it for a Riemann surface B of finite type and for holomorphic families of type $(g, 0)$ with $g \geq 2$ (cf. Faltings [7], [8], Mazur [11], Mumford [13], p. 39 and pp. 100–101, Szpiro [20]). The tool of our proof is the theory of Teichmüller spaces and Kleinian groups.

Parshin [16] showed that the above finiteness theorem implies the following finiteness theorem for holomorphic sections of a fixed holomorphic family of Riemann surfaces. But we shall show that a more direct and elementary argument makes the above finiteness theorem yield it.

FINITENESS THEOREM OF SECTIONS. *Let B be a Riemann surface of finite type. Consider a holomorphic family of Riemann surfaces of type $(g, 0)$ with $g \geq 2$ over B. If it is locally non-trivial, then there are only finitely many holomorphic sections of it. If it is locally trivial, then there are only finitely many non-constant holomorphic sections of it.*

This is called *Mordell conjecture* in the function field case. It is proved by Manin, Grauert [9] and Miwa [12] (cf. Mazur [11], Mumford [13] p. 39 and pp. 100–101, Noguchi [14,15], Samuel [17]).

In the first section, we give the definition of holomorphic families of Riemann surfaces and some examples.

In §2, we explain the terms of the theory of Teichmüller spaces which are used in this paper.

In §3, we prove a "Rigidity Theorem" for holomorphic families of Riemann surfaces.

In §4, we prove the finiteness theorem of holomorphic families of Riemann surfaces. In fact, we show that the rigidity theorem in §3 and the hyperbolic geometry of Riemann surfaces yield the theorem.

In the last section, we prove that the finiteness theorem of families implies the finiteness theorem of sections.

The authors would like to thank Professor Junjiro Noguchi for many fruitful discussions on Shafarevichi–Mordell conjecture. They also thank Professor Shigeyasu Kamiya for his help in their joint work.

§1.   Definitions and Examples.

Let $\hat{M}$ be a two-dimensional complex manifold and let C be a nonsingular one dimensional analytic subset of $\hat{M}$ or empty. Let B be a Riemann surface. Assume that there exists a holomorphic mapping $\hat{\pi} : \hat{M} \to B$ satisfying the following two conditions;

1) $\hat{\pi}$ is proper and of maximal rank at every point of $\hat{M}$, and

2) setting $M = \hat{M} - C$ and $\pi = \hat{\pi}|M$, the fiber $S_t = \pi^{-1}(t)$ of M over each t in B is an irreducible analytic subset of M and is of fixed finite type (g, n) as a Riemann surface, where g is the genus of S_t and n is the number of punctures of S_t.

We call such a triple (M, π, B) a holomorphic family of Riemann surfaces of type (g, n) over B.

We assume throughout this paper that holomorphic families (M, π, B) of Riemann surfaces are of type (g, n) with $2g - 2 + n > 0$ and that B is of finite type, that is, a Riemann surface obtained by removing at most a finite number of points from a closed Riemann surface.

Two holomorphic families (M_1, π_1, B) and (M_2, π_2, B) of Riemann surfaces over B are called *isomorphic* if there exists a biholomorphic mapping $f : M_1 \to M_2$ with $\pi_1 = \pi_1 \circ f$. A holomorphic family (M, π, B) of Riemann surfaces is *locally trivial* if for each point t_0 in B there exists a neighborhood U_0 of t_0 such that $(\pi^{-1}(U_0), \pi|\pi^{-1}(U_0), U_0)$ is isomorphic to the trivial holomorphic family $(U_0 \times \pi^{-1}(t_0), \pi_0, U_0)$, where $\pi_0 : U_0 \times \pi^{-1}(t_0) \to U_0$ is the canonical projection. A holomorphic family (M, π, B) of Riemann surfaces is locally trivial if and only if the fibers $\pi^{-1}(t)$ are all isomorphic.

We have the following typical examples of holomorphic families of Riemann surfaces:

EXAMPLE 1: Let B be a Riemann surface and S a Riemann surface of type (g,n). Let $\pi : B \times S \to B$ be the canonical projection. Then $(B \times S, \pi, B)$ is a holomorphic family of Riemann surfaces of type (g,n). Such a holomorphic family $(B \times S, \pi, B)$ is said to be globally trivial.

EXAMPLE 2: Set

$$
\begin{aligned}
B &= \mathbb{C} - \{0\}, \\
M &= \{([z_0, z_1, z_2], t) \in P^2 \times B \,|\, z_2^4 = z_1^3 z_0 - t z_0^4 \},
\end{aligned}
$$

where $[z_0, z_1, z_2]$ are homogeneous coordinates of the two dimensional complex projective space P^2. Let $\pi : M \to B$ be a holomorphic mapping with $\pi(([z_0, z_1, z_2], t)) = t$. Then (M, π, B) is a locally trivial holomorphic family of Riemann surfaces of type $(3, 0)$, but it is not globally trivial.

EXAMPLE 2': Set

$$
\begin{aligned}
B &= \mathbb{C} - \{0\}, \\
M &= \{(x, y, t) \in \mathbb{C}^2 \times B \,|\, y^4 = x^3 - t \}.
\end{aligned}
$$

Let $\pi : M \to B$ a holomorphic mapping with $\pi(x, y, t) = t$. Then (M, π, B) is a locally trivial holomorphic family of Riemann surfaces of type $(3, 1)$, but it is not globally trivial.

EXAMPLE 3: Set

$$
\begin{aligned}
B &= \mathbb{C} - \{0, 1\}, \\
M &= \{([z_0, z_1, z_2], t) \in P^2 \times B \,|\, z_2^4 = z_1(z_1 - z_0) \times (z_1 - t z_0) z_0 \}.
\end{aligned}
$$

Let $\pi : M \to B$ be a holomorphic mapping with $\pi([z_0, z_1, z_2], t) = t$. Then (M, π, B) is a locally non-trivial holomorphic family of Riemann surfaces of type $(3, 0)$.

EXAMPLE 3': Set

$$
\begin{aligned}
B &= \mathbb{C} - \{0, 1\}, \\
M &= \{(x, y, t) \in \mathbb{C}^2 \times B \,|\, y^4 = x(x - 1)(x - t) \}.
\end{aligned}
$$

Let $\pi : M \to B$ be a holomorphic mapping with $\pi(x, y, t) = t$. Then (M, π, B) is a locally non-trivial holomorphic family of Riemann surfaces of type $(3, 1)$.

§2. Representation of Families of Riemann Surfaces into Teichmüller Spaces.

1. We begin by explaining the terms of the theory of Teichmüller spaces. (See Bers [5] for more details and for references.)

Let S be a Riemann surface of type (g,n) with $2g - 2 + n > 0$. Let (S, f, S') be a marked Riemann surface, that is, S' is a Riemann surface of type (g,n) and f is a quasiconformal mapping of S onto S'. Two marked Riemann surfaces (S, f, S') and (S, g, S'') are equivalent if there exists a conformal mapping h of S' onto S'' such that $g^{-1} \circ h \circ f$ is homotopic to the identity of S. Denote by $[S, f, S']$ the equivalence class of a marked Riemann surface (S, f, S'). The Teichmüller space $T(S)$ of S is the set of all equivalence classes $[S, f, S']$.

Next, we introduce the Teichmüller space $T(G)$ of Fuchsian group G. Let U be the upper half plane and G be a finitely generated Fuchsian group of the first kind acting on U. Denote by $B_2(L, G)$ the Banach space of bounded holomorphic quadratic differentials on the lower half plane L. Namely, $B_2(L, G)$ is the set of holomorphic functions ϕ on L satisfying

$$(1) \qquad \|\phi\|_L = \sup_{z \in L} |(\operatorname{Im} z)^2 \phi(z)| < \infty,$$

and the functional equations

$$(2) \qquad \phi(g(z))g'(z)^2 = \phi(z), \quad \text{for all } g \in G.$$

For each ϕ in $B_2(L, G)$ there exists a locally schlicht meromorphic function W_ϕ on L such that

$$\{W_\phi, z\} = \phi(z), \quad z \in L,$$

and $W_\phi(-i) = 0$, $W_\phi(-2i) = 1$, $W_\phi(-3i) = \infty$, where $\{W_\phi, z\}$ is the Schwarzian derivative of W_ϕ. We define the Teichmüller space $T(G)$ of G as the set of ϕ in $B_2(L, G)$ such that W_ϕ admits a quasiconformal extension of $\hat{C}$. Then, by Nehari–Kraus' theorem $T(G)$ is bounded in $B_2(L, G)$. W_ϕ is always schlicht not only for ϕ in $T(G)$ but also for ϕ in $\overline{T(G)}$. Furthermore, we verify that for every ϕ in $\overline{T(G)}$,

$$G_\phi = W_\phi G W_\phi^{-1}$$

is a Kleinian group because of the equation (2), and the map

$$G \ni g \mapsto W_\phi \circ g \circ W_\phi^{-1}$$

is an isomorphism of G depending holomorphically on ϕ. The group G_ϕ is *quasi-Fuchsian* if ϕ is in $T(G)$. For $\phi \in \partial T(G)$, the group G_ϕ is called a boundary group. A boundary group G_ϕ is called *totally degenerate* if $W_\phi(L)$ is dense in $\hat{C}$. A parabolic element of G_ϕ is *accidental* if it is of the form $W_\phi \circ g \circ W_\phi^{-1}$ with a hyperbolic element g in G. A boundary point ϕ in $\partial T(G)$ is called a *cusp* if G_ϕ has an accidental parabolic transformation. It is known that if a boundary point ϕ is not a cusp, then G_ϕ is a totally degenerate group.

Let G be a universal cover transformation group of a Riemann surface S of type (g, n) with $2g - 2 + n > 0$. For each $[S, f, S']$ of $T(S)$, $\phi \in B_2(L, G)$ is determined as follows. We can take a quasiconformal self-mapping w of U satisfying the commutative diagram:

$$
\begin{array}{ccc}
U & \xrightarrow{\ w\ } & U \\
{\scriptstyle p}\downarrow & & \downarrow{\scriptstyle p'} \\
S & \xrightarrow{\ f\ } & S'
\end{array}
$$

where p and p' are the natural projections of U onto S and S', respectively. Set

$$
\mu(z) = \begin{cases} \mu_w(z), & z \in U, \\ 0, & z \in \hat{C} - U, \end{cases}
$$

where μ_w is the complex dilatation of w. There exists a quasiconformal mapping $\hat{w}$ of $\hat{C}$ whose complex dilatation is μ. And we set $\phi(z) = \{\hat{w}, z\}$ on L. It is known that ϕ belongs to $T(G)$ and is uniquely determined by $[S, f, S']$. Conversely, every point ϕ in $T(G)$ determines a unique point $[S, f, S']$ of $T(S)$. Since $B_2(L, G) \simeq C^{3g-3+n}$, $T(S)$ is identified with a bounded domain $T(G)$ of C^{3g-3+n}.

The Teichmüller space $T(S)$ (and $T(G)$) is a complete metric space under the *Teichmüller distance* δ defined as

$$
\delta([S, f, S'], [S, g, S'']) = \inf 2^{-1} \log K(h),
$$

where $K(h)$ is the maximal dilatation of h and h runs over all quasiconformal mappings of S' onto S'' homotopic to $g \circ f^{-1}$. It is known that the Teichmüller distance δ is the Kobayashi distance of $T(G)$ with respect to the natural complex structure of $B_2(L, G)$.

For a quasiconformal automorphism $g : S \to S$, define an automorphism $[g]_* : T(S) \to T(S)$ by

$$
[g]_*([S, f, S']) = [S, f \circ g^{-1}, S'].
$$

The automorphism $[g]_*$ is determined by the homotopy class of g. The set of all $[g]_*$ is denoted by $Mod(S)$ and is called the modular group of $T(S)$. From the identification of $T(S)$ and $T(G)$, $Mod(S)$ can be regarded as the modular group $Mod(G)$ of $T(G)$. Each element of $Mod(G)$ is an holomorphic automorphism of $T(G)$ and is an isometry with respect to the Teichmüller distance δ.

2. Let (M, π, B) be a holomorphic family of Riemann surfaces of type (g, n) with $2g - 2 + n > 0$. Let $\rho : \tilde{B} \to B$ be the universal covering of B with the cover transformation group Γ. Take a Teichmüller space $T(G)$ of G such that the quotient space $S = U/G$ is a Riemann surface of type (g, n). Then, there exists a holomorphic mapping $\Phi : \tilde{B} \to T(G)$ such that the quotient space $W_{\Phi(z)}(U)/G_{\Phi(z)}$ is conformally equivalent to $\pi^{-1}(\rho(z))$ for every z in $\tilde{B}$. This mapping $\Phi : \tilde{B} \to T(G)$ is called a representation of (M, π, B) into $T(G)$. A representation Φ induces a homomorphism $\chi_\Phi : \Gamma \to Mod(G)$ such that

$$\Phi \circ \gamma = \chi_\Phi(\gamma) \circ \Phi, \quad \text{for every } \gamma \in \Gamma.$$

This homomorphism is called the monodromy with respect to Φ. We can replace a representation Φ by $\Phi_1 = \chi \circ \Phi$ with $\chi \in Mod(G)$. Then the monodromy χ_{Φ_1} with respect to Φ_1 is given by $\chi_{\Phi_1} = \chi \circ \chi_\Phi(\gamma) \circ \chi^{-1}$ for every $\gamma \in \Gamma$.

A holomorphic family (M, π, B) is locally non-trivial if and only if a representation Φ of (M, π, B) into $T(G)$ is non-constant. If (M, π, B) is locally non-trivial, then the universal covering surface $\tilde{B}$ of B is conformally equivalent to the unit disk $(|z| < 1)$, for $T(G)$ is a bounded domain.

§3. Rigidity Theorem.

We shall prove the following rigidity theorem.

RIGIDITY THEOREM. *Let (M_1, π_1, B) and (M_2, π_2, B) be locally nontrivial holomorphic families of Riemann surfaces of type (g, n) over a Riemann surface B of finite type. If representations Φ_1 and Φ_2 of (M_1, π_1, B) and (M_2, π_2, B) into $T(G)$ induce the same monodromy $\chi_{\Phi_1} = \chi_{\Phi_2}$, then $\Phi_1 = \Phi_2$. Hence, (M_1, π_1, B) is isomorphic to (M_2, π_2, B).*

PROOF: Since (M_1, π_1, B) is locally non-trivial, the universal covering surface $\tilde{B}$ of B is the unit disk $D = (|z| < 1)$. Since B is of finite type, its cover transformation group Γ is of divergence type. Then it is known that almost all boundary points $\varsigma \in \partial D$ are angular limit points of Γ, that is, there exists a sequence $\{\gamma_j^\varsigma\}_{j=1}^\infty$ of Γ such that for each z_0 in D, $\{\gamma_j^\varsigma(z_0)\}$ converges to ς non-tangentially as $j \to \infty$ (cf. Tsuji [21], Theorem XI 20). On the other hand, since Φ_1 and Φ_2 are bounded holomorphic mappings on D, Fatou's theorem implies that for almost all $\varsigma \in \partial D$, Φ_1 and Φ_2 converge to $b_1(\varsigma)$ and $b_2(\varsigma)$, respectively, where $b_1(\varsigma)$, $b_2(\varsigma)$ are points in $\overline{T(G)}$. Thus, we can take a subset E of ∂D with mes $E = 2\pi$ such that for all ς in E and for all z_0 in D, there exists $\{\gamma_j^\varsigma\}_{j=1}^\infty$ of Γ satisfying

$$\lim_{j \to \infty} \Phi_k(\gamma_j^\varsigma(z_0)) = b_k(\varsigma), \quad k = 1, 2.$$

Let $\chi = \chi_{\Phi_1} = \chi_{\Phi_2}$. Since $T(G)$ is a bounded domain, we may assume that $\{\chi(\gamma_j^\varsigma)\}_{j=1}^\infty$ converges to a holomorphic mapping $A : T(G) \to \overline{T(G)}$ uniformly on every compact subset of $T(G)$ as $j \to \infty$. By the relations

$$\Phi_k \circ \gamma_j^\varsigma = \chi(\gamma_j^\varsigma) \circ \Phi_k, \quad k = 1, 2,$$

we have

$$b_1(\varsigma) = A \circ \Phi_1(z_0) \quad \text{and} \quad b_2(\varsigma) = A \circ \Phi_2(z_0),$$

for every z_0 in D and for every ς in E. Note that $\chi(\gamma_j^\varsigma)$ in $Mod(G)$ is an isometry with respect to the Teichmüller distance on $T(G)$. If $A \circ \Phi_1(z_0)$ is in $T(G)$, A is a non-constant mapping on D. But this is absurd because $b_1(\varsigma)$ does not depend on z_0. Therefore, both $b_1(\varsigma)$ and $b_2(\varsigma)$ belong to $\partial T(G)$. Since Φ_1 and Φ_2 have non-cusped radial limits almost everywhere in ∂D (Shiga [19], Theorem 5), we may assume that $b_1(\varsigma)$ and $b_2(\varsigma)$ are corresponding to totally degenerate groups for each ς in E. On the other hand, we have

$$\delta\big(\Phi_1 \circ \gamma_j^\varsigma(z_0), \Phi_2 \circ \gamma_j^\varsigma(z_0)\big)$$
$$= \delta\big(\chi(\gamma_j^\varsigma) \circ \Phi_1(z_0), \chi(\gamma_j^\varsigma) \circ \Phi_2(z_0)\big)$$
$$= \delta\big(\Phi_1(z_0), \Phi_2(z_0)\big),$$

because $\chi(\gamma_j^\varsigma)$ is an isometry with respect to the Teichmüller distance δ. Hence, by Bers [6] Lemma 2 we conclude that $b_1(\varsigma) = \lim_{j \to \infty} \Phi_1(\gamma_j^\varsigma(z_0)) = \lim_{j \to \infty} \Phi_2(\gamma_j^\varsigma(z_0)) = b_2(\varsigma)$ for every ς in E. Therefore, Φ_1 and Φ_2 have the same non-tangential boundary value for almost all points on ∂D, which implies that $\Phi_1 = \Phi_2$ on D. This completes the proof of the Rigidity Theorem.

REMARK: By the same proof, we see that the Rigidity Theorem holds in the case when B belongs to class O_G, i.e., B is a Riemann surface admitting no Green's functions.

§4. Proof of the Finiteness Theorem for Families.

In this section we will prove the Finiteness Theorem for holomorphic families of Riemann surfaces.

Assume that there exist infinitely many non-isomorphic and locally non-trivial holomorphic families of Riemann surfaces of fixed type (g, n) over a fixed Riemann surface B. Take an infinite sequence $\{(M_m, \pi_m, B)\}_{m=0}^{\infty}$ of such families. Let $\Phi_m : D \to T(S)$ be a representation of (M_m, π_m, B) into a fixed Teichmüller space $T(S)$.

Let $\{[P_1], [P_2], \ldots, [P_{k_0}]\}$ be the set of all congruence classes of pants decompositions of S (see Abikoff [2] for its definition). For each point $[S, f, S']$ of $T(S)$, $\{[f(P_1)], [f(P_2)], \ldots, [f(P_{k_0})]\}$ is also the set of all congruence classes of pants decompositions of S'. Furthermore, we may assume that all P_j and $f(P_j)$ $(j = 1, 2, \ldots, k_0)$ are *geodesic pants decompositions*, namely, pants decompositions whose border curves are all geodesics.

Set $\Phi_m(0) = [S, f_m, S_m]$ $(m = 1, 2, \ldots)$. We know that taking a subsequence if necessary, there exist $M > 0$ depending only on (g, n) and $P = P_j$ for some j such that $\ell(f_m(\alpha)) < M$ for every border curve α of P, where $\ell(\alpha)$ is the Poincaré length of α (Abikoff [2], Chap. II §3 Lemma 3). Hence, taking a subsequence if necessary, we may assume that $\lim_{m \to \infty} \ell(f_m(\alpha))$ exists and is finite for every border curve α of P.

Let $C = \{\alpha_1, \ldots, \alpha_k\}$ be the set of all border curves of P such that

$$(3) \qquad \lim_{m \to \infty} \ell(f_m(\alpha)) = 0.$$

Then we have the following two cases:

CASE 1. *C is empty.*

CASE 2. *C is non-empty.*

If Case 1 occurs, then $\{[S, f_m, S_m]\}_{m=1}^{\infty}$ (or their images under some modular transformations) is bounded in $T(S)$. This is verified by considering the Fenchel–Nielsen coordinates (cf. Abikoff [2] Chap. III) of $T(S)$ with respect to P. (Applying Dehn twists about border curves of P on $[S, f_m, S_m]$, we see that the "twist-coordinates" are bounded.) Hence we may assume that $\{\Phi_m(0)\}_{m=0}^{\infty}$ converges to a point w_0 in $T(S)$.

Let $\{\gamma_1, \ldots, \gamma_{N_0}\}$ be the system of generators of the universal covering transformation group Γ of the universal covering $\rho : D \to B$. Let d_D be the Poincaré distance on the unit disk D. Since the Teichmüller distance δ is the Kobayashi distance and a modular transformation is its isometry, the distance decreasing property of holomorphic mappings (cf. Kobayashi [**10**]) implies that

$$\delta(\chi_m(\gamma_j)(w_0), w_0) \leqq \delta(\chi_m(\gamma_j)(w_0), \chi_m(\gamma_j)(\Phi_m(0)))$$
$$+ \delta(\chi_m(\gamma_j)(\Phi_m(0)), \Phi_m(0)) + \delta(\Phi_m(0), w_0)$$
$$= 2\delta(\Phi_m(0), w_0) + \delta(\Phi_m(\gamma_j(0)), \Phi_m(0))$$
$$\leqq 2\delta(\Phi_m(0), w_0) + d_D(\gamma_j(0), 0),$$

where we set $\chi_m = \chi_{\Phi_m}$. Since $\Phi_m(0) \to w_0$ as $m \to \infty$ and the Teichmüller distance δ is complete, taking a subsequence if necessary, we may assume that $\{\chi_m(\gamma_j)(w_0)\}_{m=0}^{\infty}$ converges to a point of $T(S)$ for every $j = 1, \ldots, N_0$. From the discreteness of the modular group $Mod(S)$, there exists a positive integer m_0 such that $\chi_m(\gamma_j) = \chi_{m_0}(\gamma_j)$ for infinitely many m and $j = 1, \ldots, N_0$, which implies that $\chi_m = \chi_{m_0}$ for such m. Thus, from the Rigidity Theorem of §3 we conclude that $\Phi_m = \Phi_{m_0}$ for infinitely many m. In particular, (M_m, π_m, B) are isomorphic to (M_{m_0}, π_{m_0}, B) for such m, and we have a contradiction.

Next, suppose that Case 2 occurs. Then, we note the following lemmas (cf. Abikoff [**2**], Theorem 4, p. 52 and Lemma 1, p. 95).

LEMMA A. *Assume that S_1 and S_2 are Riemann surfaces of non-excluded type, and $f : S_1 \to S_2$ is a K-quasiconformal mapping. If α is a closed geodesic on S_1, then the geodesic β homotopic to $f(\alpha)$ satisfies*

$$\ell(\beta) \leqq K\ell(\alpha).$$

LEMMA B. *Let S be a Riemann surface of non-excluded type. Then, there exists a continuous function L defined on $\{x \in \mathbb{R} : x > 0\}$ such that $L(x) > 0$, $\lim_{x \to 0} L(x) = +\infty$ and for any geodesic loops α and β with $\alpha \cap \beta \neq \emptyset$*

$$\ell(\beta) \geqq L(\ell(\alpha)).$$

First, we shall prove that there exist constant $C > 0$ and $m_0 \in \mathbb{N}$ satisfying:

For every simple closed geodesic curve $\beta \notin C = \{\alpha_1, \ldots, \alpha_k\}$

$$\ell([f_m(\beta)]) \geqq C \quad \text{for all } m \geqq m_0,$$

where $\ell([f_m(\beta)])$ is the Poincaré length of the geodesic curve homotopic to $f_m(\beta)$.

If β is a border curve of $\mathcal{P}$, then it is obvious because of the assumption that $\displaystyle\lim_{m\to\infty}\ell(f_m(\alpha))$ exists and is finite for every border curve α of $\mathcal{P}$.

If $\beta \cap \alpha_j \neq \emptyset$ for some j, then we have

$$\ell([f_m(\beta)]) \geqq L(\ell(f_m(\alpha_j))) \to +\infty, \quad \text{as } m \to +\infty$$

from Lemma B.

If $\beta \cap \alpha_j = \emptyset$ for all j and β is not a border curve of $\mathcal{P}$, then β intersects a border curve $\gamma(\neq \alpha_j, j = 1, \ldots, k)$ of a pants P of $\mathcal{P}$. We may assume that $\beta \cap P$ separates other components γ' and γ'' of ∂P or it connects γ and other border curve of P. Thus, $f_m(\beta) \cap f_m(P)$ separates $f_m(\gamma')$ and $f_m(\gamma'')$, or it connects $f_m(\gamma)$ and other border curve of $f_m(P)$. Since $\ell(f_m(\gamma)), \ell(f_m(\gamma')), \ell(f_m(\gamma'')) < M$ and the triple $(\ell(f_m(\gamma)), \ell(f_m(\gamma')), \ell(f_m(\gamma'')))$ determines the conformal structure of $f_m(P)$, we verify that $\ell([f_m(\beta \cap P)]) \geqq C'$ for some constant C' (cf. Abikoff [2] Chap. II §3 Theorem and its proof). Hence, we obtained the desired result.

Take $\varepsilon(> 0)$ so small that

$$(4) \qquad\qquad \varepsilon \exp(2K_0) < C,$$

where $K_0 = \max\{d_D(0, \gamma_j(0)); j = 1, \ldots, N_0\}$. Take an integer $m(\geqq m_0)$ so large that

$$(5) \qquad\qquad \ell(f_m(\alpha_j)) < \varepsilon, \quad \text{for } j = 1, \ldots, k.$$

Since $\delta(\Phi_m(0), \chi_m(\gamma_j)(\Phi_m(0))) = \delta(\Phi_m(0), \Phi_m(\gamma_j(0))) \leqq d_D(0, \gamma_j(0)) \leqq K_0$, we can choose quasiconformal mappings $f'_m : S \to S_m$ and $g_j : S_m \to S_m$ such that

$$\Phi_m(0) = [S, f'_m, S_m],$$

$$(6) \qquad\qquad \chi_m(\gamma_j)(\Phi_m(0)) = [S, g_j f'_m, S_m] \quad \text{and}$$

$$(7) \qquad\qquad K(g_j) \leqq \exp(2K_0) \quad \text{for } j = 1, \ldots, N_0.$$

For each α in $\mathcal{C}$, from Lemma A, (4), (5) and (7) we have

$$\ell([g_j(f'_m(\alpha))]) \leqq \exp(2K_0)\ell([f'_m(\alpha)]) \leqq \varepsilon \exp(2K_0)$$
$$< C, \quad \text{for } j = 1, \ldots, N_0.$$

This implies that a simple closed curve $g_j(f'_m(\alpha))$ is not homotopic to any geodesic β on S_m which does not belong to $\{f_m(\alpha_1), \ldots, f_m(\alpha_k)\}$. Hence, the free homotopy classes of $\{f_m(\alpha_i)\}_{i=1}^k$ are preserved by $g_j : S_m \to S_m$ $(j = 1, \ldots, N_0)$. Since $\{\gamma_1, \ldots, \gamma_{N_0}\}$ is a system of generators of Γ, for each $\gamma \in \Gamma$ they are also preserved by a quasiconformal mapping $g_\gamma : S_m \to S_m$ which is corresponding to $\chi(\gamma)$ as (6). Therefore, we have

$$(8) \qquad \ell([g_\gamma(f_m(\alpha_i))]) \leqq \max\{\ell(f_m(\alpha_i)) | i = 1, \ldots, k\}.$$

As in the proof of the Rigidity Theorem in §3, we can take a sequence $\{\gamma_j\}_{j=1}^\infty$ of Γ such that $\{\chi_m(\gamma_j)(\Phi_m(0))\}_{j=1}^\infty$ converges to a point of $\partial T(G)$ which represents a totally degenerate group with no accidental parabolic transformations. Then, from Abikoff [1] Theorem 2, we have

$$\ell([g_{\gamma_j}(f_m(\alpha_i))]) \to +\infty, \quad \text{as } j \to +\infty.$$

This contradicts the inequality (8), and we complete the proof of the Finiteness Theorem for families.

§5. Proof of the Finiteness Theorem for Sections.

In this section we will give a proof of the Finiteness Theorem for holomorphic sections of a fixed holomorphic family (M, π, B) of Riemann surfaces of type (g, n) over B.

Suppose that there exist infinitely many holomorphic sections $\{s_m\}_{m=0}^\infty$ of (M, π, B). Considering an assignment

$$p(\in B) \mapsto \pi^{-1}(p) - \{s_m(p)\},$$

we see that each holomorphic section s_m induces a holomorphic family (M_m, π_m, B) of Riemann surfaces of type $(g, n + 1)$ over B. Furthermore, (M_m, π_m, B) is locally non-trivial. Indeed, when (M, π, B) is locally non-trivial, it is obvious. When (M, π, B) is locally trivial, it is also locally non-trivial because s_m is non-constant. Hence, we see from the Finiteness Theorem for families that there are only finitely many non-isomorphic families in $\{(M_m, \pi_m, B)\}_{m=0}^\infty$. We may assume that all (M_m, π_m, B) $(m = 0, 1, 2, \ldots)$ are isomorphic to each other, namely, there exist biholomorphic mappings $\Psi_m : M_0 \to M_m$ with $\pi_0 = \pi_m \circ \Psi_m$ for all m. Since $\Psi_m|\pi_m^{-1}(p)$ $(p \in B)$ is a conformal mapping of $\pi^{-1}(p) - \{s_m(p)\}$ onto $\pi^{-1}(p) - \{s_0(p)\}$, $\{s_m(p)\}_{m=0}^\infty$ is a finite set and the number of the elements is at most the number of the conformal automorphisms of $\pi^{-1}(p)$. Thus, there exists a positive integer N depending only on (g, n) such that for any p in B, $\{s_m(p)\}_{m=0}^\infty$ consists of at most N points in M. In fact, we can take $N = 84(g - 1)$ when $g > 1$. Hence, the finiteness of $\{s_m\}_{m=0}^\infty$ is deduced from the following lemma.

218

LEMMA. *Let M_1 and M_2 be complex manifolds. Suppose that holomorphic mappings f_m $(m = 0, 1, 2, \ldots)$ of M_1 to M_2 satisfy*

$$\sup_{p \in M_1} {}^\sharp \{ f_m(p); m = 1, 2, \ldots \} = N < +\infty,$$

where $^\sharp E$ is the cardinal number of a set E. Then, $\{ f_m \}_{m=0}^\infty$ consists of N holomorphic mappings.

PROOF OF LEMMA: Take a point p_0 in M_1 with

$$^\sharp \{ f_m(p_0); m = 0, 1, 2, \ldots \} = N$$

and set $\{ f_m(p_0) \}_{m=0}^\infty = \{ a_1, \ldots, a_N \}$. We can take $f_{m_1}, \ldots, f_{m_N}$ so that $f_{m_j}(p_0) = a_j$ $(j = 1, \ldots, N)$. If $\{ f_m \}_{m=0}^\infty$ consists of at least $(N+1)$ distinct mappings, then there exists a in $\{ a_1, \ldots, a_N \}$, say a_1, such that equations

$$f_m(p_0) = a_1$$

holds for at least two distinct f_m. Therefore, there exists f_{n_1} in $\{ f_m \}_{m=1}^\infty$ such that $f_{n_1}(p_0) = a_1$ and $f_{n_1} \neq f_{m_1}$.

Take a neighborhood U_0 of p_0 so small that $f_{m_j}(U_0) \cap f_{m_i}(U_0) = \emptyset$ $(i \neq j)$ and $f_{n_1}(U_0) \cap f_{m_j}(U_0) = \emptyset$ $(j \neq 1)$.

Since $f_{m_1} \neq f_{n_1}$, there exists p in U_0 such that

$$f_{m_1}(p) \neq f_{n_1}(p).$$

Therefore, $^\sharp \{ f_m(p) \} \geqq N + 1$ and, we have a contraduction.

Y. Imayoshi, Department of Mathematics, Osaka University

H. Shiga, Department of Mathematics, Kyoto University

References

1. Abikoff, W., *Two theorems on totally degenerate Kleinian groups*, Amer. J. Math. **98** (1976), 109–118.
2. Abikoff, W., "The Real Analytic Theory of Teichmüller Space," Lecture Notes in Math. **820**, Springer–Verlag, 1980.
3. Arakelov, S. Ju, *Families of curves with fixed degenerancies*, Math. USSR Izvestija **5** (1971), 1277–1302.
4. Bers, L., *An extremal problem for quasiconformal mappings and a theorem by Thurston*, Acta Math. **141** (1978), 73–98.
5. Bers, L., *Finite dimensional Teichmüller spaces and generalizations*, Bull. Amer. Math. Soc. **5** (1981), 131–172.
6. Bers, L., *On iterations of hyperbolic transformations of Teichmüller spaces*, Amer. J. of Math. **105** (1983), 1–11.
7. Faltings, G., *Arakelov's theorem for Abelian varieties*, Invent. Math. **73** (1983), 337–347.
8. Faltings, G., *Endlichkeitssäte für abelshe Varietäten über Zahlkörpern*, Invent. Math. **73** (1983), 349–366.
9. Grauert, H., *Mordells Vermutung über Punkte auf algebraischen Kurven und Funktionenkörper*, Publ. Math. I.H.E.S. **25** (1965), 131–149.
10. Kobayashi, S., "Hyperbolic Manifolds and Holomorphic Mappings," Marcel Dekker, Inc., New York, 1970.
11. Mazur, B., *Arithmetic on curves*, Bull. (new series) of Amer. Math. Soc. **14** (1986), 207–259.
12. Miwa, M., *On Mordell's conjecture for algebraic curves over function fields*, J. Math. Soc. Japan **18** (1966), 182–188.
13. Mumford, D., "Curves and Their Jacobians," The University of Michigan Press, Ann Arbor, 1975.
14. Noguchi, J., *A higher dimensional analogue of Mordell's conjecture over function fields*, Math. Ann. **258** (1981), 207–212.
15. Noguchi, J., *Hyperbolic fibre spaces and Mordell's conjecture over function fields*, Publ. RIMS, Kyoto Univ. **21** (1985), 27–46.
16. Parshin, A.N., *Algebraic curves over function fields I*, Math. USSR Izvestija **2** (1968), 1145–1170.
17. Samuel, S., *La conjecture de Mordell pour les corp fe fonctions*, Séminaire Bourbaki 17e année, 1964/65, n 287.
18. Shafarevich, I.R., *Algebraic number fields*, Amer. Math. Soc. Transl (2) **31** (1963), 25–39.
19. Shiga, H., *On analytic and geometric properties of Teichmüller spaces*, J. Math. Kyoto Univ. **24** (1984), 441–452.
20. Szpiro, L., *Sur le théorème de rigidité de Parsin et Arakelov*, Astérisque **64** (1979), 169–202.
21. Tsuji, M., "Potential Theory in Modern Function Theory," Chelsea Publishing Company, New York, 1975.

Non-variational global coordinates for Teichmüller spaces

BY IRWIN KRA

Let Γ be a terminal, torsion free, (regular) b-group of type (p, n), $2p - 2 + n > 0$. Maskit [M3] has observed that the deformation space $T(\Gamma)$ is a model for the Teichmüller space $T(p, n)$ of Riemann surfaces of finite analytic type (p, n) (because Γ represents a surface of type (p, n) on its invariant component and, in general, $2p - 2 + n$ thrice punctured spheres—the latter carry no moduli). He showed that the group Γ can be constructed from $3p - 3 + n$ terminal b-groups of type $(1, 1)$ or $(0, 4)$ (hence with a one dimensional deformation space). Each one dimensional Teichmüller space can be identified with U, the upper half plane—the Teichmüller space of the torus. Thus Maskit constructs a holomorphic injective map

$$m : T(p, n) \to U^{3p-3+n}$$

that yields coordinates (the *Maskit coordinates*) on Teichmüller space. The Maskit coordinates are *non-variational*; they *do not* depend on the choice of a base point and depend only on finitely many combinatorial choices.

In this paper we investigate concrete realizations of the Maskit coordinates. We show that we can obtain global coordinates for $T(\Gamma)$ as

(1) cross ratios of $3p - 3 + n$ quadruples of fixed points of (deformations of) elements of Γ, or as

(2) traces of $3p - 3 + n$ loxodromic elements of Γ, or as

(3) moduli of $3p - 3 + n$ marked tori canonically associated to the marked surface represented by Γ.

The cross ratio coordinates are a special case of a more general phenomenon; studied previously by Kra–Maskit [KM1]. The trace and moduli of tori coordinates are special to the particular representation of the Teichmüller space $T(p, n)$ by the deformation space $T(\Gamma)$. In contrast, S. Wolpert has observed (oral communication) that no set of $6p - 6$ traces can be used to parametrize conjugacy classes of Fuchsian groups of type $(p, 0)$, $p \geq 2$; see also [Kr7]. Similarly, no set of $3p - 3$ traces can be used to parametrize (even locally) conjugacy classes of Schottky groups of genus $p \geq 3$.

This work was supported in part by NSF grant DMS 8401280.

To each set of coordinates, we can associate bases for the space of cusp forms for Γ—usually Poincaré or relative Poincaré series of rational functions.

We proceed to describe our results. All terms are defined subsequently in this paper, and many of the results are reformulated in §6 in more general form.

To each maximal partition Σ (see §1.1) of a surface S of type (p, n) there corresponds a unique, up to conjugation in $PSL(2, \mathbb{C})$, terminal, regular, torsion free b-group Γ with the property that the loops in the partition are represented by (accidental) parabolic transformations $A_1, \ldots, A_{3p-3+n}$ (see §1.2). To each A_j we can associate a *dual* loxodromic element $C_j \in \Gamma$ (see §§4.1 and 5.2) and a marked subgroup G_j that represents a punctured torus or a four times punctured sphere. We call these groups the *modular* subgroups (see §2.1) of Γ. To the modular subgroup G_j we can *canonically* associate a cross ratio χ_j of four fixed points of elements of G_j and a modulus of a torus τ_j (even for the four punctured sphere groups). Details in §§4 and 5.

THEOREM 1. *The cross ratios $\chi_1, \ldots, \chi_{3p-3+n}$ are global coordinates for $T(p, n)$.*

To each cross ratio χ_j, we can assign a rational function r_j. Let ϕ_j be the Poincaré series of r_j (§§4.2 and 5.1).

COROLLARY 1. *The automorphic forms $\phi_1, \ldots, \phi_{3p-3+n}$ form a basis for $Q(\Gamma)$, the space of integrable holomorphic 2-forms for Γ.*

THEOREM 2. *The traces of $C_1, \ldots, C_{3p-3+n}$ are global coordinates for $T(p, n)$.*

To each loxodromic element C of Γ, there corresponds a relative Poincaré series ϕ_C (**[Kr5]**); this is an integrable holomorphic 2-form.

COROLLARY 2. *The cusp forms $\phi_{C_1}, \ldots, \phi_{C_{3p-3+n}}$ form a basis for $Q(\Gamma)$.*

Each parabolic element $A \in \Gamma$ determines an automorphic 2-form ψ_A (**[Kr3]**); in general, this form is not integrable over the entire region of discontinuity of Γ.

THEOREM 3. *The automorphic forms $\psi_{A_1}, \ldots, \psi_{A_{3p-3+n}}$ restricted to the invariant component of Γ form a basis for $Q(\Gamma)$; the same automorphic forms restricted to the union of all the non-invariant components of Γ form* *a basis for the holomorphic quadratic differentials on the Riemann surface with nodes represented by these components.*

REMARK: We will not prove the second assertion in Theorem 3 in this paper. It will be established in a forthcoming paper on surfaces with nodes.

THEOREM 4. *The $3p-3+n$ moduli $\tau_1,\ldots,\tau_{3p-3+n}$ are global coordinates for $T(p,n)$.*

REMARK: Maskit [**M3**] treated only the case of closed surfaces. We work with surfaces with punctures and ramification points not only because the generalization introduces some additional interesting problems, but because it is the natural setting. The decomposition of a closed surface into its parts (see §1) naturally leads one to consider compact surfaces with punctures.

§1. Terminal (regular) b-groups.

1.1 Let $\sigma = (p,n;\nu_1,\ldots,\nu_n)$ be a *hyperbolic signature*; that is, $p,n \in \mathbb{Z}^+ \cup \{0\}$, $\nu_j \in \mathbb{Z}^+ \cup \{\infty\}$, $\nu_j \geq 2$, $j = 1,\ldots,n$, and

$$2p - 2 + \sum_{j=1}^{n}(1 - \nu_j^{-1}) > 0.$$

Let S be a Riemann surface of signature σ; that is, S is a compact surface of genus p with n distinguished points $x_1,\ldots,x_n$, where to the point x_j we assign the ramification number ν_j (if $\nu_j = \infty$, x_j is also called a puncture). Let

$$S_0 = S - \{x_1,\ldots,x_n\}.$$

A *maximal partition* (see Figures 1 and 2) of S_0 is a collection Σ of $3p-3+n$ simple disjoint closed loops $a_1,\ldots,a_{3p-3+n}$ on S_0 satisfying:
(i) no a_j bounds a disc or punctured disc on S_0, and
(ii) no distinct pair a_j, a_k, $j \neq k$, bound a cylinder on S_0.

1.2 A *terminal (regular) b-group* is a geometrically finite non-elementary Kleinian group Γ with a simply connected invariant component Δ so that $(\Omega - \Delta)/\Gamma$ is a union of thrice punctured spheres (here, and in the rest of this paper, $\Omega(\Gamma) = \Omega$ is the entire region of discontinuity of Γ, Λ its limit set and $\Delta = \Delta(\Gamma)$ its invariant component). The group Γ *represents* the surface S with signature σ and partition Σ if
(i) $\Delta/\Gamma \cong S - \{x_j; \nu_j = \infty, j = 1,\ldots,n\}$,

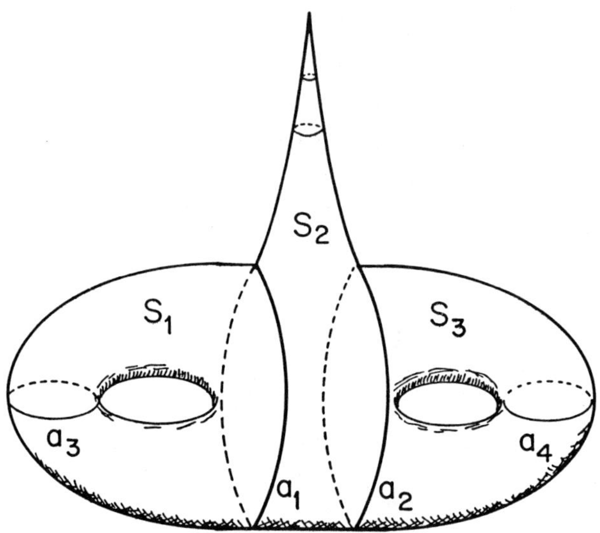

Figure 1: A maximal partition on a surface of type $(2,1)$.

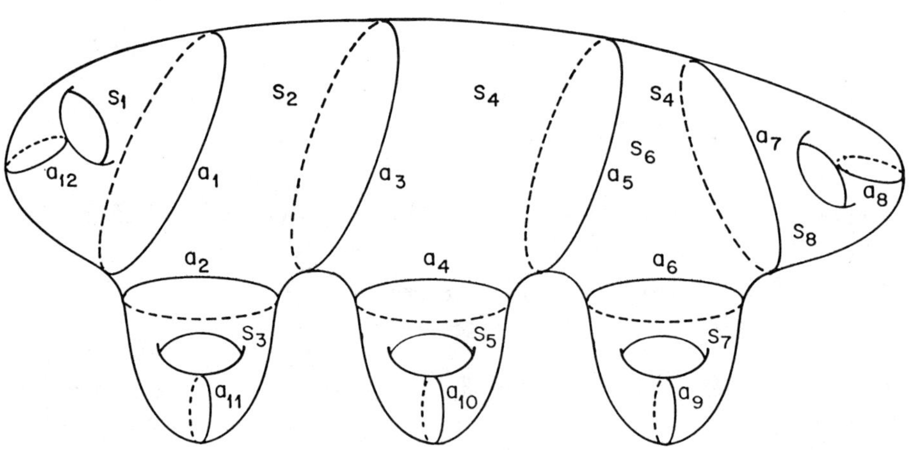

Figure 2: A maximal partition of a surface of type $(5,0)$.
See Maskit [**M3**].

(ii) the natural projection $\pi\colon \Delta \to \Delta/\Gamma$ is ν_j-to-1 over each ramification point x_j of finite order ν_j, $j = 1, \ldots, n$, (and is unramified elsewhere) and
(iii) for each curve a_j on S_0, there is a curve $\tilde{a}_j$ in Δ precisely invariant under an *accidental parabolic* cyclic subgroup $\langle A_j \rangle$ in Γ with $\pi(\tilde{a}_j) = a_j$, $j = 1, \ldots, 3p - 3 + n$.

THEOREM (MASKIT [**M1**], [**M4**]). *Let S be a Riemann surface with hyperbolic signature σ. Let Σ be a maximal partition on S. Then*
(a)   the triple (S, σ, Σ) can be represented by a terminal (regular) b-group Γ,
(b) Γ is unique up to conjugation in $PSL(2, \mathbb{C})$, and
(c) $(\Omega - \Delta)/\Gamma$ is a union of the thrice punctured spheres obtained by squeezing each curve in Σ to a puncture and discarding all surfaces of signature $(0, 3; 2, 2, \infty)$ that appear.

REMARKS: (1) If $(p, n) = (0, 3)$, then Σ is empty and Γ has two invariant components. We shall henceforth assume that $3p - 3 + n > 0$.
(2) Let $a \in \Sigma$. An element (necessarily accidental parabolic) $A \in \Gamma$ *corresponds* to a if it generates the stabilizer of a component of the preimage of a in Δ.
(3) Let x be a *distinguished* point (puncture or ramified point) on Δ/Γ. An element $E \in \Gamma$ (necessarily elliptic or parabolic) *corresponds* to x if it generates the stabilizer of a component of the preimage of x in the closure of Δ. (The preimage of a puncture is defined by continuity.)
(4) Every parabolic or elliptic element of Γ is a power of an element that corresponds to either a curve in Σ or a distinguished point on Δ/Γ.

1.3 We proceed to describe the algebraic structure of the group Γ. We start with

$$S_0 - \Sigma = S_1 \cup S_2 \cup \cdots \cup S_{2p-2+n},$$

the partition of $S_0 - \Sigma$ into its disjoint *parts* $S_1, \ldots, S_{2p-2+n}$. Each part is topologically a thrice punctured sphere (viewed as a surface with signature and boundary curves).

Let $\pi : \Delta \to \Delta/\Gamma$ be the canonical projection. The collection of curves

$$\pi^{-1}(a_1 \cup a_2 \cup \cdots \cup a_{3p-3+n}) = \bigcup_{j=1}^{3p-3+n} \Gamma \tilde{a}_j = \tilde{\Sigma}$$

divides $\Delta_0 = \Delta - \tilde{\Sigma}$ into a disjoint union of components. Each component is stabilized by a triangle group (each boundary curve of a component is stabilized by a cyclic parabolic group or a $\mathbb{Z}_2$-extension of a cyclic parabolic group); there are $2p - 2 + n$ conjugacy classes of such triangle groups; representatives $F_1, \ldots, F_{2p-2+n}$ of these conjugacy classes are the basic building blocks of Γ. The group Γ is reconstructed from these groups (known as *factor subgroups of* Γ) by two types of operations:

(i) amalgamated free products (AFP) across cyclic parabolic groups, and
(ii) HNN extensions obtained by conjugating, via a loxodromic element, one cyclic parabolic group into another.

It is convenient to pick an *assembly algorithm*: a way of reconstructing S_0 from its parts. We start with $S^1 = S_1$. Having constructed S^k, $k = 1, \ldots, 3p - 4 + n$, we construct by induction S^{k+1} as follows. Pick a boundary curve b on S^k. This boundary curve corresponds to a curve $a_j \in \Sigma$. Find the second boundary curve b' corresponding to a_j. It is either on S^k itself or on a part S_ℓ that is disjoint from S^k. In either case we obtain S^{k+1} by gluing b to b'. Then $S_0 = S^{3p-3+n}$. The assembly algorithm for S_0 leads to an assembly algorithm for Γ. We choose a component Δ_1 of Δ_0 so that $\pi(\Delta_1) = S_1$ and we let Γ_1 be the stabilizer of Δ_1. Having constructed connected sets $\Delta_1 \subset \Delta_2 \subset \cdots \subset \Delta_k$ and groups $\Gamma_1 \subset \Gamma_2 \subset \cdots \subset \Gamma_k$, $k = 1, 2, \ldots, 3p - 4 + n$, we let Δ_{k+1} be the connected component of $\pi^{-1}(S^{k+1})$ that contains Δ_k and let Γ_{k+1} be the stabilizer of Δ_{k+1} in Γ. The group Γ_k is a terminal b-group representing a Riemann surface topologically equivalent to S^k, and $\Gamma_{3p-3+n} = \Gamma$.

We proceed to describe how Γ_{k+1} is constructed from Γ_k. The boundary curve b on S^k corresponds to a curve $\tilde{a} \subset \partial \Delta_k \cap \tilde{\Sigma}$ and a parabolic element $A \in \Gamma_k$ stabilizing $\tilde{a}$. If b' is also a boundary curve of S^k, then it corresponds to another parabolic element $A' \in \Gamma_k$. The elements A and A' are not conjugate in Γ_k but are conjugate in Γ. We can find an element $B \in \Gamma$ so that

$$\Gamma_{k+1} = \text{HNN extension of } \Gamma_k \text{ by } B \text{ with } B \circ A' \circ B^{-1} = A.$$

If b' is on the boundary of a part S_ℓ disjoint from S^k, then we let U_ℓ be the component of Δ_0 disjoint from Δ_k, bounded by $\tilde{a}$, and stabilized by F_ℓ (A then belongs to F_ℓ). It follows that $\Delta_{k+1} \supset \Delta_k \cup U_\ell \cup \tilde{a}$, and in this case

$$\Gamma_{k+1} = \Gamma_k *_{\langle A \rangle} F_\ell = \text{AFP of } \Gamma_k \text{ and } F_\ell \text{ across } \langle A \rangle.$$

We also have shown how to construct the groups $F_1, \ldots, F_{2p-2+n}$. We note that the same group Γ can be reconstructed from the triangle groups $F_1, \ldots, F_{2p-2+n}$ by a finite number of distinct assembly algorithms.

§2. Modular decompositions.

2.1 We continue to use the notation of §1. For $j = 1, \ldots, 3p - 3 + n$, let T_j be the surface containing a_j obtained by cutting S_0 along all the curves a_k, $k \neq j$. It follows that T_j is a surface of type $(1,1)$ or $(0,4)$ and that the single curve a_j provides a maximal partition of T_j. We call the collection $T_1, \ldots, T_{3p-3+n}$ the *modular parts* of S_0 and the decomposition of S_0 so obtained the *modular decomposition* of the surface S_0 (see Figure 3).

2.2 If we let (for $j = 1, \ldots, 3p - 3 + n$) D_j be a component of $\pi^{-1}(T_j)$ and G_j its stabilizer, then G_j is a terminal (regular) b-group of type $(1,1)$ or $(0,4)$. We will call G_j a *modular subgroup* of Γ. We note that we can choose G_j to be an AFP of two distinct F_k if G_j has type $(0,4)$ and G_j to be an HNN extension of an F_k if G_j is of type $(1,1)$. Each group G_j contains a single conjugacy class of accidental parabolic elements, and loxodromic elements "dual" to the accidental parabolics—to be described in §§4 and 5.

§3. Maskit coordinates.

3.1 Let Γ be an arbitrary finitely generated non-elementary Kleinian group. An isomorphism $\theta : \Gamma \to PSL(2, \mathbb{C})$ is *geometric* if there exists a quasiconformal self-map w of $\hat{\mathbb{C}}$ such that

$$(3.1.1) \qquad \theta(\gamma) = w \circ \gamma \circ w^{-1}, \quad \text{all } \gamma \in \Gamma.$$

Two isomorphisms $\theta_i : \Gamma \to PSL(2, \mathbb{C})$, $i = 1, 2$, are *equivalent* provided there exists an $A \in PSL(2, \mathbb{C})$ so that

$$\theta_2(\gamma) = A \circ \theta_1(\gamma) \circ A^{-1}, \quad \text{all } \gamma \in \Gamma.$$

The deformation space $T(\Gamma)$ is the set of equivalence classes of geometric isomorphisms (see [**B**], [**M2**], [**Kr1**]).

A quasiconformal map w is Γ-*compatible* if $w \circ \gamma \circ w^{-1} \in PSL(2, \mathbb{C})$ for all $\gamma \in \Gamma$. Fix three distinct sturdy (see [**KM2**]) points x_1, x_2, x_3. (These are limit points or elliptic fixed points. In general, not all elliptic fixed points of order 2 are sturdy. However, all elliptic fixed points considered in this paper, including those of order 2, are sturdy.) The map w is *normalized* if $w(x_i) = x_i$, $i = 1, 2, 3$. The deformation space $T(\Gamma)$ can be described as the traces on Λ (restrictions to Λ) of normalized Γ-compatible quasiconformal automorphisms of $\hat{\mathbb{C}}$. In this setting, the complex structure of $T(\Gamma)$ is completely determined by decreeing that for each $x \in \Lambda$, the map

$$T(\Gamma) \ni w \mapsto w(x) \in \hat{\mathbb{C}}$$

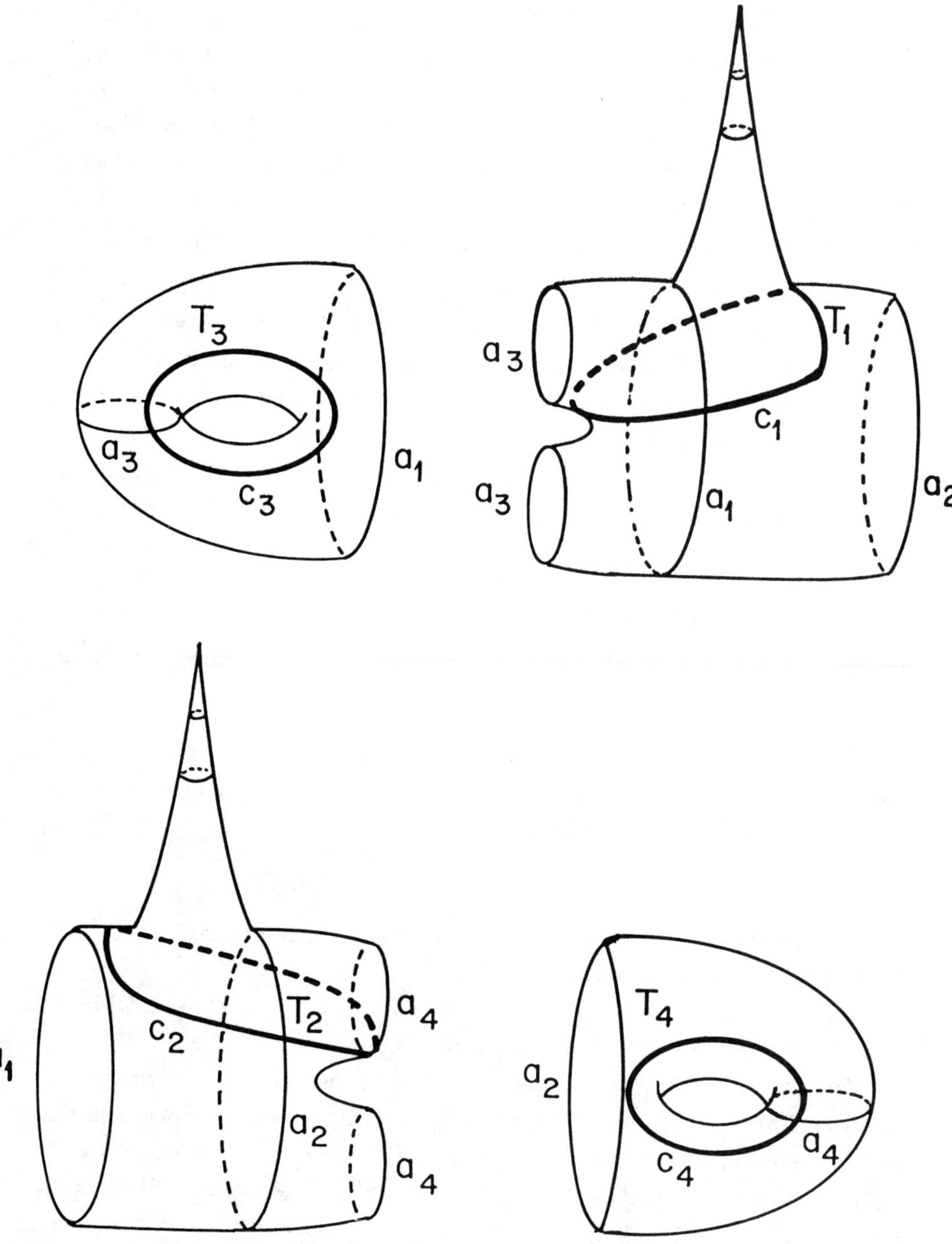

Figure 3: The modular decomposition of the surface of type (2, 1) of Figure 1. Also shown are the dual curves to the partition curves.

is holomorphic. If w is Γ-compatible, its equivalence class in $T(\Gamma)$ is denoted by $[w]$.

3.2 Let Γ now be a terminal (regular) b-group of signature σ corresponding to a maximal partition Σ. Let $G_1, \ldots, G_{3p-3+n}$ be a set of modular subgroups corresponding to the partition Σ. There is an obvious restriction map from $T(\Gamma)$ to $T(G_j)$ and hence a map

$$m : T(\Gamma) \to \prod_{j=1}^{3p-3+n} T(G_j).$$

THEOREM (MASKIT [**M3**]). *The map m is holomorphic and one-to-one. Its image is an open subset of* $\prod_{j=1}^{3p-3+n} T(G_j)$.

PROOF: It is obvious that m is holomorphic. Since the smallest subgroup of Γ containing $\cup_{j=1}^{3p-3+n} G_j$ is all of Γ, the map m is one-to-one. To be specific, if θ_1 and θ_2 are geometric isomorphisms of Γ and if for $j = 1, \ldots, 3p - 3 + n$, there exists an $A_j \in PSL(2, \mathbb{C})$ such that

$$\theta_2(\gamma) = A_j \circ \theta_1(\gamma) \circ A_j^{-1}, \quad \text{all } \gamma \in G_j,$$

then we want to conclude that $A_1 = A_2 = \cdots = A_{3p-3+n}$. Recall each T_j is the union of two distinct S_k along with their common boundary, or an S_k glued along two of its boundary curves. It follows that, after relabeling the indices, we may assume that for each $j = 2, 3, \ldots, 3p - 3 + n$, G_j intersects some G_k with $i \leq k \leq j - 1$ in the triangle group, say F_ℓ. Assume by induction that $A_1 = A_2 = \cdots = A_{j-1}$. Then

$$\theta_2(\gamma) = A_j \circ \theta_1(\gamma) \circ A_j^{-1} = A_k \circ \theta_1(\gamma) \circ A_k^{-1}$$
$$= A_1 \circ \theta_1(\gamma) \circ A_1^{-1}, \text{ all } \gamma \in F_\ell.$$

Hence $A_1^{-1} \circ A_j$ commutes with $\theta_1(F_\ell)$ elementwise. Since θ_1 is type preserving, $\theta_1(F_\ell)$ is a triangle group. If F_ℓ is non-elementary, then clearly $A_1^{-1} \circ A_j = I$. If F_ℓ is elementary, it must be of signature $(0, 3; 2, 2, \infty)$. Thus we may assume that F_ℓ is generated by $z \mapsto z + 1$ and $z \mapsto -z$. It is well known (and is easy to check) that only the identity commutes with this group.

We have shown that m is holomorphic and one-to-one. Since

$$\dim T(\Gamma) = 3p - 3 + n = \dim \left(\prod_{j=1}^{3p-3+n} T(G_j) \right),$$

invariance of domain shows that the image of m is open.

3.3 The cotangent space to $T(\Gamma)$ at the identity isomorphism may be identified with (see [**Kr2**], for example)

$$Q(\Gamma) = \{\phi \text{ holomorphic on } \Omega = \Omega(\Gamma); \ (\phi \circ \gamma)(\gamma')^2 = \phi$$
$$\text{all } \gamma \in \Gamma \text{ and } \iint_{\Omega/\Gamma} |\phi(z)dzd\bar{z}| < \infty\}.$$

The tangent space of $T(\Gamma)$ at the identity is then isomorphic to

$$L^\infty(\Gamma)/Q^\perp(\Gamma),$$

where

$$L^\infty(\Gamma) = \{\mu \text{ measurable on } \Omega; \ (\mu \circ \gamma)\bar{\gamma}'/\gamma' = \mu \text{ all } \gamma \in \Gamma \text{ and } \|\mu\|_\infty < \infty\}$$

$(\|\mu\|_\infty = \text{essential supremum of } \mu)$, and

$$Q^\perp(\Gamma) = \{\mu \in L^\infty(\Gamma); \ \iint_{\Omega/\Gamma} \mu(z)\phi(z)dzd\bar{z} = 0 \text{ all } \phi \in Q(\Gamma)\}.$$

Since every $\phi \in Q(\Gamma)$ vanishes on $\Omega - \Delta$, we may and do assume that we are considering only those $\mu \in L^\infty(\Gamma)$ that are supported on Δ.

The natural projection $T(\Gamma) \rightarrow T(G_j)$ induces the inclusion map $L^\infty(\Gamma) \rightarrow L^\infty(G_j)$ (on the tangent space level). The induced map on cotangent spaces is given by the *relative Poincaré series* operator

$$\Theta_j = \Theta_{G_j\backslash\Gamma} : Q(G_j) \rightarrow Q(\Gamma),$$

where

$$(\Theta_j\phi)(z) = \sum_{\gamma \in G_j\backslash\Gamma} \phi(\gamma z)\gamma'(z)^2, z \in \Delta, \phi \in Q(G_j).$$

(This follows by use of the identity

$$\iint_{\Delta/\Gamma} \mu(z)(\Theta_j\phi)(z)dzd\bar{z} = \iint_{\Delta_j/G_j} \mu(z)\phi(z)dzd\bar{z};$$

here $\Delta_j \supset \Delta$ is the invariant component of G_j, $\mu \in L^\infty(\Gamma)$, $\phi \in Q(G_j)$.)

Since the map m is (locally) biholomorphic, it follows that the induced map between cotangent spaces is an isomorphism. We hence have the following

COROLLARY. *Let $\phi_j \in Q(G_j)$, $\phi_j \neq 0$, $j = 1,\ldots,3p-3+n$. Then $\{\Theta_j(\phi_j); j = 1,\ldots,3p-3+n\}$ forms a basis for $Q(\Gamma)$.*

§4. Terminal (regular) b-groups of type $(1,1)$.

4.1 Let Γ be a terminal (regular) b-group of signature $(1,1;\nu)$. In this case Γ is an HNN extension (see Figure 5) of a triangle group G of signature $(0,3;\infty,\infty,\nu)$. Assume that G is generated by two parabolic elements A, B with $A \circ B$ elliptic of order ν (if $\nu < \infty$) or parabolic (if $\nu = \infty$). Then Γ is generated by G and a loxodromic element C with $C \circ A \circ C^{-1} = B^{-1}$. Note that Γ also has the presentation

$$\Gamma = \langle A, C; \ A \text{ is accidental parabolic and } |A \circ C \circ A^{-1} \circ C^{-1}| = \nu \rangle,$$

where order ∞ means that the element is parabolic. Let Δ be the invariant component of Γ.

Let the element $A \in \Gamma$ correspond to the curve a in the partition for Δ/Γ. An (necessarily loxodromic) element $C \in \Gamma$ with the property that $A \circ C \circ A^{-1} \circ C^{-1}$ corresponds to the distinguished point on Δ/Γ will be called *dual* to A. The closed curves a and c on Δ/Γ corresponding to the elements A and C then form a canonical homology basis on the torus Δ/Γ (we shall say that c is a *dual curve* to a). See Figures 3 and 4.

Let $\phi \in Q(\Gamma)$, $\phi \neq 0$. Then ϕ is non-zero except at the elliptic fixed points (if $\nu < \infty$) where it has a zero of order $2\nu - 2$. Since Δ is simply connected, we can choose a holomorphic function π on Δ such that $(\pi')^2 = \phi$. The period τ of Δ/Γ is now defined by (we call τ the *modulus of the marked group Γ*)

$$\tau = \tau(\Gamma) = \frac{\pi(Cz_0) - \pi(z_0)}{\pi(Az_0) - \pi(z_0)},$$

where $z_0 \in \Delta$ is arbitrary (τ is independent of z_0; see [**Kr6**]).

We note that π' projects to the holomorphic 1-form on the torus obtained by compactifying Δ/Γ (if $\nu < \infty$, Δ/Γ is already compact). It follows that π is an abelian integral of the first kind, and hence $\pi(Az_0) \neq \pi(z_0)$ as a consequence of the bilinear relations of Riemann. Also, τ has positive imaginary part. We note further that the modulus τ depends on the choice of partition curve a and the chocie of the dual curve c to a.

We consider

$$C_\tau = \{z \in \mathbb{C}; z \neq n + m\tau, \text{ all } n, m \in \mathbb{Z}\}.$$

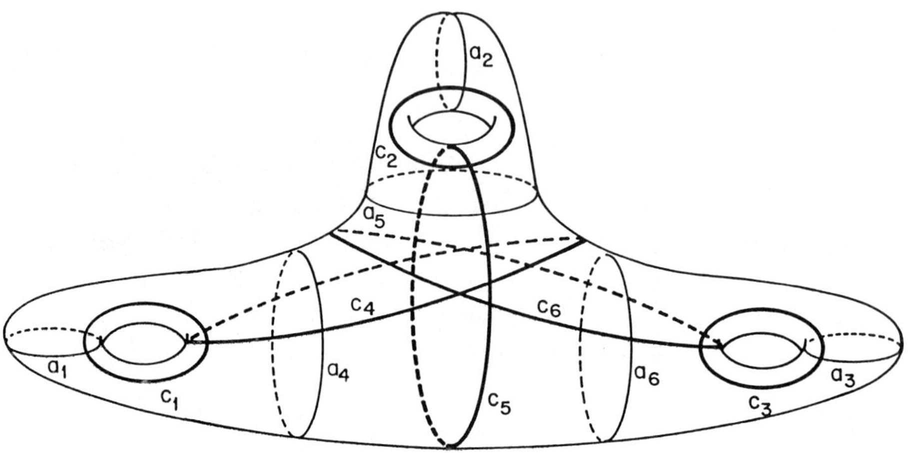

Figure 4: A maximal partition of a surface of type $(3,0)$.
Also shown are dual curves to the partition curves.

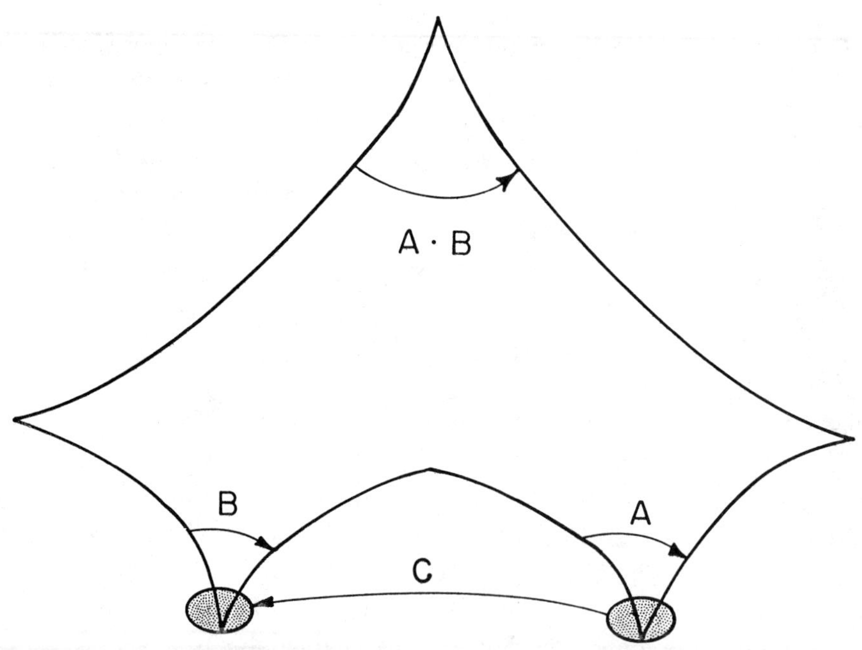

Figure 5: An HNN extension of a triangle group of
signature $(0, 3; \infty, \infty, \nu)$.

Fix $z_0 \in \Delta$, and define

$$f : \Delta \to \mathbb{C}$$

(the constant k is to be specified later) by the formula

$$(4.1.1) \qquad f(z) = \frac{\pi(z) - \pi(z_0)}{\pi(Az_0) - \pi(z_0)} + k, z \in \Delta.$$

We observe that

$$f(Az) = f(z) + 1,$$

$$f(Cz) = f(z) + \tau,$$

$$f(z_1) = f(z_2) \Leftrightarrow \left\{ \begin{array}{l} z_2 = \gamma(z_1) \text{ with } \gamma \in [\Gamma, \Gamma], \text{ the} \\ \text{commutator subgroup of } \Gamma. \end{array} \right.$$

The last equivalence follows from Abel's theorem. We also note that f is locally one-to-one except at the elliptic fixed points (if any) where it is locally ν-to-one. Since

$$\begin{array}{ccc} \Delta & \longrightarrow & f(\Delta) \\ \downarrow & & \downarrow \\ \Delta/\Gamma & \longrightarrow & f(\Delta)/\{\text{lattice generated by 1 and } \tau\} \end{array}$$

commutes, we must have $f(\Delta) = \mathbb{C}$ if $\nu < \infty$. If $\nu = \infty$, then by properly choosing $k \in \mathbb{C}$ we can achieve $f(\Delta) = \mathbb{C}_\tau$. We have shown that f is a (branched, if $\nu < \infty$) holomorphic universal cover of $\mathbb{C}_\tau$ ($\mathbb{C}$, if $\nu < \infty$).

The conjugacy class of Γ in $PSL(2, \mathbb{C})$ is determined by τ; as is shown by the following

THEOREM. *The map*

$$T(\Gamma) \ni t \mapsto \tau(\Gamma(t)) \in U = \{z \in \mathbb{C}; \, \mathrm{Im}\, z > 0\}$$

is a holomorphic bijection. Specifically, let S be a torus. Let $x_0 \in S$ and let a be a simple closed curve in $S_0 = S - \{x_0\}$. Assume that a is not contractible to a point nor the puncture in S_0. Choose a simple closed curve c on S_0 so that a and c form a canonical homology basis on S. Let τ be the period of S with respect to this canonical homology basis. There exists a unique (up to conjugation in $PSL(2, \mathbb{C})$) terminal (regular) b-group $\tilde{\Gamma}$ of signature $(1, 1; \nu)$ that represents (S_0, Σ) (where $\Sigma = \{a\}$) with $\tau(\tilde{\Gamma}) = \tau$.

PROOF: This theorem is an immediate consequence of Maskit's theorem [**M1**] as formulated in §1.2, since τ determines the complex structure of Δ/Γ as well as its marking. We outline the proof.

Let $\tau_0 = \tau(\Gamma) \in U$ and let $\tau \in U$ be arbitrary. Choose $a \in \mathbb{C}$ so that

$$a\tau_0 + (1 - a)\bar{\tau}_0 = \tau.$$

It follows that

$$(4.1.2) \qquad W(z) = az + (1 - a)\bar{z}, z \in \mathbb{C},$$

satisfies

$$W(z + 1) = W(z) + 1, W(z + \tau_0) = W(z) + \tau, \text{ all } z \in \mathbb{C};$$

and thus

$$W(\mathbb{C}_{\tau_0}) = \mathbb{C}_\tau.$$

Lift the constant Beltrami coefficient $\frac{a}{1-a}$ on $\mathbb{C}_{\tau_0}$ to Δ to obtain a Γ-invariant Beltrami coefficient μ

$$\left(\mu(z) \frac{\overline{f'(z)}}{f'(z)} = \frac{a}{1-a}, z \in \Delta \right).$$

Extend μ to be zero off Δ, and let w^μ be an appropriately normalized μ-conformal automorphism of $\hat{\mathbb{C}}$ (see [**AB**]). Let f^μ be the unique holomorphic map so that

$$
\begin{array}{ccc}
\Delta & \xrightarrow{\ w^\mu\ } & w^\mu(\Delta) \\
{\scriptstyle f}\downarrow & & \downarrow{\scriptstyle f^\mu} \\
\mathbb{C}_{\tau_0} & \xrightarrow[\ W\ ]{} & \mathbb{C}_\tau
\end{array}
$$

commutes (in the above diagram we are assuming $\nu = \infty$; if $\nu < \infty$, it should be appropriately modified). Then

$$\tau(w^\mu \Gamma (w^\mu)^{-1}) = \tau.$$

4.2 The quadratic differential ϕ needed in the proof of Theorem 4.1 can be constructed quite explicitly using several methods. Let $t \in T(\Gamma)$. Let $\theta(t)$ be an isomorphism

$$\theta(t) : \Gamma \to \Gamma(t)$$

representing the point t. For $\gamma \in \Gamma$, we let $\gamma(t) = (\theta(t))(\gamma)$. We set:

$$
\begin{aligned}
a(t) &= \text{ fixed point of } A(t), \\
b(t) &= \text{ fixed point of } B(t), \\
x(t) &= \text{ fixed point of } (A \circ B)(t) \text{ (in the invariant component} \\
&\quad\ \text{ of } \Gamma(t), \text{ if } |A \circ B| < \infty), \\
y(t) &= (C(t))(b(t)) = \text{ fixed point of } (C \circ B \circ C^{-1})(t).
\end{aligned}
$$

THEOREM (KRA–MASKIT [**KM1**]). *The map*

$$
T(\Gamma) \ni t \mapsto cr(a(t), b(t), x(t), y(t)) \in \mathbb{C} - \{0, 1\}
$$

(cr stands for cross ratio) is holomorphic and injective.

We let

$$
(4.2.1) \qquad \phi(t, z) = \sum_{\gamma \in \Gamma} r(t, (\gamma(t))(z))[(\gamma(t))'(z)]^2, z \in \Omega(\Gamma(t)),
$$

where (assuming $y(t) \neq \infty$)

$$
(4.2.2) \qquad r(t, z) = \frac{(y(t) - a(t))(y(t) - b(t))(y(t) - x(t))}{(z - a(t))(z - b(t))(z - x(t))(z - y(t))}.
$$

COROLLARY ([**KM2**], [**Kr4**]). *For each $t \in T(\Gamma)$, the function $\phi(t, \cdot)$ is not identically zero and hence forms a basis for $Q(\Gamma(t))$.*

We also define

$$
(4.2.3) \qquad \psi_A(t, z) = \sum_{\gamma \in \langle A \rangle \backslash \Gamma} \frac{[(\gamma(t))'(z)]^2}{((\gamma(t))(z) - a(t))^4}, z \in \Omega(\Gamma(t)),
$$

and

$(4.2.4)$

$$
\phi_C(t, z) = \sum_{\gamma \in \langle C \rangle \backslash \Gamma} \frac{(\alpha(t) - \beta(t))^2 [(\gamma(t))'(z)]^2}{((\gamma(t))(z) - \alpha(t))^2 ((\gamma(t))(z) - \beta(t))^2}, z \in \Omega(\Gamma(t)),
$$

where $\alpha(t)$ and $\beta(t)$ are the fixed points of $C(t)$. It was shown in [**Kr3**] that $\psi_A(t, \cdot)|\Delta(t)$ is a non-zero cusp form for all $t \in T(\Gamma)$. The arguments of [**Kr5**] (see also §4.3) can be used to show that $\phi_C(t, \cdot)$ forms a basis for $Q(\Gamma(t))$, all $t \in T(\Gamma)$. These give alternate global bases for the spaces of cusp forms for the group Γ.

4.3 The quadratic differential ϕ_C defined by (4.2.4) represents the variation of the trace of $C(t)$, $t \in T(\Gamma)$. The fact that $\phi_C(t, \cdot) \neq 0$, all $t \in T(\Gamma)$, also follows from the

THEOREM. *The holomorphic function*

$$T(\Gamma) \ni t \mapsto trC(t) \in \mathbb{C} - [-2, 2]$$

is injective.

Theorems 4.3 and 5.2 have similar proofs. In §7, we prove Theorem 5.2. We leave it to the reader to modify the arguments of §7 for groups of type $(0, 4)$ to groups of type $(1, 1)$.

§5. Terminal (regular) b-groups of type $(0, 4)$.

5.1 Let Γ be a terminal regular b-group of signature $(0, 4; \nu_1, \nu_2, \nu_3, \nu_4)$, $\sum_{j=1}^{4} \nu_j^{-1} < 2$. The group Γ is an AFP of two triangle groups (see Figure 6) G_1, G_2 across a common parabolic subgroup $\langle A \rangle$. We assume that

$$G_1 = \langle A, B_1; |A| = \infty, |B_1| = \nu_1, |A \circ B_1| = \nu_2 \rangle,$$
$$G_2 = \langle A, B_2; |A| = \infty, |B_2| = \nu_3, |A \circ B_2| = \nu_4 \rangle.$$

Note that $\Gamma = G_1 *_{\langle A \rangle} G_2$ also represents the two thrice punctured spheres of signatures $(0, 3; \infty, \nu_1, \nu_2)$ and $(0, 3; \infty, \nu_3, \nu_4)$ except when one of the two groups say G_1, is a $\mathbb{Z}_2$-extension of $\langle A \rangle$, in which case $\nu_1 = \nu_2 = 2$ and Γ represents on its non-invariant components a single sphere of signature $(0, 3; \infty, \nu_3, \nu_4)$.

We define for $t \in T(\Gamma)$:

$$a(t) = \text{ fixed point of } A(t),$$
$$b(t) = \text{ fixed point of } B_1(t),$$
$$x(t) = \text{ fixed point of } (A \circ B_1)(t),$$
$$y(t) = \text{ fixed point of } B_2(t).$$

(If any of the above elements are elliptic, then we have to choose which fixed point to use; we can always use the fixed point in $\Delta(t)$.)

Theorem 4.2 and its corollary are valid for our group Γ. Further, the function ψ defined by (4.2.3) yields a basis for the quadratic differentials on $\Gamma(t)$ for all $t \in T(\Gamma)$. The presentation for Γ is

$$\Gamma = \langle A \circ B_2, B_2^{-1}, B_1, (A \circ B_1)^{-1}; |A \circ B_2| = \nu_4, |B_2^{-1}| = \nu_3,$$
$$|B_1| = \nu_1, |(A \circ B_1)^{-1}| = \nu_2, (A \circ B_2) \circ B_2^{-1} \circ B_1 \circ (A \circ B_1)^{-1} = I,$$
$$(A \circ B_2) \circ B_2^{-1} \text{ accidental parabolic} \rangle.$$

REMARK: We have some freedom in selecting the presentations of G_1 and G_2 (and hence of Γ). We shall henceforth assume that $\nu_1 \geq \nu_2$, $\nu_3 \geq \nu_4$ and $\nu_1 \leq \nu_3$. This involves no loss of generality.

5.2 Again, the trace of an appropriately chosen element of Γ can serve as a coordinate for Teichmüller space. The element $C = B_2^{-1} \circ B_1$ of Γ will be called *dual* to $A = (A \circ B_2) \circ B_2^{-1}$.

THEOREM. *The function*

$$T(\Gamma) \ni t \mapsto trC(t) \in \mathbb{C} - [2,2]$$

is holomorphic and locally injective. It is injective if either
(a) both G_1 and G_2 are generated by two parabolic elements, or
(b) either G_1 or G_2 is elementary.

The theorem is proven in §7.

5.3 Assume now that Γ is torsion free. Let $\phi \in Q(\Gamma)$, $\phi \neq 0$. Then ϕ has no zeros on Δ and we can choose a holomorphic square root ψ. It follows that for each $\gamma \in \Gamma$, there is a $c_\gamma \in \{\pm 1\}$ such that

$$\psi(\gamma z)\gamma'(z)c_\gamma = \psi(z), \text{ all } z \in \Delta.$$

The map $\Gamma \ni \gamma \mapsto c_\gamma \in \{\pm 1\}$ is a surjective homomorphism. As a matter of fact, if $\gamma \in \Gamma$ is parabolic and corresponds to a puncture on Δ/Γ, then $c_\gamma = -1$. To see this, we may assume by conjugation that $\gamma(z) = z + 1$, γ is primitive, and Δ contains the half plane

$$U_1 = \{z \in \mathbb{C}; \text{ Im } z > 1\}.$$

It is well known that ϕ has a Fourier series expansion in U_1:

$$\phi(z) = e^{2\pi i z} \sum_{n=0}^{\infty} a_n e^{2\pi i n z}, a_0 \neq 0.$$

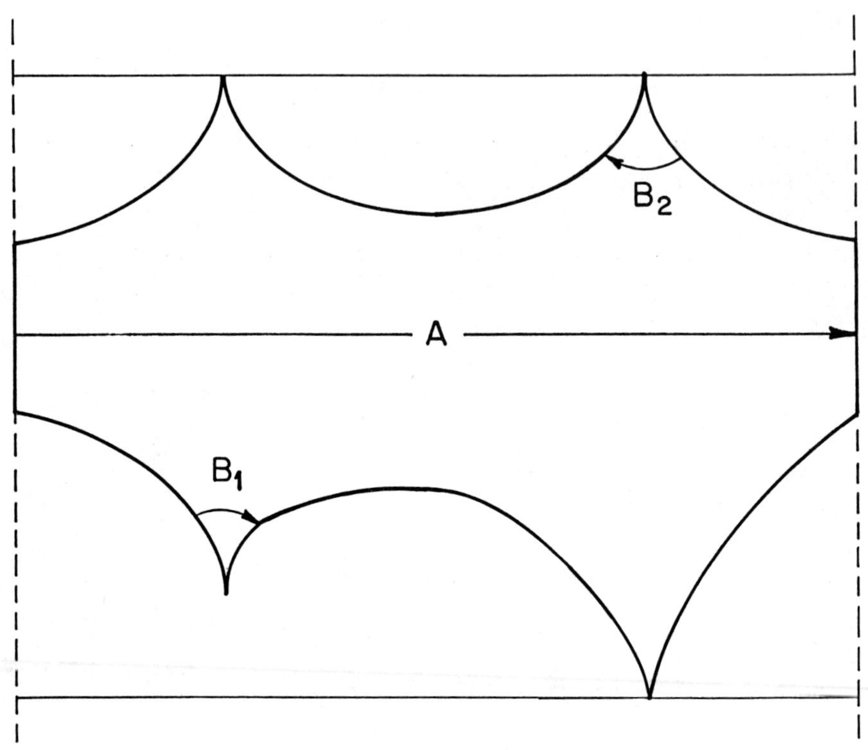

Figure 6: An AFP of two triangle groups.

It follows that

$$\psi(z) = e^{\pi i z} \sum_{n=0}^{\infty} b_n e^{2\pi i n z}, b_0^2 = a_0.$$

Hence $\psi(z+1) = -\psi(z)$, all $z \in \Delta$.

It is convenient at this point to relabel the generators of Γ. We rewrite the presentation (5.1.1) as

$$\Gamma = \langle A_j, j = 1, 2, 3, 4; |A_j| = \infty, j = 1, 2, 3, 4, A_1 \circ A_2 \circ A_3 \circ A_4 = I,$$

$$A = A_1 \circ A_2 \text{ accidental parabolic} \rangle.$$

Let Γ_0 be the kernel of the homomorphism c. Then $\Gamma_0 \lhd \Gamma$, $[\Gamma : \Gamma_0] = 2$. Since $c_{A_j} = -1$, $A_j \notin \Gamma_0$ for $j = 1, 2, 3, 4$; however, $A \in \Gamma_0$. Since Γ_0 is of finite index in Γ, both groups have the same limit set and Δ is also the invariant component of Γ_0.

The Möbius transformation A_1 induces an involution J on $\Delta/\Gamma_0 = X$ to produce $\Delta/\Gamma = Y$. (See Figure 7.) The involution J has no fixed points on X. If $B \in \Gamma_0$ is parabolic, then it is conjugate in Γ to an even power of A_j, $j = 1, 2, 3, 4$, or to a power of A. Let $C \in \Gamma$ and $k \in \mathbb{Z}$. Assume that

$$B = C \circ A_j^{2k} \circ C^{-1}, j = 1, 2, 3, 4.$$

If $C \notin \Gamma_0$, then $\tilde{C} = C \circ A_j \in \Gamma_0$ and

$$B = \tilde{C} \circ A_j^{2k} \circ \tilde{C}^{-1}.$$

We conclude that any element in Γ_0 conjugate in Γ to a parabolic representing a puncture, is already conjugate in Γ_0 to that element.

Now $A = A_1 \circ A_2$ is conjugate in Γ to $A_2 \circ A_1(= A_2 \circ (A_1 \circ A_2) \circ A_2^{-1})$. However, these two elements are not conjugate in Γ_0. (For if $A_2 \circ A_1 = C \circ A_1 \circ A_2 \circ C^{-1}$ for some $C \in \Gamma_0$, then $C^{-1} \circ A_2$ commmutes with $A_1 \circ A_2$. Thus $C^{-1} \circ A_2 = (A_1 \circ A_2)^k$ for some $k \in \mathbb{Z}$, and $C^{-1} = (A_1 \circ A_2)^k \circ A_2^{-1} \notin \Gamma_0$.) Again, let $C \in \Gamma$ and $k \in \mathbb{Z}$. Assume that

$$B = C \circ (A_1 \circ A_2)^k \circ C^{-1}.$$

If $C \notin \Gamma_0$, then $C \circ A_2^{-1} \in \Gamma_0$ and

$$B = C \circ A_2^{-1} \circ A_2 \circ (A_1 \circ A_2)^k \circ A_2^{-1} \circ A_2 \circ C^{-1}$$
$$= (C \circ A_2^{-1}) \circ (A_2 \circ A_1)^k \circ (C \circ A_2^{-1})^{-1}.$$

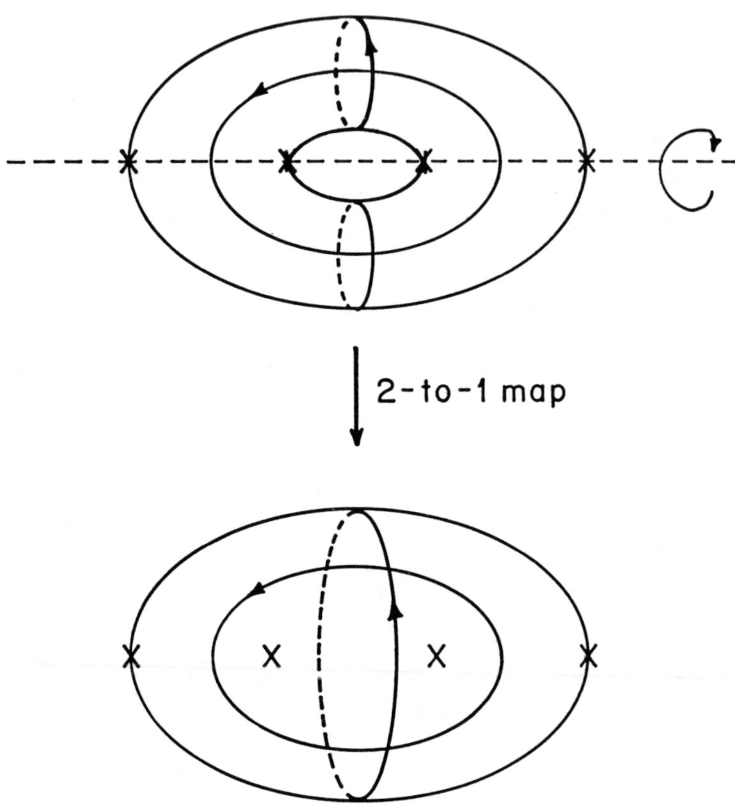

Figure 7: The torus canonically associated to a four–punctured sphere. The "hyperelliptic involution" on the torus is the rotation by π radians through the axis shown.

Thus Γ_0 has two primitive accidental parabolic conjugacy classes. Each puncture on Δ/Γ_0 is fixed by J. If x is the fixed point of A_j, then $A_1(x)$ is the fixed point of $A_1 \circ A_j \circ A_1^{-1}$. It follows that J has 4 fixed points on $\overline{\Delta/\Gamma_0}$, the compactification of Δ/Γ_0. Since ψ projects to an abelian differential on $\overline{\Delta/\Gamma_0}$ without zeros and without poles, $\overline{\Delta/\Gamma_0}$ has genus 1 (this also follows from Riemann–Hurwitz). We conclude that J is the "hyperelliptic involution" on Δ/Γ_0 fixing the four half-periods of the torus.

Now let

$$G_{0j} = G_j \cap \Gamma_0, j = 1, 2.$$

Recall that G_1 is generated by A_3 and A_4 and has 3 conjugacy classes of parabolic cyclic subgroups $(\langle A_3 \rangle, \langle A_4 \rangle, \langle A_3 \circ A_4 = A^{-1} \rangle)$. Similarly, G_2 is generated by A_1 and A_2 (its parabolic cyclic subgroups are conjugates of $\langle A_1 \rangle, \langle A_2 \rangle, \langle A_1 \circ A_2 = A \rangle)$. We have already remarked that Γ_0 has 6 non-conjugate cyclic parabolic groups:

$$\langle A_1^2 \rangle, \ldots, \langle A_4^2 \rangle, \langle A_1 \circ A_2 = A_4^{-1} \circ A_3^{-1} \rangle,$$

$$\langle A_2 \circ A_1 = A_2 \circ (A_4^{-1} \circ A_3^{-1}) \circ A_2^{-1} = A_2 \circ A_3 \circ (A_3^{-1} \circ A_4^{-1}) \circ A_3^{-1} \circ A_2^{-1} \rangle.$$

Clearly G_{01} contains $\langle A_3^2 \rangle, \langle A_4^2 \rangle, \langle A_4^{-1} \circ A_3^{-1} \rangle, \langle A_3^{-1} \circ A_4^{-1} \rangle$; while G_{02} contains $\langle A_1^2 \rangle, \langle A_2^2 \rangle, \langle A_1 \circ A_2 \rangle, \langle A_2 \circ A_1 \rangle$. We conclude that each of these groups represents a four times punctured sphere and the involution J fixes exactly two of the four punctures and interchanges two others. We can write down presentations:

$$G_{02} = \langle A_2 \circ A_1, A_1^{-2}, A_1 \circ A_2, A_2^{-2}; \text{ all four generators parabolic,}$$
$$(A_2 \circ A_1) \circ A_1^{-2} \circ (A_1 \circ A_2) \circ A_2^{-2} = I \rangle$$
$$G_{01} = \langle A_3^{-1} \circ A_4^{-1}, A_4^2, A_4^{-1} \circ A_3^{-1}, A_3^2; \text{ all four generators parabolic,}$$
$$(A_3^{-1} \circ A_4^{-1}) \circ A_4^2 \circ (A_4^{-1} \circ A_3^{-1}) \circ A_3^2 = I \rangle.$$

The group Γ_0 is an HNN extension of an AFP of G_{01} and G_{02}. To be specific, first one forms the AFP of G_{01} with G_{02} across the element $A_1 \circ A_2 = A_4^{-1} \circ A_3^{-1}$. This group represents a sphere with six punctures on its invariant component. This is a free group with standard presentation

$$\langle A_1^2, A_1^{-1} \circ A_2^{-1}, A_2^2, A_3^2, A_3^{-1} \circ A_4^{-1}, A_4^2; \text{ all six generators parabolic,}$$
$$A_1^2 \circ (A_1^{-1} \circ A_2^{-1}) \circ A_2^2 \circ A_3^2 \circ (A_3^{-1} \circ A_4^{-1}) \circ A_4^2 = I,$$
$$A_1 \circ A_2 = A_1^2 \circ (A_1^{-1} \circ A_2^{-1}) \circ A_2^2 \text{ accidental parabolic} \rangle.$$

Next one observes that $A_4 \circ A_1$ conjugates $A_2 \circ A_1$ onto

$$(A_4 \circ A_1) \circ (A_2 \circ A_1) \circ (A_4 \circ A_1)^{-1} = A_3^{-1} \circ A_4^{-1}.$$

We conclude that

$$
\begin{aligned}
\Gamma_0 = \langle &A_4 \circ A_1, A_2 \circ A_1, A_2^2, A_2^{-1} \circ A_1^2 \circ A_2, A_2^{-1} \circ A_1^{-1} \circ A_4^2 \circ A_1 \circ A_2, \\
&A_2^{-1} \circ A_1^{-1} \circ A_4^{-1} \circ A_3^2 \circ A_4 \circ A_1 \circ A_2; \text{ last four generators parabolic,} \\
&(A_4 \circ A_1) \circ (A_2 \circ A_1) \circ (A_4 \circ A_1)^{-1} \circ (A_2 \circ A_1)^{-1} \circ \\
&\circ (A_2^2) \circ (A_2^{-1} \circ A_1^2 \circ A_2) \circ (A_2^{-1} \circ A_1^{-1} \circ A_4^2 \circ A_1 \circ A_2) \circ \\
&\circ (A_2^{-1} \circ A_1^{-1} \circ A_4^{-1} \circ A_3^2 \circ A_4 \circ A_1 \circ A_2) = I, \\
&A_2 \circ A_1 \text{ and } A_1 \circ A_2 \text{ accidental parabolic} \rangle;
\end{aligned}
$$

which yields the standard presentation for our group. It follows that the closed curves on Δ/Γ_0 corresponding to $A_4 \circ A_1$ and $A_2 \circ A_1$ form a canonical homology basis for the (compact) torus $\overline{\Delta/\Gamma_0}$. From which it follows easily that the curves corresponding to the elements $A = A_1 \circ A_2$ and $C = A_2 \circ A_3 = B_2^{-1} \circ B_1$ also form a canonical homology basis for $\overline{\Delta/\Gamma_0}$. Let $\tau(\Gamma) = \tau(\Gamma_0)$ be the period of this marked torus. (We shall call $\tau(\Gamma_0)$ the *modulus of the marked group* Γ.)

THEOREM. *The function*

$$T(\Gamma) \ni t \mapsto \tau(\Gamma_0(t)) \in U$$

is biholomorphic.

PROOF: The proof is similar to the proof of Theorem 4.1. One observes that the function f defined by (4.1.1) and the function W defined by (4.1.2) satisfy

$$
\begin{aligned}
f(A_1 z) &= -f(z), z \in \Delta, \\
W(z + 1/2) &= W(z) + 1/2, z \in \mathbb{C}, \\
W\left(z + \frac{\tau_0}{2}\right) &= W(z) + \frac{\tau}{2}, z \in \mathbb{C}.
\end{aligned}
$$

§6. Coordinates (non-variational) for Teichmüller spaces.

6.1 Before describing our results, we introduce some (standard) terminology. If G and Γ are Kleinian groups, and $G \subset \Gamma$, then for every G-automorphic form ϕ on $\Omega(G)$, we define the *relative Poincaré series* (see §3.3)

$$\Theta_{G\backslash\Gamma}\phi = \sum_{\gamma \in G\backslash\Gamma} (\phi \circ \gamma)(\gamma')^2,$$

whenever the series converges absolutely and uniformly on compact subsets of $\Omega(\Gamma)$. As usual $\Theta_{\{I\}\backslash\Gamma} = \Theta_\Gamma$. If $G_1 \subset G_2 \subset \Gamma$, then

$$\Theta_{G_1\backslash\Gamma} = \Theta_{G_2\backslash\Gamma} \circ \Theta_{G_1\backslash G_2}.$$

If a, b, x, y are four distinct points of $\hat{\mathbb{C}}$, then (4.2.2) defines a rational function r (we are considering only the case $t = 0$) whose Poincaré series ϕ of (4.2.1) is $\Theta_\Gamma r$ (according to the above notation). It follows that

$$\Theta_\Gamma r = \Theta_{G\backslash\Gamma}(\Theta_G r).$$

Similarly, the function ψ_A (respectively, ϕ_C) associated to the parabolic (loxodromic) element $A(C)$ by formula (4.2.3) ((4.2.4)) for $t = 0$ should be denoted $\psi_{A,\Gamma}(\phi_{C,\Gamma})$. It follows that if $A \in G$,

$$\psi_{A,\Gamma} = \Theta_{G\backslash\Gamma}(\psi_{A,G}), \phi_{C,\Gamma} = \Theta_{G\backslash\Gamma}(\phi_{C,G}).$$

6.2 Let S be a Riemann surface of hyperbolic signature σ of type (p, n). Let Σ be a maximal partition of S, $\Sigma = \{a_1, \ldots, a_{3p-3+n}\}$ as in §1.1. Let Γ be a terminal (regular) b-group that represents the triple (S, σ, Σ) as in Theorem 1.2. Let A_j $(j = 1, 2, \ldots, 3p - 3 + n)$ be an accidental parabolic element of Γ corresponding to the curve a_j. Let G_j be the modular subgroup of Γ containing the accidental parabolic element A_j. Let $C_j \in G_j$ be a loxodromic element dual to A_j.

THEOREM. *The map*

$$cr : T(\Gamma) \to \mathbb{C}^{3p-3+n}$$

defined by

$$cr(t) = (cr(G_1(t)), \ldots, cr(G_{3p-3+n}(t))), t \in T(\Gamma),$$

is holomorphic and one to one (hence its image is open).

The cross ratio of a modular subgroup of Γ is defined in Theorems 4.2 and §5.1.

PROOF: This result is a consequence of Theorem 4.2, the version of Theorem 4.2 for terminal (regular) b-groups of type $(0, 4)$, and Theorem 3.2.

Let r_j be the rational function associated to $cr(G_j)$ by formula (4.2.2), and let

$$\phi_j = \Theta_\Gamma r_j.$$

COROLLARY. *The automorphic forms $\phi_1, \ldots, \phi_{3p-3+n}$ form a basis for* $Q(\Gamma)$.

PROOF: Use the Corollary to Theorem 4.2, the Corollary to Theorem 3.2, and the remarks of §6.1, or use Theorem 6.2 and the description of the cotangent space of $T(\Gamma)$ given in §3.3.

6.3 Similar results apply to traces.

THEOREM. *The map*

$$tr : T(\Gamma) \to \mathbb{C}^{3p-3+n}$$

defined by

$$trt = (trC_1(t), \ldots, trC_{3p-3+n}(t)), t \in T(\Gamma),$$

is holomorphic and locally injective. It is (globally) injective provided each factor subgroup is generated by either two parabolics or two elliptics of order 2.

PROOF: Use Theorems 3.2, 4.3 and 5.2.

COROLLARY. *The cusp forms*

$$\phi_{C_1, \Gamma}, \ldots, \phi_{C_{3p-3+n}, \Gamma}$$

form a basis for $Q(\Gamma)$.

6.4 The accidental parabolic elements lead to interesting automorphic forms. Let Ω be the region of discontinuity of Γ and Δ its invariant component.

THEOREM. *The automorphic forms*

$$\psi_{A_1,\Gamma},\ldots,\psi_{A_{3p-3+n},\Gamma}$$

restricted to Δ *form a basis for* $Q(\Gamma)$.

PROOF: This theorem is a consequence of our work in [**Kr3**].

REMARK: See Theorem 3 of the Introduction for properties of $\psi_{A_j,\Gamma}$ restricted to $\Omega - \Delta$.

6.5 THEOREM. *If* Γ *is torsion free, then the period map*

$$\tau : T(\Gamma) \to \mathbb{C}^{3p-3+n}$$

defined by

$$\tau(t) = (\tau(G_1(t)),\ldots,\tau(G_{3p-3+n}))), t \in T(\Gamma),$$

is holomorphic and injective.

PROOF: Use Theorems 3.2, 4.1 and 5.3.

6.6 Theorems 6.3 (since traces are equivalent to multipliers) and 6.5 show that $T(\Gamma)$ is a bounded domain in $\mathbb{C}^{3p-3+n}$. These trace and period coordinates hence yield new compactifications of Teichmüller space. We expect to study these compactifications in the future.

§7. Appendix: Trace calculations.

7.1 In this section we prove Theorem 5.2. The reader will note that we are varying the arguments of [**KM1**]. As with all such calculations, the only problem is to eliminate certain ambiguities. Usually it is easily shown that candidates for parameters determine at most finitely many groups. Some special techniques are needed to make sure that the parameters determine at most one group.

7.2 We shall consider three cases:

   (I) $\nu_1 = \infty = \nu_3$,
   (II) $3 \leq \nu_1 < \infty$,
  (III) $\nu_1 = 2$.

Let us choose three distinct sturdy points in $\hat{\mathbb{C}}$ and normalize all Γ-compatible quasiconformal maps w at these points. If $x \in \Lambda$, then $w \mapsto w(x)$ defines a holomorphic function on $T(\Gamma)$. Let $\gamma \in \Gamma$. Then $\gamma(t) = w \circ \gamma \circ w^{-1} \in PSL(2, \mathbb{C})$ varies holomorphically with $t = [w] =$ equivalence class of w in $T(\Gamma)$. We can choose lifts of $\gamma(t)$ to $SL(2, \mathbb{C})$ whose entries vary holomorphically. Consider

$$cr(z, z_1, z_2, z_3) = \frac{z - z_2}{z - z_3} \div \frac{z_1 - z_2}{z_1 - z_3},$$

the cross ratio (of four distinct points in $\hat{\mathbb{C}}$)). Choose three distinct limit points $x_1, x_2, x_3 \in \Lambda$. Then

$$y_i = \gamma(t)(x_i) = w \circ \gamma \circ w^{-1}(x_i), i = 1, 2, 3,$$
$$cr(\gamma(t)(z), y_1, y_2, y_3) = cr(z, x_1, x_2, x_3), \text{ all } z \in \hat{\mathbb{C}}.$$

It follows that $\gamma(t)$ can be lifted to $GL(2, \mathbb{C})$ so that its entries

$$\begin{pmatrix} a(t) & b(t) \\ c(t) & d(t) \end{pmatrix}$$

vary holomorphically with t. Since $\gamma(t)$ is a Möbius transformation, the holomorphic function

$$T(\Gamma) \ni t \mapsto a(t)d(t) - b(t)c(t) \in \mathbb{C}$$

never vanishes, and hence has a holomorphic square root. It follows that we can lift $\gamma(t)$ to be a holomorphic function from $T(\Gamma)$ to $SL(2, \mathbb{C})$.

We have shown that $t \mapsto trC(t)$ is a holomorphic function on $T(\Gamma)$. It remains to show that this function is (locally) injective. Throughout, $a(t), x_j(t), \ldots$ will denote holomorphic functions on $T(\Gamma)$ with $a, x_j, \ldots$ denoting constant functions on $T(\Gamma)$.

7.3 (Case I) We normalize so that for all $t \in T(\Gamma)$, $A(t)$ fixes ∞, $B_1(t)$ fixes 0, and $B_2(t)$ fixes 1. (We conjugate Γ so that $A(0) = A$, $B_1(0) = B_1$ and $B_2(0) = B_2$ fix ∞, 0, and 1, respectively. We then consider only Γ-compatible w that are normalized at these three points.) It follows that

$$A(t) = \begin{pmatrix} 1 & a(t) \\ 0 & 1 \end{pmatrix}, B_1(t) = \begin{pmatrix} 1 & 0 \\ b(t) & 1 \end{pmatrix}, B_2(t) = \begin{pmatrix} 1 + c(t) & -c(t) \\ c(t) & 1 - c(t) \end{pmatrix}.$$

We claim that $x_3(t) = tr(B_2^{-1} \circ B_1)(t)$ is enough to determine the group $\Gamma(t)$. We have (straightforward calculation):

$$(7.3.1) \qquad 2 + a(t)b(t) = trA \circ B_1 = x_1,$$

$$(7.3.2) \qquad 2 + a(t)c(t) = trA \circ B_2 = x_2,$$

$$(7.3.3) \qquad 2 + b(t)c(t) = tr(B_2^{-1} \circ B_1)(t) = x_3(t).$$

From these equations it follows that

$$a(t)^2 = \frac{(x_1 - 2)(x_2 - 2)}{x_3(t) - 2},$$

hence $(x_1 \neq 2, x_2 \neq 2, x_3(t) \neq 2$ for all t, and$)$ $a^2(t)$ depends only on the values $x_1, x_2, x_3(t)$. We have shown that

$$t \mapsto \sqrt{x_3(t) - 2} = x_4(t)$$

is a one-to-one holomorphic function on $T(\Gamma)$ (since $a(t), b(t), c(t)$ are computable from $x_4(t)$). We claim that x_3 is already one-to-one. A given value of x_3 determines two solutions to the equations (7.3.1)–(7.3.3). If (a, b, c) is one solution, then $(-a, -b, -c)$ is the other. Without loss of generality, we may assume that the first solution corresponds to the group Γ ($t = 0$ in $T(\Gamma)$). If the second solution corresponds to a point $t = [w]$ in $T(\Gamma)$, then $\Gamma(t) = \Gamma$ and the isomorphism $\theta : \Gamma \to \Gamma$ corresponding to the Γ-compatible normalized quasiconformal automorphism w of $\hat{C}$ sends A, B_1, B_2 to $A^{-1}, B_1^{-1}, B_2^{-1}$. The automorphism θ induces an element of the modular group (see [**Kr2**] for definition) of order 2. By Kerckhoff's theorem [**Ke**] (actually we need a special case that was known much earlier) θ has a fixed point. Since w is normalized, θ induces a translation of $T(\Gamma)$. Since a translation has no fixed points, we have arrived at a contradiction.

7.4 (Case II) We normalize so that A fixes 1, B_1 fixes 0, and B_2 fixes ∞. Hence

$$A(t) = \begin{pmatrix} 1 + a(t) & -a(t) \\ a(t) & 1 - a(t) \end{pmatrix}, B_1(t) = \begin{pmatrix} \alpha & 0 \\ b(t) & \alpha^{-1} \end{pmatrix}, B_2(t) = \begin{pmatrix} \beta & c(t) \\ 0 & \beta^{-1} \end{pmatrix}.$$

The constants α, β are known. We need to determine the unknown functions a, b, c. Again we compute

$$(7.4.1) \qquad (\alpha + \alpha^{-1}) + (\alpha - \alpha^{-1})a(t) - a(t)b(t) = trA \circ B_1,$$

$$(7.4.2) \qquad (\beta + \beta^{-1}) + (\beta - \beta^{-1})a(t) + a(t)c(t) = trA \circ B_2,$$

$$(7.4.3) \qquad \alpha\beta^{-1} + \beta\alpha^{-1} - b(t)c(t) = tr(B_2^{-1} \circ B_1)(t).$$

248

We note that $\alpha + \alpha^{-1} = trB_1$, $\beta + \beta^{-1} = trB_2$. Routine calculations show that

$$(7.4.4) \qquad a(t) = \frac{-\chi_2 \pm \sqrt{\chi_2^2 - 4\chi_1\chi_3(t)}}{2\chi_3(t)},$$

where

$$\chi_1 = (trB_1 - trA \circ B_1)(trB_2 - trA \circ B_2),$$
$$\chi_2 = (trB_1 trA \circ B_1)(\beta - \beta^{-1}) + (trB_2 - trA \circ B_2)(\alpha - \alpha^{-1}),$$
$$\chi_3(t) = \alpha\beta + \alpha^{-1}\beta^{-1} - tr(B_2^{-1} \circ B_1)(t).$$

We observe that $\chi_3(t) \neq 0$ all t; because $(B_2^{-1} \circ B_1)(t)$ is loxodromic while $\alpha\beta + \alpha^{-1}\beta^{-1}$ is the trace of an elliptic element of finite order. Since $a(t)$ is well defined on $T(\Gamma)$ we are able to conclude that $t \mapsto \sqrt{\chi_2^2 - 4\chi_1\chi_3(t)}$ is a (well defined) holomorphic function on $T(\Gamma)$, and we may assume that the plus sign holds in (7.4.4). Clearly, $t \mapsto a(t)$ is one-to-one (since $b(t)$ and $c(t)$ are computed in terms of $a(t)$). From

$$\frac{da}{dt} = [(1/2)\chi_2\chi_3(t)^{-2} + (1/2)((1/4)\chi_2\chi_3(t)^{-2} - \chi_1\chi_3(t)^{-1})^{-1/2} \times$$
$$\times ((-1/2)\chi_2\chi_3(t)^{-3} + \chi_1\chi_3(t)^{-2})]\frac{d\chi_3}{dt}$$

it follows that $t \mapsto tr(B_2^{-1} \circ B_1)(t)$ is locally one-to-one.

7.5 (Case III) If $trB_1 = trA \circ B_1$ or $trB_2 = trA \circ B_2$ (for example, if $\nu_1 = \nu_2 = 2$), then (7.4.4) shows that χ_3 is one-to-one on $T(\Gamma)$.

REMARK: I do not know whether χ_3 can ever fail to be injective.

State University of New York, Stony Brook, NY 11794-3651
and Mathematical Sciences Research Institute, Berkeley, CA 94720

REFERENCES

[AB] Ahlfors, L.V. and Bers, L., *Riemann's mapping theorem for variable metrics*, Ann. of Math. **72** (1960), 385–404.

[B] Bers, L., *Spaces of Kleinian groups*, in "Several Complex Variables, Maryland 1970", Lecture Notes in Mathematics **155** (1970); Springer, Berlin, 9–34.

[Ke] Kerckhoff, S.P., *The Nielsen realization problem*, Ann. of Math. **117** (1983), 235–265.

[Kr1] Kra, I., *On spaces of Kleinian groups*, Comment. Math. Helv. **47** (1972), 53–69.

[Kr2] Kra, I., *Canonical mappings between Teichmüller spaces*, Bull. Amer. Math. Soc. **4** (1981), 143–179.

[Kr3] Kra, I., *On cohomology of Kleinian groups IV. The Ahlfors–Sullivan construction of holomorphic Eichler integrals*, J. d'Analyse Math. **43** (1983/84), 51–87.

[Kr4] Kra, I., *On the vanishing of and spanning sets for Poincaré series for cusp forms*, Acta Math. **153** (1984), 47–116.

[Kr5] Kra, I., *Cusp forms associated to loxodromic elements of Kleinian groups*, Duke Math. J. **52** (1985), 587–625.

[Kr6] Kra, I., *On algebraic curves (of low genus) defined by Kleinian groups*, Ann. Polonici Math. **46** (1985), 147–156.

[Kr7] Kra, I., *Uniformization, automorphic forms and accessory parameters*, RIMS (Kyoto Univ.) Kokyuroku **571** (1985), 54–84.

[KM1] Kra, I. and Maskit, B., *The deformation space of a Kleinian group*, Amer. J. of Math. **103** (1981), 1065–1102.

[KM2] Kra, I. and Maskit, B., *Bases for quadratic differentials*, Comment. Math. Helv. **57** (1982), 603–626.

[M1] Maskit, B., *On boundaries of Teichmüller spaces and on Kleinian groups: II*, Ann. of Math. **91** (1970), 607–639.

[M2] Maskit, B., *Self-maps of Kleinian groups*, Amer. J. Math. **93** (1971), 840–856.

[M3] Maskit, B., *Moduli of marked Riemann surfaces*, Bull. Amer. Math. Soc. **80** (1974), 773–777.

[M4] Maskit, B., *On the classification of Kleinian groups: I - Koebe groups*, Acta Math. **135** (1975), 249–270.

Parameters for Fuchsian Groups I: Signature $(0,4)$

BY BERNARD MASKIT

This is the first of a series of notes presenting new parameters for certain torsion-free finitely generated Fuchsian and quasifuchsian groups. In this note we consider signature $(0,4)$. Other low signatures, as well as the general case, will be dealt with elsewhere. Every Fuchsian group of signature $(0,4)$, acting on the upper half-plane U, can be generated by four parabolic transformations, A,B,C,D, where the product $ABCD = 1$. Normalize so that AB has its attracting fixed point at ∞, its repelling fixed point at 0, and so that the fixed point of C is at 1. Let x be the fixed point of D, and let y be the fixed point of B. Then $x > 1$ and $y < 0$. We show that x and y serve as parameters for the deformation space of these groups (this is really two results, one having to do with Fuchsian, and the other with quasifuchsian groups). We also explicitly write the matrices A,B,C,D in $PGL(2,\mathbb{R})^+$ (these are 2×2 real matrices with positive determinant) as functions of x and y; this gives an explicit example of a stratification (see [**K-M**]). We also construct an explicit fundamental domain for the Teichmüller modular group for signature $(0,4)$, and we identify the side pairing transformations.

1. Our computations will all be done in $PGL(2,\mathbb{R})$. There is a canonical isomorphism between this group and the group of all isometries of the hyperbolic plane, or equivalently, the group of all conformal homeomorphisms of the upper half-plane, including orientation reversing ones, obtained as follows. Regard hyperbolic 3-space as being the upper half-space $\mathsf{H}^3 = \{(z,t) \mid z \in \mathbb{C}, t \in \mathbb{R}, t > 0\}$, and consider the hyperbolic plane as being the subset $\mathsf{H}^2 = \{(z,t) \in \mathsf{H}^3 \mid \mathrm{Im}(z) = 0\}$. Regard $A \in PGL(2,\mathbb{R}) \subset PGL(2,\mathbb{C}) = PSL(2,\mathbb{C})$ as an orientation preserving isometry of H^3. Since the coefficients of A are real, A preserves the real line; hence it preserves H^2, the hyperbolic plane supported by the real line. If $\det(A) > 0$, then A preserves both half-spaces bounded by H^2; hence it preserves orientation on H^2. If $\det(A) < 0$, then A reverses these two half-spaces; hence it reverses orientation on H^2.

This work was supported in part by NSF grant DMS-8401280.

We will make considerable use of the following. If $a \neq b$ are real, then they determine a hyperbolic line, C in the upper half-plane. As an element of $PGL(2,\mathbb{R})$, the reflection in C is

$$R = \begin{pmatrix} a + b & -2ab \\ 2 & -a - b \end{pmatrix}$$

We remark that the matrix above, as an element of $PGL(2,\mathbb{C})$, i.e., a Möbius transformation, is an elliptic element of order 2, with fixed points at a and b.

2. Consider a Fuchsian group F representing a sphere with two (parabolic) punctures, and one (hyperbolic) hole. Such a group can be generated by two parabolic elements, A and B, where the product AB is boundary hyperbolic (i.e., the axis L of AB bounds a hyperbolic half-space H, where H is precisely invariant under AB in F). Let M_1' be the hyperbolic line with one endpoint at the fixed point of A, and orthogonal to the axis L of AB; let M_2' be the hyperbolic line with one endpoint at the fixed point of B, also orthogonal to L; and let M' be the hyperbolic line with one endpoint at the fixed point of A, and the other at the fixed point of B (in Figure 1, the fixed point of A is labelled "xy", and the fixed point of B is labelled "y").

We know that we can write $A = R_1 R$, and $B = R R_2$, where R is reflection in M, R_1 is reflection in some line ending at the fixed point of A, R_2 is reflection in some line ending at the fixed point of B, and $AB = R_1 R_2$ is hyperbolic with axis L if and only if the lines of fixed points of both R_1 and R_2 are both orthogonal to L. We conclude that R_m is reflection in M_m.

3. We start with a Fuchsian group F of signature $(0,4)$. Pick parabolic generators A, B, C, D, where $ABCD = 1$, as above, and normalize so that the attracting fixed point of AB is at ∞, and the repelling fixed point is at 0. Then the fixed points of A and B are either both positive or both negative. We further normalize (this is actually a choice of orientation), so that the fixed points of A and B are both negative. In this case, it follows that the fixed point of B greater than that of A, and the fixed points of C and D are both positive, with the fixed point of D greater than that of C. This last choice (of orientation) is equivalent to an orientation requirement on the corresponding homotopy basis; this is, our requirement is that we have chosen a basis A, B, C, D for $\pi_1(\mathsf{U}/F)$, where $ABCD = 1$, so that each of A, B, C, D is a simple loop separating one of the four punctures

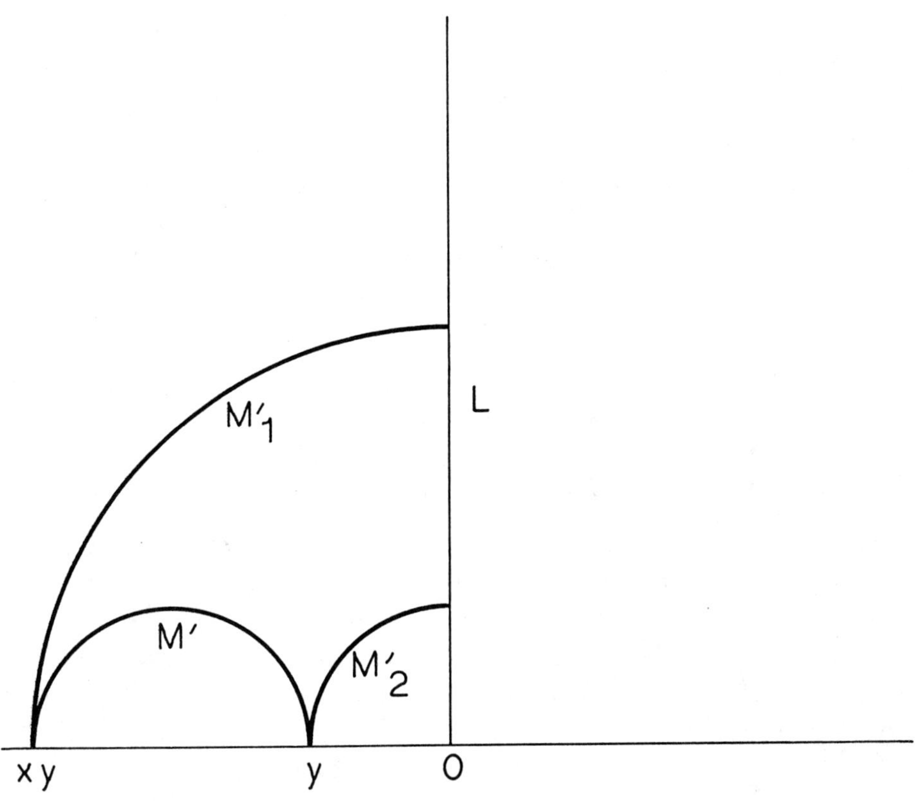

Figure 1

from the other three, and traversing a loop about that puncture in the positive direction.

A set of generators for a Fuchsian group of signature $(0,4)$, as described above, is called a *good* set of generators.

Having made the choice as above, we further normalize so that the fixed point of C is at 1. Let x be the fixed point of D, and let y be the fixed point of B. We then have the inequalities:

$$x > 1, \text{ and } y < 0.$$

4.   THEOREM. *Let $x > 1$, and $y < 0$, be real numbers. Then there is a unique set of good generators A, B, C, D, for a Fuchsian group F of signature $(0,4)$, where $ABCD = 1$, so that D has its fixed point at x, C has its fixed point at 1, and B has its fixed point at y.*

PROOF: We first construct F_1, the group generated by C and D. Draw the (hyperbolic) line M with one endpoint at x and the other at 1. Let M_1 be the hyperbolic line with one endpoint at x and orthogonal to the imaginary axis (i.e., M_1 is the Euclidean circle: $|z| = x$). Similarly, let M_2 be the line with one endpoint at 1, and orthogonal to the imaginary axis (see Figure 2, where the fixed points of C and D are at 1 and x, respectively). Let R, R_m denote reflection in M, M_m, respectively. Set $D = RR_1$, and $C = R_2R$. Then D is parabolic with fixed point x, C is parabolic with fixed point 1, and $CD = R_2R_1$ is hyperbolic with attracting fixed point at 0, and repelling fixed point at ∞.

To prove the uniqueness of F_1, suppose there are other good generators, $\tilde{A}$, $\tilde{B}$, $\tilde{C}$, $\tilde{D}$, with the same normalization, where $\tilde{D}$ also has its fixed point at x. Then $\tilde{D} = R\tilde{R}_1$, and $\tilde{C} = \tilde{R}_2R$, where, as above, R is reflection in the line joining 1 to x, $\tilde{R}_1$ is reflection in some line $\tilde{M}_1$, with one endpoint at x, and $\tilde{R}_2$ is reflection in some line $\tilde{M}_2$ with one endpoint at 1. Since the axis of $\tilde{C}\tilde{D}$ is the common orthogonal of $\tilde{M}_1$ and $\tilde{M}_2$, $\tilde{M}_m = M_m$, and $\tilde{R}_m = R_m$.

We write:

$$R = \begin{pmatrix} 1+x & -2x \\ 2 & -1-x \end{pmatrix}, \quad R_1 = \begin{pmatrix} 0 & x^2 \\ 1 & 0 \end{pmatrix}, \quad R_2 = \begin{pmatrix} 0 & 1 \\ 1 & 0 \end{pmatrix},$$

and compute

$$D = \begin{pmatrix} -2x & x^2(1+x) \\ -1-x & 2x^2 \end{pmatrix}, \quad C = \begin{pmatrix} 2 & -1-x \\ 1+x & -2x \end{pmatrix}.$$

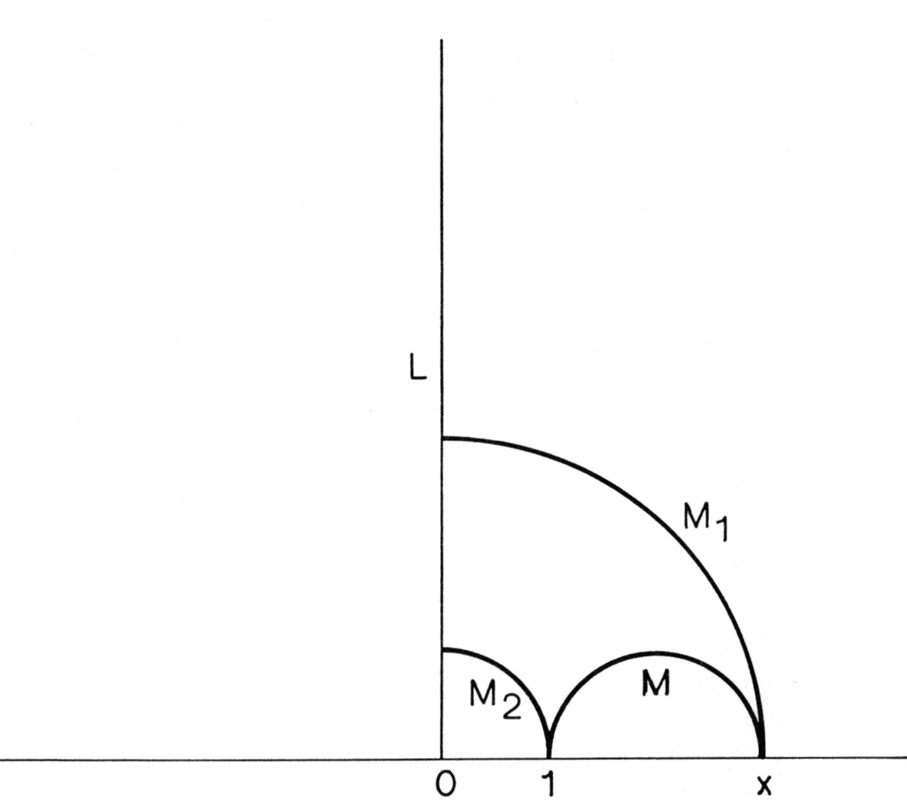

Figure 2

The three lines M, M_1 and M_2 bound a common region, $\hat{D}_1$; by Poincaré's polygon theorem $\hat{D}_1$ is a fundamental polygon for $\hat{F}_1 = \langle R, R_1, R_2 \rangle$, the group generated by R, R_1 and R_2. Using this fact, together with the fact that M_1 and M_2 are both orthogonal to L, it is easy to see that the closed left half-plane is precisely invariant under $\langle CD \rangle$ in F_1, from which it follows that the closed left half-plane is a $(\langle CD \rangle, F_1)$-block (see [**M**, section VII.B]). Also, since $\hat{D}_1$ is a fundamental polygon for $\hat{F}_1$, it is easy to see that there are points in the first quadrant that are not F_1-translates of any point of the second quadrant.

5. We next construct F_2, the group generated by A and B. We know that the fixed point of B is at y; let M_1' be the hyperbolic line with one endpoint at y, and orthogonal to the imaginary axis. We also know that

$$AB = (CD)^{-1} = \begin{pmatrix} x^2 & 0 \\ 0 & 1 \end{pmatrix}.$$

Hence the hyperbolic distance along the positive imaginary axis, from the point of intersection of M_1' to that of M_2', must equal the distance from the point of intersection of M_1 to that of M_2, where M_2' is the line with one endpoint at the fixed point of A, and orthogonal to the imaginary axis (see Figure 1). We conclude that the fixed point of A is at xy. As above, A and B are uniquely determined by the location of their fixed points, and those of the product, AB.

Let M' be the hyperbolic line with endpoints at the fixed points of A and B; let R' be reflection in M', and let R_m' be reflection in M_m'. Then

$$R' = \begin{pmatrix} y(1+x) & -2xy^2 \\ 2 & -y(1+x) \end{pmatrix}, \quad R_1' = \begin{pmatrix} 0 & y^2 \\ 1 & 0 \end{pmatrix}, \quad R_2' = \begin{pmatrix} 0 & x^2y^2 \\ 1 & 0 \end{pmatrix},$$

from which we compute,

$$B = R'R_1' = \begin{pmatrix} -2xy & y^2(1+x) \\ -1-x & 2y \end{pmatrix}, \quad A = R_2'R' = \begin{pmatrix} 2x^2y & -x^2y^2(1+x) \\ 1+x & -2xy \end{pmatrix}.$$

Observe that A and B are parabolic, B has its fixed point at y, A has its fixed point at xy, and $AB = (CD)^{-1}$.

We saw above that since $AB = (CD)^{-1}$, given that the fixed point of B is at y, the fixed point of A is necessarily at xy. As above, we conclude that $F_2 = \langle A, B \rangle$ is uniquely determined by the location of the fixed points of A, B, and AB.

By Poincaré's polygon theorem, $\hat{D}_2$, the region bounded by M', M_1', and M_2' is a fundamental polygon for $\hat{F}_2 = \langle R', R_1', R_2' \rangle$. As above, one now easily sees that the closed right half-plane is precisely invariant under $\langle AB \rangle$ in F_2, and that it is an $(\langle AB \rangle, F_2)$-block.

6. We now have that $AB = (CD)^{-1}$, so $J = \langle AB \rangle = \langle CD \rangle$ is a common subgroup of F_1 and F_2. We also know that B_1, the closed left half-plane, is a (J, F_1)-block; B_2, the closed right half-plane is a (J, F_2)-block, and these two sets form a proper interactive pair. Hence (see [**M**, Section VII.C.2]) $F = \langle F_1, F_2 \rangle$ is discrete, and $\Omega(F)/F$ can be obtained from $\Omega(F_1)/F_1$ and $\Omega(F_2)/F_2$ by deleting the projection of B_1 from the first, and the projection of B_2 from the second, and gluing these two together along the projection of their common boundary.

Since F_1, F_2, and F all keep the upper half-plane U invariant, it is easy to see that U/F can be realized as follows. The surface U/F_m is a sphere with two parabolic punctures, and one hole. The geodesic w about the hole divides this surface into two subsurfaces; one of these subsurfaces is again a sphere with two parabolic punctures and a hole, called the *core*, and the other is an annulus, called the *boundary annulus*.

Let T_1 be the complement of the closed second quadrant, and all its F_1-translates, and let T_2 be the complement of the closed first quadrant and all its F_2-translates. Then T_1/F_1 is the core of U/F_1, and T_2/F_2 is the core of U/F_2. Then U/F is $T_1/F_1 \cup T_2/F_2$, where these two surfaces are glued along their common boundary, w. We have shown that U/F is a surface of signature $(0,4)$, and that A, B, C, D are a set of good generators for F. $\qquad\qquad\square$

7. We turn next to quasifuchsian groups. It is clear that if A, B, C, D are parabolic generators of a quasifuchsian group G, of signature $(0,4)$, where $ABCD = 1$, then we can normalize so that ∞ is the attracting fixed point of AB, 0 is the repelling fixed point of AB, and 1 is the fixed point of C. Then we can read off complex parameters x and y, where x is the fixed point of D, and y is the fixed point of B. The exact meaning of a good set of generators is unclear; we make the following convention. We start with a Fuchsian group F having good generators A, B, C, D and consider the deformation space $T(F)$.

A deformation of a non-elementary Fuchsian or Kleinian group F is an orientation preserving quasiconformal homeomorphism $w : \hat{\mathsf{C}} \to \hat{\mathsf{C}}$, which induces an isomorphism ϕ of F into $PGL(2, \mathsf{C})$; that is, $\phi(f)(z) =$

$w \circ f \circ w^{-1}(z)$. Two deformations w_1 and w_2 are equivalent if they agree on the limit set of F, or, equivalently for non-elementary groups, if they induce the same isomorphism.

The set of equivalence classes of deformations is called the deformation space of F; the basic facts about deformation spaces of Fuchsian and Kleinian groups can be found in Bers [B], and the references cited there. The deformation space here has a natural structure as a complex manifold; it can be identified with $T \times T$, where T is the usual Teichmüller space of Fuchsian groups of signature $(0,4)$. We will identify a point in $T(F)$ either as a representative quasiconformal mapping, or as the induced homomorphism.

We now can state that, by definition, if ϕ is a point in $T(F)$, then $\phi(A)$, $\phi(B)$, $\phi(C)$, $\phi(D)$ is a good set of generators for $\phi(F)$.

We now define the map $\Psi : T(F) \to \mathbb{C}^2$, by $\Psi(\phi) = (x, y)$. We know that Ψ is well defined, and complex analytic.

PROPOSITION. $\Psi : T(F) \to \mathbb{C}^2$ *is a complex analytic embedding.*

PROOF: It suffices to show that Ψ is injective. The proof is similar to that in the Fuchsian case, except that here, instead of writing each generator as a product of reflections, we write it as a product of half-turns; that is, elliptic elements of order 2. Our proof takes place in the context of H^3, so that "line" refers to a hyperbolic line in H^3.

Let M be the line with one endpoint at x and the other at 1. Then there is a line M_1, with one endpoint at x, so that $D = RR_1$, where R is the half-turn about M (that is, M is its fixed point set), and R_1 the half-turn about M_1. Similarly, there is a line M_2, with one endpoint at 1, so that $C = R_2R$, where R is the half-turn about M_2. The product CD is loxodromic if and only if the lines M_1 and M_2 do not meet, even at the sphere at infinity; in this case, the axis of CD is the common orthogonal of M_1 and M_2. Since CD is required to have fixed points at 0 and ∞, M_1 and M_2 are both orthogonal to L, the line joining 0 to ∞. We have shown that x, together with our normalization, uniquely determines C and D.

Every loxodromic transformation has two square roots. Both square roots of CD map M_1 onto M_2; choose $(CD)^{1/2}$ so that it maps x to 1.

The fixed point of B is at y; let z be the fixed point of A. We need to show that z is determined by x and y, together with the normalization; as above, once we know the fixed points of A, B, and AB, we know A and B. Let M_1' be the line orthogonal to L, with one endpoint at z. Then

$AB = R'_2 R'_1$, where R'_m is the half-turn about M'_m. Since $AB = (CD)^{-1}$, $M'_2 = (CD)^{-1/2}(M'_1)$. This gives us two possible choices for z; we write these as $z = \pm xy$. Since x, y, and z are holomorphic functions on $T(F)$, and $z - xy = 0$ for Fuchsian groups, while $z + xy \neq 0$ for Fuchsian groups, we conclude that $z - xy = 0$ throughout $T(F)$. Now that we have located z, we know R'_2, and hence we know both A and B. $\qquad\square$

8. In section 4, we wrote down matrices for A, B, C, and D as functions of x and y, for Fuchsian groups. It is easy to see that these same matrices remain valid for quasifuchsian groups.

9. In order to find a fundamental domain for the modular group, we start with a Riemann surface S of signature $(0,4)$; i.e., S is a sphere with four punctures. Let w be the shortest hyperbolic geodesic on S. Then w is a simple loop separating two of the four punctures from the other two. Next look at the set of all simple geodesics on S, each crossing w at exactly two points; let v be the shortest such geodesic. It is clear that generically w and v are unique; if one or both are not unique on the particular surface S, then we make some choice of shortest geodesics.

The loop w divides S into two subsurfaces; call one of them the inside. Likewise, v divides S into two subsurfaces; call one of them the inside. Let w_1 be the arc of w lying inside v, and let w_2 be the other arc of w. Similarly, let v_1 be the arc of v lying inside w, and let v_2 be the other arc.

We next orient w and v as follows. We orient both loops so that $w_1 \cdot v_1$ is a well defined (simple) loop. This loop divides S into two subsurfaces, one of which is a punctured disc. We orient w and v so that $w_1 \cdot v_1$ describes the boundary of this punctured disc in the positive direction. We set $A = w_1 \cdot v_1$, $B = v_1^{-1} \cdot w_2$, $C = w_2^{-1} \cdot v_2^{-1}$, and $D = v_2 \cdot w_1^{-1}$. Then each of these loops surrounds a distinct puncture on S in the positive direction. It follows that in the Fuchsian group F representing S, the elements corresponding to these generators of the fundamental group, which we call by the same name, are parabolic, and they satisfy the one relation: $ABCD = 1$.

The construction above involves some choices: namely, we could reverse the inside and outside of either w or v. If we reverse the inside and outside of w, while keeping the inside and outside of v fixed, then we replace (A, B, C, D) by $(CDC^{-1}, C, B, B^{-1}AB)$; an easy computation shows that this has no effect on the moduli x and y. Similarly, if we reverse the inside and outside of v, while keeping the inside and outside of w fixed, then we replace (A, B, C, D) by $(B, B^{-1}AB, CDC^{-1}, C)$, which also has no effect on the moduli x and y. (In signature $(0,4)$, the Teichmüller modular group does not act effectively; there is a subgroup of order four, generated by the above two motions, that fixes every point of the Teichmüller space $T_{(0,4)}$). We have shown that our conditions, that w be the shortest geodesic, and that v be the shortest simple geodesic crossing w at exactly two points, defines a fundamental domain for the modular group.

We have chosen A, B, C, D so that $w = w_1 \cdot w_2 = AB$, and, after changing the base point and direction of v, so that $v = v_1^{-1} \cdot v_2^{-1} = BC$. That is, in our moduli, a point (x, y) lies in the fundamental domain for the modular group if $|tr(AB)|$ is minimal among all hyperbolic elements of the Fuchsian group, and $|tr(BC)|$ is minimal among all primitive simple hyperbolic elements whose projected axes cross the projected axis of AB exactly twice.

The statement above defines the fundamental domain in terms of a set of inequalities, some of which are obscure. Our next goal is to clarify this set of inequalities, and also, to reduce it to a finite number.

10. The lemma below is stated in some generality for future use. We use $|\cdot|$ to denote the length of a curve.

LEMMA. *Let S be a hyperbolic planar Riemann surface with at least four boundary components, where S is given with the Poincaré metric. Let w and v be simple closed geodesics on S, where neither w nor v is a boundary geodesic, and w and v intersect at exactly two points. If $|w| \leq |v|$, and if for any simple closed geodesic u crossing w exactly twice, $|u| \geq |v|$, then no simple closed geodesic crossing w is shorter than w.*

PROOF: Since w is a simple non-boundary geodesic, it divides S into two subsurfaces, and each of these is a sphere with at least three boundary components. Let t be a simple closed geodesic, where t crosses w at least four times. Let the total number of points of intersection of t with w be $2n$. These $2n$ points divide both t and w into $2n$ arcs; we label these arcs, in their natural order on both t and w, as $t_1, \ldots, t_{2n}$, and $w_1, \ldots, w_{2n}$. Consider the arc $t_1 \cdot t_2$. This arc starts at a point of w, crosses w and ends at some other point of w. It is easy to see tha there is a unique arc $\hat{w}_1$ of w, between the starting and ending points of $t_1 \cdot t_2$, so that $\hat{t}_1 = t_1 \cdot t_2 \cdot \hat{w}_1$ is a simple loop. In essentially the same manner, we form $\hat{t}_2, \ldots, \hat{t}_{2n}$, where $\hat{t}_{2n}$ is formed from $t_{2n} \cdot t_1$.

The loop w divides S into two discs, each with at least three boundary components. Looking at homotopy relative to the boundary component w, each of the arcs t_m is homotopically non-trivial. It follows that each $\hat{t}_m$ divides S into two discs, where each disc contains at least two boundary components of S; hence, each $\hat{t}_m$ is homotopically non-trivial and hyperbolic. Let $\tilde{t}_m$ be the unique geodesic freely homotopic to $\hat{t}_m$. It is also easy to see that each $\tilde{t}_m$ crosses w at exactly two points.

Since each $\tilde{t}_m$ crosses w at exactly two points, $|\tilde{t}_m| \geq |v|$.

Now consider

$$(1) \qquad \Sigma|\hat{t}_m| = \Sigma|t_m \cdot t_{m+1} \cdot \hat{w}_m| = 2|t| + \Sigma|\hat{w}_m|.$$

Fix m, and let x be one of the endpoints of w_m. Then x is an endpoints of two of the arcs of t; for some k, these are t_k and t_{k+1}. It is clear that $\hat{w}_k$ and w_m are disjoint. Since this is true for both endpoints of w_m, each w_m is disjoint from at least two of the $\hat{w}_k$. We conclude that

$$(2) \qquad \Sigma|\hat{w}_m| \leq (2n - 2)|w|.$$

We combine (1) and (2), together with

$$(3) \qquad |w| \leq |v| \leq |\tilde{t}_m| \leq |\hat{t}_m|$$

to obtain

$$(4) \qquad 2n|w| \leq \Sigma|\hat{t}_m| \leq 2|t| + \Sigma|\hat{w}_m| \leq 2|t| + (2n - 2)|w|.$$

$\square$

11. In signature $(0,4)$ any two non-boundary geodesics necessarily cross, and of course the shortest geodesic is necessarily simple. Hence, the lemma above, when applied to our situation, has the following consequence. A point (x,y) in our space of Fuchsian groups of signature $(0,4)$ lies in the fundamental domain defined above if

$$(5) \qquad |tr(AB)| \leq |tr(BC)|,$$

and if, for every primitive hyperbolic element T, where the corresponding geodesic (i.e., the projection of the axis of T), which we call by the same name T, is a simple loop crossing AB exactly twice,

$$(6) \qquad |tr(T)| \geq |tr(BC)|.$$

We next explore the set of simple geodesics crossing $w = AB$ exactly twice. Each such geodesic has one arc inside w and one arc outside. Except for w, the inside of w has exactly two boundary components; hence for any two such geodesics, the arcs inside w are homotopic modulo the boundary (the boundary of course is w); similarly, the arcs outside w are homotopic, in the outside of w, modulo the boundary.

On the sphere with three boundary components, there are essentially only three simple loops. If we contract w to a point, and make it the base point, then there is only one simple loop inside w, and one simple loop outside w. Hence any loop crossing w exactly twice is of the form $w^m A w^{-m} w^n C w^{-n}$, or $w^m B w^{-m} w^n C w^{-n}$. Since we are only interested in free homotopy, the two forms are $A(AB)^m C(AB)^{-m}$, and $B(AB)^m C(AB)^{-m}$.

We can restate the above as follows. Let α denote the Dehn twist about $w = AB$. Then, up to free homotopy, every simple loop crossing AB exactly twice is of the form $\alpha^m(AC)$, or $\alpha^m(BC)$.

12. It was observed by Kerchkoff [**K**] that the length of a geodesic is a convex function of the twist parameter, in performing Dehn twists about a crossing geodesic. In our case, it is a simple observation that if we set $E_\lambda(z) = \lambda z$, then, for any matrices B and C with fixed points different from 0 and ∞, $|tr(E_\lambda B E_\lambda^{-1} C)|$ is a strictly convex function of λ; so it has a unique minimum. Hence, once we know

$$(7) \qquad |tr(BC)| \le |tr((AB)B(AB)^{-1}C)|,$$

and

$$(8) \qquad |tr(BC)| \le |tr((AB)^{-1}B(AB)C)|,$$

we know that $|tr(BC)| \le |tr(AB)^m(AB)^{-m}C|$, for all m.

We have shown that a fundamental domain D for the modular group is defined by the inequalities (5), (7), (8), and

$$(9) \qquad |tr(BC)| \le |tr((AB)^n A(AB)^{-n}C)|, \quad n \in \mathbb{Z}.$$

13. We now compute, using the matrices A, B, C, D above.

$$(10) \qquad |tr(AB)| = (1 + x^2)/x;$$

$$(11) \qquad |tr(BC)| = [8xy - (1 + y^2)(1 + x)^2]/y(x - 1)^2;$$

$$(12) \quad |tr((AB)B(AB)^{-1}C)| = [8x^3y - (1 + x^4y^2)(1 + x)^2]/x^2y(x - 1)^2;$$

$$(13) \quad |tr((AB)^{-1}B(AB)C)| = [8x^3y - (x^4 + y^2)(1 + x)^2]/x^2y(x - 1)^2;$$

$$(14) \quad \begin{aligned} |tr((AB)^n A(AB)^{-n}C)| = \\ = [8x^{2n+2}y - (1 + x^{4n+2}y^2)(1 + x)^2]/x^{2n+1}y(x - 1)^2; \end{aligned}$$

Inserting (10) and (11) in (5) yields

$$(15) \quad xy^2 + (x^2 - 4x + 1)y + x \geq 0.$$

Solving for y, we obtain:

$$(16) \quad \begin{aligned} y \leq -[(x^2 - 4x + 1) - (x - 1)(x^2 - 6x + 1)^{1/2}]/2x, \text{ or} \\ y \geq -[(x^2 - 4x + 1) + (x - 1)(x^2 - 6x + 1)^{1/2}]/2x. \end{aligned}$$

Inserting (11) and (12) into (7), we obtain:

$$(17) \quad |y| \geq 1/x.$$

Similarly, using (8), (11), and (13), we obtain:

$$(18) \quad |y| \leq x.$$

Inserting (11) and (14) into (9) splits into two cases, according as n is negative or nonnegative. If $n \geq 0$, we obtain the inequality

$$(19) \quad y^2 \geq x^{-(2n+1)}.$$

Since $x > 1$, it suffices to assume $n = 0$, and we obtain

$$(20) \quad |y| \geq 1/\sqrt{x}.$$

Similarly, for $n < 0$, we obtain

$$(21) \quad y^2 \leq x^{-(2n+1)};$$

again, it suffices to consider only $n = -1$, and we obtain

$$(22) \quad |y| \leq \sqrt{x}.$$

We have shown that the region, in the fourth quadrant, defined by the inequalities (16), (20), and (22) is a fundamental domain D for the action of the Teichmüller modular group, for signature $(0, 4)$.

14. One can regard the fundamental domain D described by the above inequalities as being bounded by three sides as follows. The first side s_1 is defined by the equation: $xy^2 + (x^2 - 4x + 1)y + x = 0$, the second side s_2 is defined by $y = -1/\sqrt{x}$, and the third side s_2' is defined by $y = -\sqrt{x}$. The sides s_2 and s_2' meet at the vertex $v_1 = (1, -1)$; this point is not in the domain. The sides s_1 and s_2 meet at the vertex $v_2 = 1/2(7+3\sqrt{5}, -3+\sqrt{5})$, while s_2' and s_1 meet at $v_2' = 1/2(7 + 3\sqrt{5}, -3 - \sqrt{5})$. It might be noted that the first coordinate of v_2 and v_2' satisfies: $x^2 - 7x + 1 = 0$, while the second coordinate of both points satisfies: $y^2 + 3y + 1 = 0$.

Let M^* be the Teichmüller modular group factored by the subgroup that acts as the identity. One expects that in the case of signature $(0, 4)$, M^* is isomorphic to the elliptic modular group. We will make this isomorphism explicit.

The side s_1 is mapped onto itself by the transformation α, defined by $\alpha(A, B, C, D) = (B, C, D, A)$. Since the transformation $(A, B, C, D) \to (C, D, A, B)$ acts trivially, α has order two in M^*.

We can write α in terms of coordinates:

$$\alpha(x, y) = \frac{(1 + x)(y - 1) - A}{(1 + x)(y - 1) + A}, \frac{(y + 1)(x - 1) + A}{(y + 1)(x - 1) - A}$$

where $A^2 = (x + 1)^2(y + 1)^2 - 16xy$.

The transformation α has a fixed point at $(3+2\sqrt{2}, -1)$, and interchanges the two arcs of s_1 on either side of this point.

The side s_2 is mapped onto the side s_2' by the transformation β, defined by $\beta(A, B, C, D) = (ABA^{-1}, A, C, D)$, which has infinite order. In our coordinates, this is given by:

$$\beta(x, y) = (x, xy).$$

An easy computation shows, as one expects, that the composition $\alpha \circ \beta$ has order 3 in M^*.

Since the Teichmüller space $T_{(0,4)}$ is connected, the transformations α and β generate M^*. Also, since $T_{(0,4)}$ is simply connected, M^* has no relations other than those at the vertices: $\alpha^2 = (\alpha\beta)^3 = 1$.

Department of Mathematics, State University of New York at Stony Brook, Stony Brook, NY 11794

References

[**B**] Bers, L., *Uniformization, Moduli, and Kleinian groups*, Bull. London Math. Soc. **4** (1972), 257–300.

[**K**] Kerchkoff, S., *The Nielsen realization problem*, Ann. of Math. (2) **117** (1983), 235–265.

[**K-M**] Kra, I. and Maskit, B., *The deformation space of a Kleinian group*, Amer. J. Math. **103** (1981), 1065–1102.

[**M**] Maskit, B., "Kleinian Groups," Springer–Verlag, to appear.

Parametrization of Teichmüller spaces by geodesic length functions

BY MIKA SEPPÄLÄ AND TUOMAS SORVALI

Introduction.

The Teichmüller space $T(\Sigma)$ of a compact C^∞-surface Σ can be parametrized by geodesic length functions. More precisely, we can find a set $\{\alpha_1, \ldots, \alpha_n\}$ of closed curves α_j on Σ such that the isotopy class of a hyperbolic metric d on Σ (i.e. the point $[d] \in T(\Sigma)$) is determined by the lengths of geodesic curves homotopic to the curves α_j on (Σ, d). However, since the fundamental group of Σ is not freely generated there is a quite complicated relation among these geodesic length function.

If the genus of Σ is p, $p > 1$, and Σ is oriented then $\dim_{\mathbb{R}} T(\Sigma) = 6p - 6$. Consequently we need at least $6p - 6$ geodesic length functions to parametrize $T(\Sigma)$. Scott Wolpert has, however, observed that no set of fixed $6p - 6$ curves can ever parametrize $T(\Sigma)$ even locally. Hence we need more curves. In this work we give a set of $6p - 4$ curves such that the associated geodesic length functions give a global parametrization for $T(\Sigma)$. The relation between the generators of $\pi_1(\Sigma)$ implies an algebraic relation between the so called trace parameters of $T(\Sigma)$ which correspond to these geodesic length functions.

We consider also non-orientable surfaces Σ. The relation between the generators of $\pi_1(\Sigma)$ is more simple than in the case of oriented surfaces. For our considerations this situation is quite good since, in §6, we can give a global parametrization for $T(\Sigma)$ in terms of $3p - 3$ geodesic length functions where $3p - 3 = \dim_{\mathbb{R}} T(\Sigma)$.

These considerations are actually based on elementary computations concerning groups of Möbius-transformations. The details are in [S-S]. The main point which makes these computations successful is the introduction of the classes $\mathcal{H}$, $\mathcal{P}$ and $\mathcal{E}$ of pairs of hyperbolic Möbius-transformations. This is actually a topological classification of pairs of closed curves of Σ as is explained in §4. This classification allows us to simplify formulas concerning multipliers of Möbius-transformations. Hence the computations become manageable.

This work was supported in part by the Academy of Finland.

As a byproduct of these considerations we compute, in §2, two inequalities. The first one, Theorem 2.1, shows that short simple closed geodesics on a Riemann surface have to be rather far apart from each other and that the distance between them grows toward infinity as the lengths of the curves tend to zero. The second inequality relates the lengths of intersecting simple closed geodesics. This is only a fresh version of a well known inequality (cf. [**Abikoff**]). The interesting point here is that these facts concerning the hyperbolic geometry of Riemann surfaces are consequences of elementary computations concerning Möbius-transformations. The discontinuity of the corresponding Fuchsian group plays no role in our considerations.

1.   The classes $\mathcal{H}$, $\mathcal{P}$ and $\mathcal{E}$.

A *hyperbolic* Möbius transformation g is determined by three parameters: the *attracting* fixed point $a(g)$, the *repelling* fixed point $r(g)$ and the *multiplier* $k(g) > 1$. Then, for any $z \in \hat{C}$ not fixed by g,

$$(1.1) \qquad k(g) = (g(z), z, r(g), a(g)),$$

the cross-ratio being defined such that $(t, 1, 0, \infty) = t$.

Define $f(k) = \sqrt{k} + 1/\sqrt{k} = \sqrt{k + 1/k + 2}$ for $k > 0$ and denote $f(g) = f(k(g))$. Then $f(g) > 2$ for all hyperbolic transformations g. For a *parabolic* g, set $k(g) = 1$ and $f(g) = 2$.

Since the fixed points of an *elliptic* transformation g are of the same type, equation (1.1) gives two values $k = e^{i\theta}$ and $k' = e^{-i\theta}$ for the multiplier $k(g)$. However, since $k(g) + 1/k(g) + 2$ is uniquely determined and non-negative, $f(g)$ is well-defined and $0 \le f(g) < 2$ for all elliptic transformations g.

If $z \mapsto (az + b)/(cz + d)$, $ad - bc = 1$, is any transformation conjugate to g, then, regardless of the type of g, $f(g) = |a + d|$.

Let (g, h) be a pair of hyperbolic Möbius transformations mapping the unit disk D onto itself. Suppose that g and h have no common fixed points and denote

$$k_1 = k(g)$$

$$k_2 = k(h)$$

$$t = (r(g), r(h), a(h), a(g)).$$

An elementary calculation shows that

$$(1.2) \qquad f(g \circ h) = |tf(k_1 k_2) + (1 - t)f(k_1/k_2)|.$$

If k_1 and k_2 are given, then to every $f(g \circ h) \neq 0$ there exist two values of t satisfying (1.2). For our purposes it is necessary to limit ourselves to cases where $f(g \circ h)$ determines t uniquely. To this end, we divide the pairs (g, h) into disjoint classes $\mathcal{H}$ (for handle), $\mathcal{P}$ (for pants) and $\mathcal{E}$ (for elliptic) as follows:

$$(g,h) \in \mathcal{H} \Leftrightarrow f(g \circ h) = tf(k_1 k_2) + (1 - t)f(k_1/k_2) \geq 2$$
$$\Leftrightarrow t \geq t_1 \quad = \frac{2 - f(k_1/k_2)}{f(k_1 k_2) - f(k_1/k_2)},$$
$$(g,h) \in \mathcal{P} \Leftrightarrow f(g \circ h) = -tf(k_1 k_2) - (1 - t)f(k_1/k_2) \geq 2$$
$$\Leftrightarrow t \leq t_2 \quad = \frac{-2 - f(k_1/k_2)}{f(k_1 k_2) - f(k_1/k_2)},$$
$$(g,h) \in \mathcal{E} \Leftrightarrow t_2 < t < t_1 \quad .$$

We restate Theorem 1.1 in [S-S] in the following form:

THEOREM 1.1. *For any $k_1 > 1$, $k_2 > 1$ and $k_3 \geq 1$ there exists an up to conjugation unique pair $(g, h) \in \mathcal{P}$ such that*

$$(1.3) \qquad k(g) = k_1, k(h) = k_2, k(g \circ h) = k_3.$$

For a pair $(g, h) \in \mathcal{H}$, $(k_3 - k_1 k_2)(k_3 - k_1/k_2)(k_3 - k_2/k_1) \neq 0$. If $k_1 > 1$, $k_2 > 1$ and $k_3 \geq 1$ satisfy this condition then there exists an up to conjugation unique pair $(g, h) \in \mathcal{H}$ satisfying (1.3).

PROOF: Let k_1, k_2 and k_3 be given and normalize by conjugation in such a way that $r(h) = 1$, $a(h) = 0$ and $a(g) = \infty$. Then $t = r(g)$. In Figure 1, the axes of g and h are drawn in the same $z = t + iu$ plane with the graph of

$$(1.4) \qquad u(t) = |tf(k_1 k_2) + (1 - t)f(k_1/k_2)|.$$

Then $f(g \circ h)$ is the ordinate of the intersection point of (1.4) and the axis of g whereas the abscissa t gives the class of (g, h). Since $t \neq 0, 1, f(g \circ h)$ omits the values $f(k_1 k_2)$ and $f(k_1/k_2) = f(k_2/k_1)$ in $\mathcal{H}$. Since f maps the interval $[1, \infty]$ injectively onto $[2, \infty]$, the assertions follow. $\qquad \square$

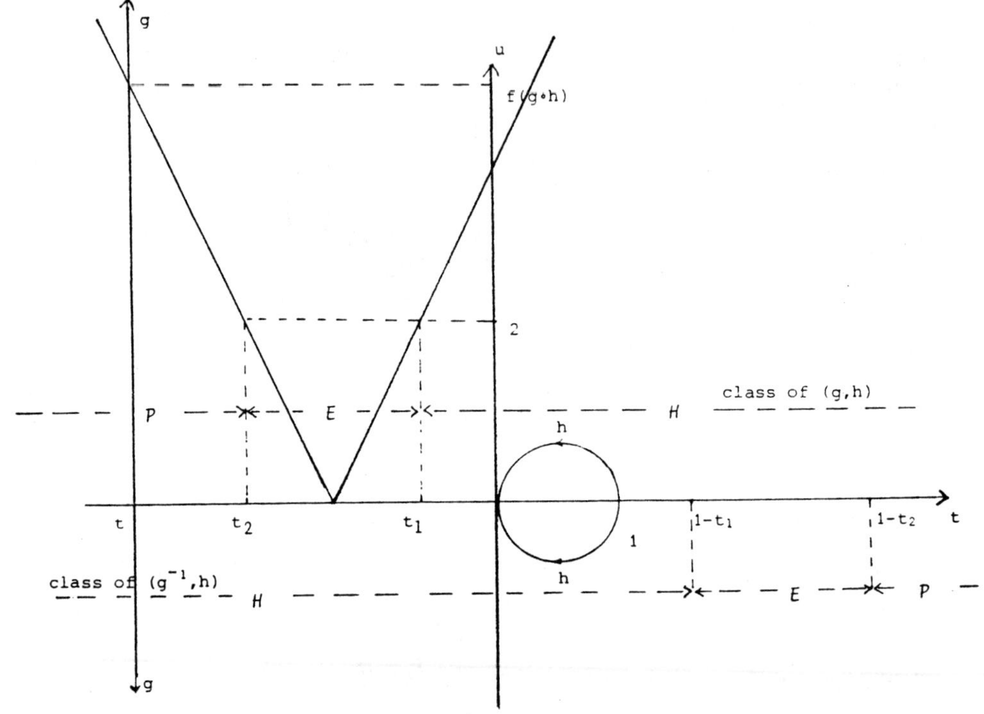

Figure 1

REMARK: The pairs (g,h) and (h,g) are in the same class. Similarly, (g^{-1},h) and (g,h^{-1}) are in the same class. In Figure 1 also the class of (g^{-1},h) is indicated.

2. Short geodesics.

Let (g,h) be a pair of hyperbolic Möbius transformations mapping the unit disk D onto itself. Suppose that g and h have no common fixed points. Then we may normalize by conjugating in such a way that $a(h) = 1, r(h) = -1, a(g) = e^{i\theta}$ and $r(g) = -e^{\pm i\theta}$, $0 < \theta < \pi$. There are three topologically different cases for different values of t (Figure 2):

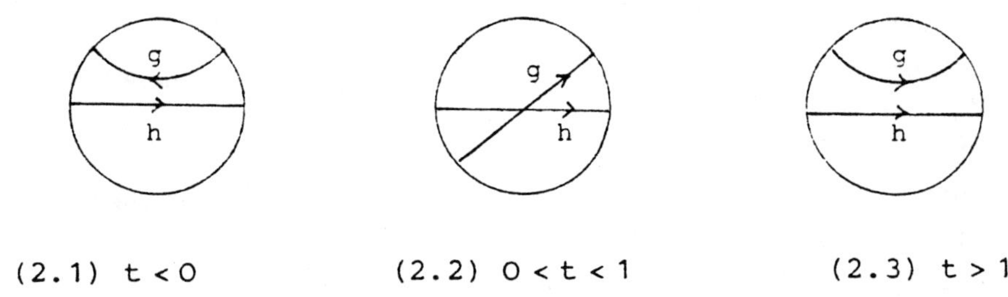

(2.1) $t < 0$ (2.2) $0 < t < 1$ (2.3) $t > 1$

Figure 2

The case (2.1) has five different subcases which are found by drawing the isometric circles of g, h, g^{-1} and h^{-1}.

(2.1.1) If $t < t_2$, the isometric circles $I(h)$ and $I(g^{-1})$ are exterior to each other and $(g, h) \in \mathcal{P}$ (Figure 3).

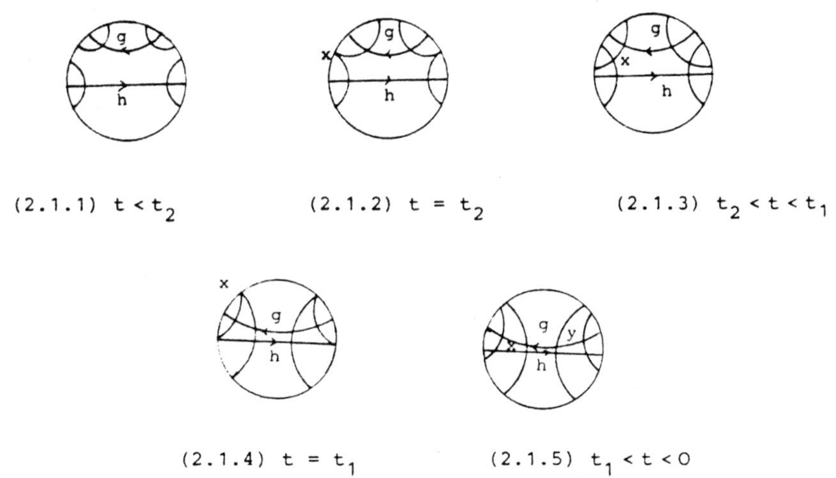

(2.1.1) $t < t_2$ (2.1.2) $t = t_2$ (2.1.3) $t_2 < t < t_1$

(2.1.4) $t = t_1$ (2.1.5) $t_1 < t < 0$

Figure 3

(2.1.2) If $t = t_2$, $I(h)$ is externally tangent to $I(g^{-1})$, the tangential point x is fixed by the parabolic transformation $g \circ h$, and $(g, h) \in \mathcal{P}$.

(2.1.3) If $t_2 < t < t_1$, $I(h)$ and $I(g^{-1})$ intersect, the intersection point x is fixed by the elliptic transformation $g \circ h$, and $(g, h) \in \mathcal{E}$.

(2.1.4) If $t = t_1$, $I(h)$ and $I(g^{-1})$ are internally tangent to each other, the tangential point x is fixed by the parabolic transformation $g \circ h$, and $(g,h) \in \mathcal{H}$.

(2.1.5) If $t_1 < t < 0$, then one of the circles $I(h)$ and $I(g^{-1})$ is interior to another. If e.g. $I(g^{-1})$ is interior to $I(h)$, then there are points x and y on the axis of g such that $h(x) = y$. Moreover $(g,h) \in \mathcal{H}$.

In the case (2.2), $0 < t < 1$, the axes of g and h intersect and $(g,h) \in \mathcal{H}$.

In the case (2.3), $t > 1$ and $(g,h) \in \mathcal{H}$. If we consider the pair (g^{-1}, h) we have (see Figure 1) five subcases:

(2.3.1) If $t > 1 - t_2$, $I(h)$ and $I(g)$ are exterior to each other and $(g^{-1}, h) \in \mathcal{P}$.

(2.3.2) If $t = 1 - t_2$, $I(h)$ and $I(g)$ are externally tangent to each other, $g^{-1} \circ h$ is parabolic and $(g^{-1}, h) \in \mathcal{P}$.

(2.3.3) If $1 - t_1 < t < 1 - t_2$, $I(h)$ and $I(g)$ intersect, $g^{-1} \circ h$ is elliptic and $(g^{-1}, h) \in \mathcal{E}$.

(2.3.4) If $t = 1 - t_1$, $I(h)$ and $I(g)$ are internally tangent to each other $g \circ h$ is parabolic and $(g^{-1}, h) \in \mathcal{H}$.

(2.3.5) If $1 < t < 1 - t_1$, then one of the circles $I(h)$ and $I(g)$ is interior to another. If e.g. $I(g)$ is interior to $I(h)$ then there are points x and y on the axis of g such that $h(x) = y$. Moreover $(g^{-1}, h) \in \mathcal{H}$.

After this detailed listing of different possibilities we can prove an inequality concerning the lengths of non-intersecting closed geodesics on a compact Riemann surface R of genus $p > 1$.

Let R be represented as a quotient space D/G where G is a group of hyperbolic Möbius transformations $g : D \to D$. Equip R with the metric induced by the hyperbolic metric $2|dz|/(1 - |z|^2)$ of D.

THEOREM 2.1. *Suppose that $g \in G$ and $h \in G$ correspond to simple closed geodesics α and β, respectively, on D/G. If α and β do not intersect, then either $(g,h) \in \mathcal{P}$ or $(g^{-1}, h) \in \mathcal{P}$ depending on the cyclic order of the fixed points of g and h. If $\log k > 0$ is the shortest distance between α and β, $\log k_1$ is the length of α and $\log k_2$ is the length of β, then*

$$k > \frac{f(\sqrt{k_1 k_2}) + f(\sqrt{k_1/k_2})}{f(\sqrt{k_1 k_2}) - f(\sqrt{k_1/k_2})}.$$

PROOF: The pair (g, h) satisfies exactly one of the cases (2.1.1)–(2.1.5) and (2.3.1)–(2.3.5). Since $g \circ h$ and $g^{-1} \circ h$ are hyperbolic, only cases (2.1.1), (2.3.1), (2.1.5) and (2.3.5) remain. Since the axes of g and h are projected onto α and β, respectively, it follows in cases (2.1.5) and (2.3.5) that at least one of the curves α and β is self-intersecting. Hence, since α and β are simple closed curves, (g, h) satisfies either (2.1.1) or (2.3.1) and the first assertion is proved.

We can choose g and h in such a way that the distance between the axes of g and h is $\log k$. We may suppose that $(g, h) \in \mathcal{P}$. We may further suppose that $a(h) = 1$, $r(h) = -1$, $a(g) = e^{i\theta}$ and $r(g) = -e^{-i\theta}$ (Figure 4). Let iy be the point, where the axis of g intersects the imaginary-axis. Since

$$t < t_2 = \frac{-2 - f(k_1/k_2)}{f(k_1 k_2) - f(k_1/k_2)},$$

we have

$$-\cos\theta = \frac{1}{1 - 2t},$$

$$y = \sqrt{\frac{1 + \cos\theta}{1 - \cos\theta}} = \sqrt{\frac{-t}{1 - t}}.$$

Hence

$$y > \sqrt{\frac{-t_2}{1 - t_2}} = \frac{f(\sqrt{k_1/k_2})}{f(\sqrt{k_1 k_2})}.$$

Since $k = (1 + y)/(1 - y)$, the assertion follows. $\qquad\square$

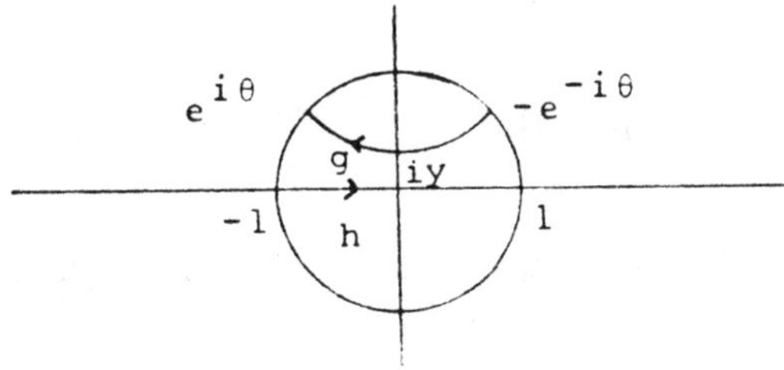

Figure 4

COROLLARY. *If, under the hypotheses of Theorem 2.1, $k_j < 1 + \varepsilon < 2$, $j = 1, 2$, then $k > 8/\varepsilon^2$.*

PROOF: For all $\varepsilon \in (0, 1)$ we have

$$(1 + \varepsilon)^{\frac{1}{2}} < 1 + \frac{1}{2}\varepsilon$$

274

and

$$(1+\varepsilon)^{-\frac{1}{2}} < 1 - \frac{1}{2}\varepsilon + \frac{3}{8}\varepsilon^2.$$

Hence

$$f(1+\varepsilon) < 2 + \frac{3}{8}\varepsilon^2.$$

Consequently

$$k > \frac{f(\sqrt{k_1 k_2}) + f(\sqrt{k_1/k_2})}{f(\sqrt{k_1 k_2}) - f(k_1/k_2)} > \frac{f(\sqrt{k_1 k_2}) + 2}{f(\sqrt{k_1 k_2}) - 2}$$

$$\geq \frac{f(1+\varepsilon) + 2}{f(1+\varepsilon) - 2} > \frac{4 + \frac{3}{8}\varepsilon^2}{\frac{3}{8}\varepsilon^2} > \frac{8}{\varepsilon^2},$$

because $x \mapsto \frac{x+2}{x-2}$ is strictly decreasing for $x > 2$. $\qquad\square$

We restate an inequality about the lengths of intersecting geodesics (cf. [S-S] (1.5)):

THEOREM 2.2. *Suppose that $g \in G$ and $h \in G$ correspond to simple closed geodesics α and β of lengths $\log k_1$ and $\log k_2$ on D/G, respectively. If α and β intersect, then*

$$(2.4) \qquad\qquad \frac{16\, k_2}{(k_2 - 1)^2} < \frac{(k_1 - 1)^2}{k_1}.$$

PROOF: A direct computation shows that

$$f(c) = \left| t(1-t)\frac{(k_1 - 1)^2 (k_2 - 1)^2}{k_1 k_2} - 2 \right| > 2$$

for $c = h \circ g^{-1} \circ h^{-1} \circ g$. Since the axes of g and h intersect, we have, by Figure 1, $0 < t(1-t) \leq \frac{1}{4}$. Hence

$$2 < f(c) = t(1-t)\frac{(k_1 - 1)^2 (k_2 - 1)^2}{k_1 k_2} - 2$$

$$\leq \frac{1}{4}\frac{(k_1 - 1)^2 (k_2 - 1)^2}{k_1 k_2} - 2$$

from which the assertion follows. $\qquad\square$

For practical purposes, the rather sharp inequality (2.4) can be modified e.g. as follows (cf. [S-S], Theorem 1.3):

COROLLARY. *If, under the hypotheses of Theorem 2.2, $k_2 < 1 + \varepsilon$, then $k_1 > 4/\varepsilon$.* $\square$

3. Geometric isomorphisms.

Let G and G' be Fuchsian groups of the first kind, acting in D. An isomorphism $j : G \to G'$ is geometric if there exists a homeomorphism $\varphi : D \to D$ such that

$$(3.1) \qquad\qquad j(g) = \varphi \circ g \circ \varphi^{-1}$$

for all $g \in G$. Then φ can be extended homeomorphically to the closure of D such that (3.1) remains valid. Hence j preserves the cyclic order of the fixed points of the elements of G.

THEOREM 3.1. *A geometric isomorphism $j : G \to G'$ preserves the type of the pairs (g, h) of hyperbolic transformations of G.*

PROOF: Denote $g' = j(g)$ and $h' = j(h)$ and let t', t'_1 and t'_2 be the parameters associated with the pair (g', h').

Suppose that $t < t_2$. Since j preserves the cyclic order of the fixed points and the type of the transformations, we have, by Figures 2 and 3, either $t' < t'_2$ or $t'_1 < t' < 0$. In case (2.1.1) $a(g)$ and $r(g)$ bound an arc of the unit circle not containing points x' and y' such that $h(x') = y'$. Hence cases (2.1.1) and (2.1.5) cannot be geometrically isomorphic, and it follows that $t' < t'_2$.

Suppose that $t = t_2$. Since j is typepreserving and $g \circ h$ is parabolic we have either $t' = t'_2$ or $t' = t'_1$. In cases (2.1.2) and (2.1.4) the cyclic order of the fixed points of g, h and $g \circ h$ is different. Hence $t' = t'_2$. We have proved, by Figure 1, that $(g, h) \in \mathcal{P}$ if and only if $(g', h') \in \mathcal{P}$. The remaining cases follow similarly. $\square$

4. Hyperbolic surfaces.

Let us now consider the geometric meaning of the classification of pairs of Möbius-transformations. Assume that Σ is an oriented compact C^∞-differentiable surface with negative Euler characteristic $\chi(\Sigma)$. Then Σ carries metrics of constant negative curvature -1. The surface Σ together with such a metric d is a Riemann surface $R = (\Sigma, d)$.

Let $R = D/G$ where G is a Fuchsian group of the first kind. There is an isomorphism between the fundamental group of Σ, $\pi_1(\Sigma)$, and G. This isomorphism is defined by lifting the closed curves α, $[\alpha] \in \pi_1(\Sigma)$, to D.

The isomorphism is not quite canonical; it depends on the choice of the point of D to which we lift the curves. Hence the isomorphism $\pi_1(\Sigma) \to G$ is canonically defined up to an inner automorphism of G. Let us now fix an isomorphism $\pi_1(\Sigma) \to G$.

Let α' and β' be simple closed curves on Σ representing elements of $\pi_1(\Sigma)$ and let α and β be the geodesic curves freely homotopic to α' and β' on $R = (\Sigma, d)$. Let g and h be the hyperbolic Möbius-transformations corresponding to $[\alpha']$ and $[\beta']$ under the isomorphism $\pi_1(\Sigma) \to G$. Then the axis of g covers α and that of h covers β in the projection $D \to R = D/G$.

Assume that α^n is not homotopic to any β^m for all $m, n \in \mathbb{Z}$. This implies that the axes of g and h are not the same. Consequently, g and h can have at most one common fixed point. If they do have exactly one common fixed point then their commutator is parbolic. But this is not possible since G contains only hyperbolic elements. Hence g and h do not have any common fixed points. Consequently, the pair (g, h) belongs either to $\mathcal{P}$ or to $\mathcal{H}$.

Assume now that the geodesic curves α and β intersect. Then the axes of g and h cross each other. Hence, in this case, $(g, h) \in \mathcal{H}$.

The pair (g, h) may belong to the class $\mathcal{H}$ even if the geodesic curves α and β do not intersect. That depends on the orientation of the curves α' and β'. Possibly replacing $[\alpha']$ by its inverse in $\pi_1(\Sigma)$ we may suppose, by Theorem 2.1, that, in this case, $(g, h) \in \mathcal{P}$.

The classification of the pairs of Möbius-transformations of G gives now a classification of all pairs $([\alpha'], [\beta'])$, $[\alpha'], [\beta'] \in \pi_1(\Sigma)$, for which $[\alpha']^m \neq [\beta']^n$ for all $m, n \in \mathbb{Z}$. And this gives actually a classification of pairs of oriented geodesic curves α and β on (Σ, d).

This classification of pairs of elements of $\pi_1(\Sigma)$ depends neither on the choice of the metric d of Σ nor on the choice of the uniformization $(\Sigma, d) = U/G$. This follows immediately from Theorem 3.1. Hence the classification is purely topological. In subsequent application to Teichmüller spaces we now assume that all these classes of pairs of elements of $\pi_1(\Sigma)$ are known. If $(\Sigma, d) = U/G$ then also the classes of all pairs (g, h), $g, h \in G$, are known for any metric d.

For later reference let us recall here that the fundamental group $\pi_1(\Sigma)$ of a compact oriented surface of genus p has a standard set of generators $[\alpha_1], [\beta_1], \ldots, [\alpha_p], [\beta_p]$ such that any curve homotopic to α_j intersects β_j

and there is one relation namely

$$(4.1) \qquad \prod_{j=1}^{p} [\alpha_j][\beta_j]^{-1}[\alpha_j]^{-1}[\beta_j] = 1.$$

Let $g_j \in G$ be the Möbius-transformation corresponding to $[\alpha_j] \in \pi_1(\Sigma)$ under the isomorphism $\pi_1(\Sigma) \to G$, and let $h_j \in G$ be that corresponding to $[\beta_j]$. Then $\{g_1, h_1, \ldots, g_p, h_p\}$ generates G and

$$(4.2) \qquad \prod_{j=1}^{p} g_j h_j^{-1} g_j^{-1} h_j = \text{Id}.$$

The axes of g_j and h_j cross each other for all j. Hence $(g_j, h_j) \in \mathcal{H}$. We may similarly assume that the classes of all other pairs $(\gamma, \gamma'), \gamma, \gamma' \in \{g_1, h_1, \ldots, g_p, h_p\}$ are known.

Non-orientable surfaces have to be treated separately. Assume now that Σ is a non-orientable compact C^∞-surface with $\chi(\Sigma) < 0$. Σ together with a metric d of constant curvature -1 is a Klein surface $K = (\Sigma, d)$. Such surfaces can always be represented as $K = D/G$ where G is a non-Euclidean crystallographic group (a NEC group). The NEC group G is a properly discontinuous group of isometries of the hyperbolic unit disk. Orientation preserving elements of G are hyperbolic Möbius-transformations and orientation reversing elements are glide reflections. Such a glide reflection s can always be written as $s = \sigma \circ g$ where g is a hyperbolic Möbius-transformation and σ is the reflection in the axis of g.

Let G_+ be the subgroup of G consisting of orientation preserving elements only. Then G_+ is a Fuchsian group of the first kind, $K_o = D/G_+$ is a compact Riemann surface and the projection $K_o \to K$ is an unramified double covering map. K_o is the orientable double covering of K. The genus or the arithmetic genus of K is defined to be the genus of the Riemann surface K_o.

If the genus p of K is even then the fundamental group $\pi_1(\Sigma)$ of Σ can be generated by p curves $\alpha_1, \beta_1, \ldots, \alpha_{p/2}, \beta_{p/2}$ going around the handles of Σ and by an additional curve γ going through the cross cap of Σ. The relation in this case is

$$(4.3) \qquad \left(\prod_{j=1}^{p/2} [\alpha_j][\beta_j]^{-1}[\alpha_j]^{-1}[\beta_j] \right) [\gamma]^2 = 1.$$

If the genus p is odd, we need two curves γ_1 and γ_2 going through cross caps. The relation is then

$$(4.4) \qquad \left(\prod_{j=1}^{p-1/2} [\alpha_j][\beta_j]^{-1}[\alpha_j]^{-1}[\beta_j] \right) [\gamma_1]^2[\gamma_2]^2 = 1.$$

Consider again the isomorphism $\pi_1(\Sigma) \to G$. Let g_j correspond to $[\alpha_j]$, h_j to $[\beta_j]$, s to $[\gamma]$ and s_1 and s_2 to $[\gamma_1]$ and to $[\gamma_2]$, respectively. The genus p being even, G can be generated by $\{g_1, h_1, \ldots, g_{p/2}, h_{p/2}, s\}$ satisfying

$$(4.5) \qquad s^2 \prod_{j=1}^{p/2} g_j h_j^{-1} g_j^{-1} h_j = \mathrm{Id}.$$

The genus being odd the generators are $\{g_1, h_1, \ldots, g_{p-1/2}, h_{p-1/2}, s_1, s_2\}$ and the relation is

$$(4.6) \qquad s_1^2 s_2^2 \prod_{j=1}^{p-1/2} g_j h_j^{-1} g_j^{-1} h_j = \mathrm{Id}.$$

All pairs (g_j, h_j) belong to the class $\mathcal{H}$.

The classes of other pairs of Möbius-transformations in $\{g_1, h_1, \ldots, g_{p/2}, h_{p/2}, s^2\}$ or in $\{g_1, h_1, \ldots, g_{p-1/2}, h_{p-1/2}, s_1^2, s_2^2\}$ are likewise determined by the fundamental group of Σ (or actually by that of the orientable double covering). These generators for the fundamental group of a non-orientable surface Σ can easily be derived from the considerations in [**Sibner**] and in [**Wilkie**].

5. Geometric parametrizations of Teichmüller spaces of oriented surfaces.

The Teichmüller space $T(\Sigma)$ of a surface Σ, orientable or not, is defined in the usual way. Let $M(\Sigma)$ denote the set of hyperbolic metrics on Σ. The group $\mathrm{Diff}(\Sigma)$ of diffeomorphisms of Σ acts on $M(\Sigma)$ by pull back. More precisely, the mapping $\alpha \in \mathrm{Diff}(\Sigma)$ induces $\alpha^* : M(\Sigma) \to M(\Sigma)$ where $\alpha^*(d)$ for any $d \in M(\Sigma)$ is defined by requiring $\alpha : (\Sigma, \alpha^*(D)) \to (\Sigma, d)$ be an isometry. Let $\mathrm{Diff}_o(\Sigma)$ denote the subgroup of $\mathrm{Diff}(\Sigma)$ consisting of mappings homotopic to the identity. The Teichmüller space of Σ is, as usual $T(\Sigma) = M(\Sigma)/\mathrm{Diff}_o(\Sigma)$. In particular, two metrics $d, d' \in M(\Sigma)$ correspond to the same point in $T(\Sigma)$ if and only if there is an isometry $(\Sigma, d) \to (\Sigma, d')$ homotopic to the identity mapping of Σ. Observe that in the case of an orientable surface Σ with $\chi(\Sigma) < 0$ this definition for the Teichmüller space agrees with the usual one since for such surfaces Σ, all mappings $\Sigma \to \Sigma$, which are homotopic to the identity, are also sense preserving.

If the surface Σ is oriented, $T(\Sigma)$ is a complex manifold. For non-orientable surfaces $T(\Sigma)$ is a real analytic manifold. For the constructions of these analytic structures see for instance [**Lehto**] Chapter 5. From now on we understand that the Teichmüller spaces $T(\Sigma)$ are equipped with the corresponding analytic structures.

For any closed curve α on Σ we may define the geodesic length function $\ell_\alpha : T(\Sigma) \to \mathbb{R}_+$ setting

$$\ell_\alpha([d]) = \inf\{d - \text{length of } \alpha' | \alpha' \approx \alpha\}.$$

The geodesic length functions are then real analytic. Our aim is to find the smallest possible set $\{\alpha_1, \ldots, \alpha_n\}$ of closed curves on Σ such that the corresponding geodesic length functions give an injective mapping

$$T(\Sigma) \to (\mathbb{R}_+)^n, \ [d] \mapsto (\ell_{\alpha_1}([d]), \ldots, \ell_{\alpha_n}([d])).$$

To that end we will apply the considerations of the preceding chapters and the results of [**S-S**].

We will handle orientable surfaces first. For the reader's convenience, let us restate a result from [**S-S**] (Corollary to Theorem (2.1)).

In this theorem we consider the presentation $R = D/G$ of a Riemann surface R of genus p. Let $K = \{g_1, h_1, \ldots, g_p, h_p\}$ the standard set of generators of G satisfying the relation (4.2).

THEOREM 5.1. *The multipliers of the following* $6p - 4$ *Möbius-transformations determine* K *uniquely up to a conjugation by a Möbius-transformation:*

$$(5.1) \quad \begin{aligned} &g_j, h_j, \ j = 1, \ldots, p \\ &g_j \circ h_1, \ j = 1, \ldots, p \\ &h_j \circ h_1, \ j = 2, \ldots, p \\ &g_j \circ g_1, \ h_j \circ g_1, \ j = 2, \ldots, p-1, \\ &g_p \circ h_p \ . \end{aligned}$$

Note, in particular, that in the above theorem we did allow conjugation by a Möbius-transformation mapping the inside of the unit disk onto its outside. Allowing conjugation also by orientation reversing isometries Theorem 5 can be stated in the following form:

LEMMA 5.1. *The multipliers of the Möbius-transformations (5.1) determine K uniquely up to a conjugation by an isometry of the hyperbolic unit disk.*

PROOF: Let K_1 and K_2 be two sets of Möbius-transformations for which the corresponding mappings (5.1) have same multipliers. Then there exists a Möbius-transformation f which conjugates K_1 to K_2.

If f maps the unit disk onto itself, we are done. If not, then let $\sigma(z) = \frac{1}{\bar{z}}$. On the boundary of the unit disk σ is the identity mapping. Hence, if g is a Möbius-transformation mapping D onto itself, $\sigma \circ g \circ \sigma^{-1} = g$. Consequently, since f conjugates K_1 to K_2 also $\sigma \circ f$ does the same. But $\sigma \circ f$ is now an isometry of D onto itself. $\qquad\square$

Consider again the representation $R = D/G$ for a Riemann surface $R = (\Sigma, d)$. Recall that if $g \in G$ is a Möbius-transformation corresponding to a closed curve α under the isomorphism $\pi_1(\Sigma) \to G$ then $\ell_\alpha([d]) = \log k(g)$, so that the multipliers determine and are determined by the lengths of geodesic curves. Let now $\alpha_1, \ldots, \alpha_{6p-4}$ be closed curves on Σ corresponding to the Möbius-transformations (5.1).

THEOREM 5.2. *The mapping*

$$L : T(\Sigma) \to (\mathbb{R}_+)^{6p-4}, \quad [d] \mapsto (\ell_{\alpha_1}([d]), \ldots, \ell_{\alpha_{6p-4}}([d])),$$

is injective.

PROOF: Assume that for metrics d and d' we have $L([d]) = L([d'])$. We have to show then that there exists an isometry $(\Sigma, d) \to (\Sigma, d')$ homotopic to the identity.

Let $(\Sigma, d) = D/G$, $(\Sigma, d') = D/G'$ and let $K = \{g_1, h_1, \ldots, g_p, h_p\}$ (resp. $K' = \{g_1', h_1', \ldots, g_p', h_p'\}$) be the standard set of generators for G (for G', resp.). Since $L([d]) = L([d'])$, the multipliers of the corresponding Möbius-transformations (5.1) are the same. Hence, by Lemma 5.1, there exists an isometry $f : D \to D$ which conjugates K to K'. In particular,

$$(5.2) \qquad f \circ g_j \circ f^{-1} = g_j' \text{ and } f \circ h_j \circ f^{-1} = h_j'$$

for all indices $j = 1, \ldots, p$.

This isometry $f : D \to D$ induces an isometry $f : (\Sigma, d) \to (\Sigma, d')$. By (5.2) the homeomorphism $f : \Sigma \to \Sigma$ induces the trivial mapping $\pi_1(\Sigma) \to \pi_1(\Sigma)$. Hence, by a theorem of Baer [**ZVC**, 5.14], f is homotopic to the identity mapping of Σ. $\qquad\square$

6. Teichmüller spaces of non-orientable surfaces.

We start with an almost obvious lemma.

LEMMA 6.1. *Let* $s : D \to D$ *be a glide-reflection for which* s^2 *is a hyperbolic Möbius-transformation. Then* s *is uniquely determined by* s^2.

PROOF: The attractive and the repelling fixed points of s are those of s^2 and s does not have any other fixed points. The multipliers satisfy $k(s^2) = k(s)^2$. Hence s is uniquely determined by s^2. $\qquad\square$

Let g_1 and h_1 be given hyperbolic Möbius-transformations with intersecting axes. Let f be a hyperbolic Möbius-transformation whose axis does not intersect those of g_1 and h_1. Assume also that the classes of the pairs (f, g_1) and (f, h_1) are known.

LEMMA 6.2. *Under the above assumptions the numbers* $k(f)$, $k(f \circ g_1)$ *and* $k(f \circ h_1)$ *determine* f *uniquely.*

PROOF: Lemma 2.1 in [S-S]. $\qquad\square$

Let Σ be a non-orientable compact C^∞-surface of genus $p > 1$. Because the generators and relations given in §4 for $\pi_1(\Sigma)$ depend on the parity of p, let us first assume that p is even. Let $p = 2n$. Then $\pi_1(\Sigma)$ can be generated by the curves $\alpha_1, \beta_1, \ldots, \alpha_n, \beta_n$ and γ satisfying the relation (4.3). Let d be a hyperbolic metric on Σ. Then $K = (\Sigma, d)$ is a Klein surface. Consider the representation $K = D/G$ where G is a NEC group which is generated by $K = \{g_1, h_1, \ldots, g_n, h_n, s\}$ satisfying the relation (4.5). Here g_j and h_j are hyperbolic Möbius-transformations with intersecting axes and there are no other intersections between the axes of the elements of K. Consider the $6n - 3$ Möbius-transformations

$$(6.1) \qquad \begin{aligned} &g_j, h_j, g_j \circ h_1, \quad j = 1, \ldots, n \\ &g_j \circ g_1, h_j \circ g_1, h_j \circ h_1, \quad j = 2, \ldots, n. \end{aligned}$$

PROPOSITION 6.1. *The multipliers of the Möbius-transformations* (6.1) *determine* K *uniquely up to a conjugation by a Möbius-transformation.*

PROOF: By a conjugation we may normalize K in such a way that $a(g_1) = 1$, $r(g_1) = -1$ and $a(h_1) = \sqrt{-1}$. By the considerations of §4 all the classes of pairs (g_j, h_i) can be assumed known.

By Theorem 1.1 the Möbius-transformations g_1 and h_1 are then determined by the multipliers $k(g_1 \circ h_1)$, $k(g_1)$ and $k(h_1)$. By Lemma 6.2 the transformations g_j and h_j, $j > 2$, are now determined by $k(g_j)$, $k(g_j \circ g_1)$,

$k(g_j \circ h_1)$ and $k(h_j)$, $k(h_j \circ g_1)$, $k(h_j \circ h_1)$. Hence the set $\{g_1, h_1, \ldots, g_n, h_n\}$ is determined. By the relation (4.5) s^2 is then determined. Finally, by Lemma 6.1, also s is now determined. $\qquad\square$

Let now $\delta_1, \ldots, \delta_{6n-3}$ be closed curves on Σ corresponding to the Möbius-transformations (6.1) under an isomorphism $\pi_1(\Sigma) \to G$.

Repeating, word by word, the argument of Theorem 5.2 we get:

PROPOSITION 6.2. *For a non-orientable surface Σ of genus $2n$, the mapping*

$$L : T(\Sigma) \to (\mathbb{R}_+)^{6n-3}, \quad [d] \mapsto (\ell_{\delta_1}([d]), \ldots, \ell_{\delta_{6n-3}}([d])),$$

is injective.

Surfaces of odd genus $p = 2n + 1$ can be treated similarly. The set K is then replaced by the set $\{g_1, h_1, \ldots, g_n, h_n, s_1, s_2\}$ satisfying (4.6). Instead of the multipliers of the Möbius-transformations (6.1) we have to consider the multipliers of the following $6n$ transformations:

$$
\begin{aligned}
&g_j, h_j, g_j \circ h_1, \quad j = 1, \ldots, n \\
(6.2) \qquad &g_j \circ g_1, h_j \circ g_1, h_j \circ h_1, \quad j = 2, \ldots, n, \\
&s_1^2, s_1^2 \circ g_1, s_1^2 \circ h_1 \,.
\end{aligned}
$$

Let $\delta_1, \ldots, \delta_{6n}$ be the closed curves on Σ corresponding to the Möbius-transformations (6.1). Repeating the preceding considerations we get now:

PROPOSITION 6.3. *For non-orientable surfaces Σ of genus $p = 2n+1$, the mapping $L : T(\Sigma) \to (\mathbb{R}_+)^{6n}$, $[d] \mapsto (\ell_{\delta_1}([d]), \ldots, \ell_{\delta_{6n}}([d]))$, is injective.*

$\qquad\square$

In conclusion let us combine Propositions 6.2 and 6.3:

THEOREM 6.1. *Let Σ be a non-orientable surface of genus p, $p > 1$. There is a set $\{\delta_1, \ldots, \delta_{3p-3}\}$ of $3p-3$ closed curves on Σ such that the associated geodesic length functions define an injective mapping*

$$L : T(\Sigma) \to (\mathbb{R}_+)^{3p-3}, \quad [d] \to (\ell_{\delta_1}([d]), \ldots, \ell_{\delta_{3p-3}}([d])).$$

The mapping L is real analytic and open.

PROOF: By Propositions 6.2 and 6.3 we can find curves $\delta_1, \ldots, \delta_{3p-3}$ for which the mapping L is injective. It is well known that the geodesic length functions are real analytic (cf. [**Abikoff**]). It remains to show that L is open. But this follows from the invariance of domain, since $T(\Sigma)$ is a manifold and $\dim_{\mathbb{R}} T(\Sigma) = 3p - 3$. $\qquad\qquad\square$

We wish to conclude these considerations by the following conjecture.

CONJECTURE: The smallest number of geodesic length functions parametrizing the Teichmüller space of a compact and oriented surface of genus p, $p > 1$, is $6p - 4$.

M. Seppälä, Universität Regensburg, Fakultät für Mathematik, BRD–8400 Regensburg and University of Helsinki, Department of Mathematics, Hallituskatu 15, SF–00100 Helsinki, Finland

T. Sorvali, University of Joensuu, Department of Mathematics, SF–80100 Joensuu, Finland

References

[**Abikoff**] Abikoff, W., "The real analytic theory of Teichmüller space," Lecture Notes in Mathematics **820**, Springer, 1980.

[**Lehto**] Lehto, O., "Univalent functions and Teichmüller spaces," Graduate Texts in Mathematics, Springer, 1986.

[**S-S**] Seppälä, M. and Sorvali, T., *Parametrization of Möbius groups acting in a disk*, Comment. Math. Helvetici **61** (1986), 149–160.

[**Sibner**] Sibner, R.J., *Symmetric Fuchsian groups*, Amer. J. Math. **90** (1968), 1237–1259.

[**ZVC**] Zieschang, H., Vogt, E. and Coldewey, H.-D., "Surfaces and planar discontinuous groups," 2nd transl. ed. Springer, 1980.

[**Wilkie**] Wilkie, W.C., *On the non-euclidean crystallographic groups*, Math. Z. **91** (1966), 87–102.

Families of compact Riemann surfaces
which do not admit n^{th} roots

BY PATRICIA L. SIPE

Let $\pi : V \to B$ be a holomorphic family of compact Riemann surfaces of genus $p \geq 2$ (to be defined in section 1). For any $t \in B$, the fiber $X_t = \pi^{-1}(t)$ is a closed Riemann surface; the canonical line bundle $K(X_t)$ is the holomorphic cotangent bundle of X_t. A standard construction (see section 1 for details) produces a line bundle $K_{\mathrm{rel}}(V) \to V$, called the relative canonical bundle, whose restriction to each Riemann surface $X_t \subseteq V$ is equivalent to the canonical bundle $K(X_t)$. (Throughout this paper, all line bundles will be holomorphic complex line bundles and equivalence will be holomorphic equivalence.)

For any given closed Riemann surface X of genus $p \geq 2$, an nth root of $K(X)$ is a line bundle $L \to X$ such that $L^{\otimes n} \simeq K(X)$. These n^{th} roots do not exist unless n divides $2p - 2$, in which case there are n^{2p} different n^{th} roots. The case $n = 2$ is exceptional; for the remainder of this paper we assume that n divides $2p - 2$ and that n (and therefore p) is ≥ 3.

If $V \to B$ is a holomorphic family of compact Riemann surfaces of genus p, an nth root of the relative canonical bundle is a line bundle $L \to V$ such that $L^{\otimes n} \simeq K_{\mathrm{rel}}(V)$. It is natural to ask whether n^{th} roots of the relative canonical bundle exist; such an nth root can be thought of as a family of nth roots corresponding to the given family of Riemann surfaces.

The purpose of this paper is to present some simple and natural examples of families whose relative canonical bundles admit no n^{th} roots for $n \geq 3$. This is in marked contrast to the situation for the universal Teichmüller curve $\pi_p : V_p \to T_p$ over the Teichmüller space T_p of closed surfaces of genus p. The author studied n^{th} roots of $K_{\mathrm{rel}}(V_p)$ in [2] and [3] and found that there are n^{2p} roots; each nth root on any given fiber extends to an nth root of $K_{\mathrm{rel}}(V_p)$. She also studied the action of the Teichmüller modular group on the finite set of nth roots. The results from [2] and [3] will play a central role in the construction of our examples.

The author thanks the referee for helpful comments.

1. Families of Riemann Surfaces; the Relative Canonical Bundle.

DEFINITION: A *holomorphic family of compact complex manifolds* is a pair of connected complex manifolds (V, B) and a mapping $\pi : V \to B$ which is a proper holomorphic submersion.

$\pi : V \to B$ is a *family of compact Riemann surfaces* if $\pi^{-1}(t)$ is a closed Riemann surface for every $t \in B$.

DEFINITION: A *holomorphic family of line bundles over B* is given by $L \to V \to B$ where $V \to B$ is a holomorphic family of Riemann surfaces and $L \to V$ is a holomorphic complex line bundle.

Note that the projection $L \to B$ has the property that the fibre over $t \in B$ is a line bundle L_t over X_t.

Given a family $\pi : V \to B$, the differential of the projection π induces a mapping between the tangent spaces; $\ker(d\pi) \subseteq T(V)$ is also a bundle over V, consisting of the vertical tangent vectors. The fibers of the mapping $\ker(d\pi) \to B$ are the tangent spaces $T(X_t)$, so $\ker(d\pi)$ is the family of tangent bundles associated with the family $V \to B$. The dual $[\ker(d\pi)]^*$ is a family of canonical bundles associated with $V \to B$, since the fiber over $t \in B$ is the canonical bundle $K(X_t)$. The bundle $[\ker(d\pi)]^*$ is called the *relative canonical bundle* and is denoted by $K_{\mathrm{rel}}(V)$.

PROPOSITION 1. *Suppose $\pi_1 : W \to D$ and $\pi_2 : V \to B$ are holomorphic families of compact Riemann surfaces, such that there are holomorphic mappings f and g making the diagram*

$$
\begin{array}{ccc}
W & \xrightarrow{\ g\ } & V \\
\downarrow & & \downarrow \\
D & \xrightarrow{\ f\ } & B
\end{array}
$$

commute. Suppose also that the restriction $g_t : \pi_1^{-1}(t) \to \pi_2^{-1}(f(t))$ is a biholomorphism for every $t \in D$. Then $K_{\mathrm{rel}}(W) \simeq g^*(K_{\mathrm{rel}}(V))$.

PROOF: The differential $dg : T(W) \to T(V)$ maps $\ker(d\pi_1)$ into $\ker(d\pi_2)$ and is an isomorphism on fibers because g_t is a biholomorphism for each $t \in D$. This implies that $\ker(d\pi_1) = g^*(\ker(d\pi_2))$, so $K_{\mathrm{rel}}(W)$ is the pullback of $K_{\mathrm{rel}}(V)$ as claimed. $\qquad\square$

2. Families of nth Roots.

We continue with $V \to B$ a holomorphic family of compact Riemann surfaces of genus $p \geq 3$ and n an integer, $n > 2$ and $n | 2p - 2$.

DEFINITION: A *holomorphic family of nth roots of the canonical bundle associated with the holomorphic family of compact Riemann surfaces $V \to B$ is a holomorphic family of line bundles $L \to V \to B$ with $L^{\otimes n} \simeq K_{\mathrm{rel}}(V)$.*

REMARK: When the meaning is clear from the context, we speak of a family of nth roots over B or simply a family of nth roots.

In the case of the universal Teichmüller curve, $K_{\mathrm{rel}}(V_p)$ and $K(V_p)$ are the same. Therefore, these families of nth roots are the same as the nth roots of the canonical bundle studied in [2].

Let Mod_p denote the Teichmüller modular group. It is well known that Mod_p acts as a group of biholomorphic mappings on the complex manifolds V_p and T_p. Also, any element of Mod_p induces a mapping on $H_1(X, \mathbb{Z})$ and also on $H_1(X, \mathbb{Z}_n)$. Let $Sp(p, \mathbb{Z})$ and $Sp(p, \mathbb{Z}_n)$ be the symplectic groups of $2p \times 2p$ matrices with coefficients in $\mathbb{Z}$ and $\mathbb{Z}_n$ respectively. Since the induced maps on homology are represented by symplectic matrices, we have homomorphisms $\rho : \mathrm{Mod}_p \to Sp(p, \mathbb{Z})$ and $\rho_n : \mathrm{Mod}_p \to Sp(p, \mathbb{Z}_n)$.

Denote $\ker(\rho)$ by I and $\ker(\rho_n)$ by I_n. Then as subgroups of Mod_p, I and I_n act on T_p and V_p. In fact, I and I_n (for $n > 2$) act on T_p and therefore on V_p without fixed points. For if f has a fixed point $s \in T_p$, it induces an automorphism $f_s : X_s \to X_s$ on the fiber, which by definition acts as the identity at the level of homology (with coefficients in $\mathbb{Z}$ or $\mathbb{Z}_n$ depending on the case). This implies that f is the identity (see for example, [1], pages 269–276). Therefore the quotients V_p/I_n and T_p/I_n are complex manifolds. The action of Mod_p on V_p induces an action on the set of families of nth roots over T_p by the rule

$$f \cdot L = (f^{-1})^*(L)$$

for all $f \in \mathrm{Mod}_p$ and L a family of nth roots over T_p.

PROPOSITION 2. *If $p \geq 3$, there is an $f \in I$ such that for any $n \geq 3$ and any n^{th} root $L \to V_p$ of $K_{\mathrm{rel}}(V_p)$, $f \cdot L \neq L$.*

PROOF: In previous work, the author defined a one to one correspondence between the set of families of nth roots over T_p and a certain set Φ_n whose elements can be represented by vectors in $(\mathbb{Z}_n)^{2p}$. That correspondence respects the actions defined and studied in [2,3]. This proposition follows directly from Lemma 1, page 520 of [3]. Indeed, given any non-zero vector $v = (v_1, v_2) \in (2\mathbb{Z}_n)^{2p}$, using that lemma, we can construct an element $f \in I$ (and therefore $f \in I_n$ for any $n|2p-2$) that acts on elements of Φ_n by $f \cdot \varphi = \varphi + v$. By choosing v so that all entries are 0 or 2, and recalling that $n > 2$, we see that $f \cdot \varphi \neq \varphi$ for any $\varphi \in \Phi_n$. Therefore no family of nth roots over T_p is left fixed by this element f. $\qquad\square$

THEOREM. *Suppose $p \geq 3$, $f \in I$ satisfies the conclusion of Proposition 2, and $\langle f \rangle$ is the cyclic subgroup of Mod_p generated by f. Then for $n \geq 3$, there are no holomorphic families of nth roots associated with the holomorphic family $V_p/\langle f \rangle \to T_p/\langle f \rangle$ of compact Riemann surfaces.*

PROOF: Suppose $\mathcal{L}$ is a holomorphic family of nth roots associated with the holomorphic family $V_p/\langle f \rangle \to T_p/\langle f \rangle$. Then if L is the pullback of $\mathcal{L}$, we have the commutative diagram

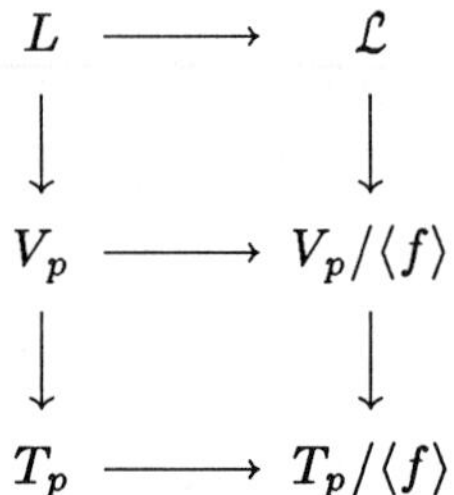

and it follows that the diagram

$$
\begin{array}{ccc}
L^{\otimes n} & \longrightarrow & \mathcal{L}^{\otimes n} \\
\downarrow & & \downarrow \\
V_p & \longrightarrow & V_p/\langle f \rangle \\
\downarrow & & \downarrow \\
T_p & \longrightarrow & T_p/\langle f \rangle
\end{array}
$$

also commutes.

Since $\mathcal{L}^{\otimes n} \simeq K_{\mathrm{rel}}(V_p/\langle f \rangle)$ by definition, $L^{\otimes n} \simeq K_{\mathrm{rel}}(V_p) \simeq K(V_p)$. That is, L is a family of nth roots over $V_p \to T_p$. Moreover, studying the action of $\langle f \rangle$ on these families gives the commutative diagram

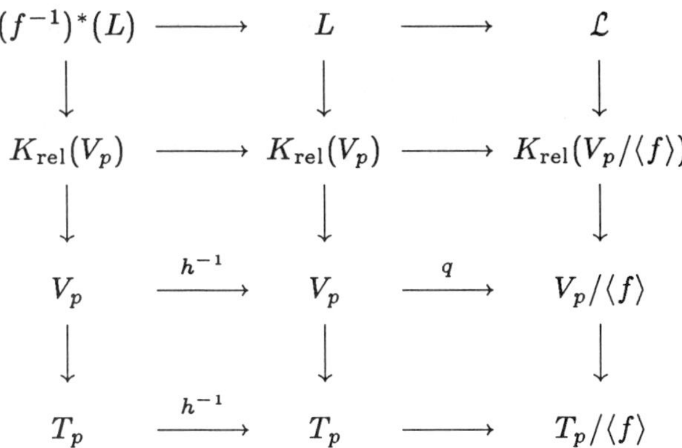

for any $h \in \langle f \rangle$. Now since $L \simeq q^*(\mathcal{L})$ and $(q \circ h^{-1}) = q$, this means

$$L \simeq q^*(\mathcal{L}) \simeq (q \circ h^{-1})^*(\mathcal{L}) \simeq (h^{-1})^*(q^*(\mathcal{L})) \simeq (h^{-1})^*(L).$$

Thus, L is invariant under the action of $\langle f \rangle$. But this is impossible by proposition 2, contradicting the assumption of the existence of $\mathcal{L}$. $\square$

COROLLARY 1. *There do not exist families of nth roots $(n \geq 3)$ of the canonical bundle associated with the family $V_p/I \to T_p/I$.*

PROOF: A similar argument shows that such a family of nth roots would pull back to a family over $V_p \to T_p$ which is invariant under the action of I. But this is impossible, since proposition 2 gives an element $f \in I$ which moves any family of nth roots. $\square$

Since $I \subseteq I_n$ for every n, the same proof establishes the following result.

COROLLARY 2. *If $p \geq 3$ and $n \geq 3$, there do not exist families of nth roots of the canonical bundle associated with the family $V_p/I_n \to T_p/I_n$.*

REMARK: In a forthcoming paper, the author and C.J. Earle construct related examples of families over the punctured disk which admit no families of nth roots for $n \geq 3$.

Department of Mathematics, Smith College, Northampton MA 01063